Augsburg College
George Sverdrup Library
Minneapolis, Minnesota 55404

c.1

POLITICS
IN
EUROPE

Comparisons and Interpretations

POLITICS IN EUROPE

Comparisons and Interpretations

edited by

AREND LIJPHART

University of California, Berkeley

PRENTICE-HALL, INC., *Englewood Cliffs, New Jersey*

© 1969 by Prentice-Hall, Inc., Englewood Cliffs, N.J.

All rights reserved. No part of this book may be reproduced in any form or by any means without permission in writing from the publisher.

13-685743-4

Library of Congress Catalog Card Number 70-76296

Printed in the United States of America

CURRENT PRINTING (LAST DIGIT):

12 11 10 9 8 7 6 5 4 3 2 1

PRENTICE-HALL INTERNATIONAL, INC., *London*
PRENTICE-HALL OF AUSTRALIA, PTY. LTD., *Sydney*
PRENTICE-HALL OF CANADA, LTD., *Toronto*
PRENTICE-HALL OF INDIA PRIVATE LTD., *New Delhi*
PRENTICE-HALL OF JAPAN, INC., *Tokyo*

for

Anna Margaretha

PREFACE

Is comparative government "essentially noncomparative" and "essentially descriptive," as Roy C. Macridis argues in his 1955 indictment of this field of political science?[1] In particular, is the basic undergraduate course in comparative government and politics, which usually covers four countries—Great Britain, France, Germany, and the Soviet Union—essentially noncomparative and descriptive? The charge is undoubtedly not completely true, and less so today than in 1955, but most instructors do feel the need to strengthen the comparative and interpretative character of the course.

This book of readings focuses, as its subtitle indicates, on *comparisons* and *interpretations*. Its aim is not just to provide further, more detailed, and more specialized information not covered in the usual textbook or series of individual-country studies, but also to stimulate and enrich the students' thinking about that information by focusing on three kinds of materials: (1) explicitly comparative analyses, drawing parallels or contrasts between the four countries studied in the course, between these four and other countries such as the United States, and between present systems and previous regimes; (2) general analyses of broad types of political systems, placing the four countries in relationship to the rest of the world's political systems and providing a global perspective to the course; (3) thoughtful and thought-provoking interpretations and speculations about future developments by acknowledged experts.

This book is based on the assumption that the traditional four-country approach to the basic comparative government course will persist, although with minor variations, and on the conviction that this tradition, like so many traditions, makes a lot of sense. In the first place, political science is still an underdeveloped discipline, and our knowledge of foreign political systems is still extremely limited; in this relative sense, the British, French, German, and Soviet systems are the most thoroughly known and understood. Second, the four countries are also great powers in international politics. Although from a comparative politics point of view large and small countries are equally significant, an analysis of the great

[1] Roy C. Macridis, *The Study of Comparative Government* (New York: Random House, Inc., 1955).

powers serves not only to increase our understanding of the internal processes of government, but also to assess the role of some of the major actors in world politics. The traditional basic course in comparative politics thus performs a dual function. And third, one has to keep in mind that this course is not, and does not claim to be, more than an introduction to comparative politics; it is not an exhaustive treatment.

The chapters in this book may be coordinated with readings from textbooks in several ways. Part One (types and varieties of political systems) precedes Parts Two to Five (the four individual countries), and within each of the book's parts the material is arranged from the more general to the more specific. There is no need, however, to adhere rigidly to this sequence. The instructor may assign some selections in conjunction with text sections, while other selections may be more profitably read for review at the end of the study of each country. In particular, many instructors will find that some or all of the readings in Part One should be reserved for the general review at the end of the course, when they can provide an excellent basis for the final lectures and/or well-informed class discussion.

This book also meets four other widespread preferences of instructors teaching the basic comparative politics course: (1) Its length has been determined with the student's reading load kept in view; (2) the "snippets" approach has been deliberately avoided—articles, chapters, or sections are reprinted in their entirety; (3) within the limitations of the first two objectives, the coverage of the subject matter is as comprehensive as possible; (4) the introductions to the five parts of the book have been kept brief. This book is intended as a book of *readings*, not summaries of or commentaries on these readings, which tend to arrogate the tasks of both instructor and student.

Berkeley, California A.L.
June, 1968

CONTENTS

Part One

TYPES AND VARIETIES OF POLITICAL SYSTEMS
Introduction 3

1. *The Non-Western Political Process*
   LUCIAN W. PYE   5

2. *Is There a Non-Western Political Process?*
   ALFRED DIAMANT   18

3. *The General Characteristics of Totalitarian Dictatorship*
   CARL J. FRIEDRICH AND
   ZBIGNIEW BRZEZINSKI   22

4. *Comparative Political Systems*
   GABRIEL A. ALMOND   32

5. *Typologies of Democratic Systems*
   AREND LIJPHART   46

Part Two

THE BRITISH POLITICAL SYSTEM
Introduction   83

6. *Rebuilding the House of Commons*
   WINSTON S. CHURCHILL   85

7. *Group Representation in Britain and the United States*
   SAMUEL H. BEER   88

8. *The Americanisation of British Politics*
   THOMAS J. CARBERY   100

9. *Party Systems in the United Kingdom and the Older Commonwealth: Causes, Resemblances, and Variations*
   LESLIE LIPSON   104

10. *Party Affiliation and International Opinions in Britain and France, 1947–1956*
    MORRIS DAVIS AND SIDNEY VERBA   121

Part Three

THE FRENCH POLITICAL SYSTEM
Introduction   137

11. *The Bayeux Manifesto*
    CHARLES DE GAULLE   139

12. *The Separation of Powers in the Constitution of the Fifth Republic*
    MAX BELOFF   143

13. *Succession and Stability in France*
    STANLEY H. HOFFMANN   150

14. *Democratic Stability and Instability:
    The French Case*
    ERIC A. NORDLINGER   164

15. *Politicization of the Electorate in France
    and the United States*
    PHILIP E. CONVERSE AND GEORGES DUPEUX   178

Part Four

THE POLITICAL SYSTEM
OF THE GERMAN FEDERAL REPUBLIC
Introduction 199

16. *The Foundations of a Democratic Future
    for Germany*
    JAMES B. CONANT   201

17. *Parties and Pressure Groups in Weimar
    and Bonn*
    CHARLES E. FRYE   209

18. *The New Germanies: Restoration,
    Revolution, Reconstruction*
    RALF DAHRENDORF   226

19. *Three Constitutional Courts: A Comparison*
    TAYLOR COLE   237

20. *Disaffection and Participation in Western
    Democracies: The Role of Political
    Oppositions*
    GIUSEPPE DI PALMA   255

Part Five

THE POLITICAL SYSTEM OF THE SOVIET UNION
Introduction   281

21. *The Nature of Communist Society*
    V.I. LENIN   282

22. *The Essence of Contemporary Communism*
    MILOVAN DJILAS   293

23. *Russia and the West: A Comparison and Contrast*
    HENRY L. ROBERTS   299

24. *Soviet Russia as a Model for Underdeveloped Areas*
    W. DONALD BOWLES   309

25. *The Comparison of Soviet and American Law*
    HAROLD J. BERMAN   326

POLITICS

IN

EUROPE

Comparisons and Interpretations

Part One

TYPES AND VARIETIES OF POLITICAL SYSTEMS

INTRODUCTION

Typologies of political systems entail the broadest possible kind of comparisons in political science. Each of the world's more than a hundred political systems is, of course, unique, but this variety may be reduced to a smaller number of basic types into which these systems may be classified. The chapters in Part I concern general types of political systems, and thus provide a wide context for the analysis of individual systems. For the student of the British, French, German, and Soviet political systems, it is instructive to consider questions such as: Where do these four major systems fit into these general types? Do they fit a type perfectly, only to some extent, or not at all? If they can be considered examples of a particular type, do they fit so well that they can be considered representative examples?

These discussions of types and varieties of political systems differ from the many prevalent "analytical frameworks," which are usually no more than recommended approaches and strategies for comparative analysis, or systematic vocabularies designed to facilitate comparisons across all political systems. However useful these approaches, strategies, and vocabularies may be, they are primarily methodological in nature. In contrast, the selections which follow contain significant substantive generalizations. These generalizations are propositions concerning the substance of politics and the characteristics of actual political systems, rather than just potentially useful methods for studying them.

Lucian W. Pye discusses the characteristics of the non-Western political process and also, by implication, states the characteristics of the "Western" political process, thus setting up two broad types. Do Britain, France, and Germany fit the Western type? Where does the Soviet Union belong in this twofold typology? Are the two types mutually exclusive? Alfred Diamant, in Chapter 2, challenges Pye's analysis on a number of important points.

In Chapter 3, a selection from their well-known book Totalitarian Dictatorship and Autocracy, *Carl J. Friedrich and Zbigniew Brzezinski discuss the general characteristics of totalitarianism and, therefore, by implication, the characteristics of non-totalitarian systems. How well does the contemporary Soviet Union fit the totalitarian type? What about the Soviet Union in the 1930's? Do the six characteristics of Friedrich and Brzezinski's totalitarian "syndrome" adequately distinguish this type of system from democracies like Britain and Germany, and from France under de Gaulle?*

Gabriel A. Almond *distinguishes four types of political systems in his famous article* "Comparative Political Systems," *first published in 1956. This article (Chapter 4) represents the first explicit attempt to use the concept of political culture in comparative analysis. Almond's* "Pre-industrial" *and Totalitarian types correspond to a large extent with the types proposed by Pye and by Friedrich and Brzezinski. His two other types, the Anglo-American and the Continental European, are of particular significance in comparing the three major European democracies. Arend Lijphart develops Almond's classification of democracies further in Chapter 5. He also discusses and criticizes the traditional classification of democracies into two-party and multiparty systems. Can the three major European democracies—Britain, France, and Germany—be classified adequately in the fourfold typology he proposes?*

ABOUT THE AUTHORS

LUCIAN W. PYE *is professor of political science at the Massachusetts Institute of Technology and chairman of the Social Science Research Council's Committee on Comparative Politics. He has written both on general comparative politics and on Southeast Asian politics. Among his many works are* Guerrilla Communism in Malaya *and* Politics, Personality, and Nation-Building, *a study of Burma.*

ALFRED DIAMANT *is professor of political science at the University of Indiana. His area specialty is Western Europe, and he has written* Austrian Catholics and the First Republic *and other works.*

CARL J. FRIEDRICH *is professor of government at Harvard University and the University of Heidelberg, president of the International Political Science Association, and a former president of the American Political Science Association. Among his many writings is* Constitutional Government and Democracy, *published in 1951, a major landmark in the development of modern comparative politics.*

ZBIGNIEW BRZEZINSKI *is professor of public law and government at Columbia University and director of Columbia's Research Institute on Communist Affairs. He has written* The Permanent Purge, The Soviet Bloc: Unity and Conflict, *and other works on the Soviet Union and Soviet bloc politics.*

GABRIEL A. ALMOND *is professor of political science and chairman of the department at Stanford University. He was president of the American Political Science Association in 1965–1966. He is the author of* The American People and Foreign Policy *and* The Appeals of Communism, *and co-author of* The Politics of the Developing Areas *and* The Civic Culture.

AREND LIJPHART, *formerly associate professor of political science at the University of California in Berkeley, is now professor of international relations at the University of Leiden in the Netherlands. He is the author of* The Trauma of Decolonization, The Politics of Accommodation, *and other books and articles.*

CHAPTER 1

The Non-Western Political Process

LUCIAN W. PYE

The purpose of this article is to outline some of the dominant and distinctive characteristics of the non-Western political process. In recent years, both the student of comparative politics and the field worker in the newly emergent and economically underdeveloped countries have found it helpful to think in terms of a general category of non-Western politics.[1]

There are, of course, great differences among the non-Western societies. Indeed, in the past, comparative analysis was impeded by an appreciation of the rich diversity in the cultural traditions and the historical circumstances of the Western impact; students and researchers found it necessary to concentrate on particular cultures, and as a consequence attention was generally directed to the unique features of each society. Recently, however, attempts to set forth some of the characteristics common to the political life of countries experiencing profound social change have stimulated fruitful discussions among specialists on the different non-Western regions as well as among general students of comparative politics.

For this discussion to continue, it is necessary for specialists on the different areas to advance, in the form of rather bold and unqualified statements, generalized models of the political process common in non-Western societies.[2] Then, by examining the ways in which particular non-Western countries differ from the generalized models, it becomes possible to engage in significant comparative analysis.

---

[1] For two excellent discussions of the implications for comparative politics of the current interest in non-Western political systems, see: Sigmund Neumann, "Comparative Politics: A Half-Century Appraisal," *Journal of Politics*, XIX (August, 1957), 269–290; and Dankwart A. Rustow, "New Horizons for Comparative Politics," *World Politics*, IX (July, 1957), 530–549.

[2] The picture of the non-Western political process contained in the following pages was strongly influenced by: George McT. Kahin, Guy J. Pauker, and Lucian W. Pye, "Comparative Politics in Non-Western Countries," *American Political Science Review*, XLIX (December, 1955), 1022–41; Gabriel A. Almond, "Comparative Political Systems," *Journal of Politics*, XVIII (August, 1956), 391–409; Rustow, *op. cit.*, and also his *Politics and Westernization in the Near East*, Center of International Studies (Princeton, 1956).

Reprinted from Lucian W. Pye, "The Non-Western Political Process," JOURNAL OF POLITICS, *XX, No. 3 (August, 1958), 468–86, by permission of the publisher. Revised version of a paper presented at the annual meeting of the American Political Science Association on September 5–7, 1957.*

1. In non-Western societies the political sphere is not sharply differentiated from the spheres of social and personal relations. Among the most powerful influences of the traditional order in any society in transition are those forces which impede the development of a distinct sphere of politics. In most non-Western societies, just as in traditional societies, the pattern of political relationships is largely determined by the pattern of social and personal relations. Power, prestige, and influence are based largely on social status. The political struggle tends to revolve around issues of prestige, influence, and even of personalities, and not primarily around questions of alternative courses of policy action.

The elite who dominate the national politics of most non-Western countries generally represent a remarkably homogeneous group in terms of educational experience and social background. Indeed, the path by which individuals are recruited into their political roles, where not dependent upon ascriptive considerations, is essentially an acculturation process. It is those who have become urbanized, have received the appropriate forms of education, and have demonstrated skill in establishing the necessary personal relations who are admitted to the ranks of the elite. Thus, there is in most non-Western societies a distinctive elite culture in which the criteria of performance are based largely on non-political considerations. To be politically effective in national politics, one must effectively pass through such a process of acculturation.

At the village level it is even more difficult to distinguish a distinct political sphere. The social status of the individual and his personal ties largely determine his political behavior and the range of his influence. The lack of a clear political sphere in such communities places severe limits on the effectiveness of those who come from the outside to perform a political role, be it that of an administrative agent of the national government or of a representative of a national party. Indeed, the success of such agents generally depends more on the manner in which they relate themselves to the social structure of the community than on the substance of their political views.

The fundamental framework of non-Western politics is a communal one, and all political behavior is strongly colored by considerations of communal identification. In the more conspicuous cases the larger communal groupings follow ethnic or religious lines. But behind these divisions there lie the smaller but often more tightly knit social groupings that range from the powerful community of Westernized leaders to the social structure of each individual village.

This essentially communal framework of politics makes it extremely difficult for ideas to command influence in themselves. The response to any advocate of a particular point of view tends to be attuned more to his social position than to the content of his views. Under these conditions it is inappropriate to conceive of an open market place where political ideas can freely compete on their own merits for support. Political discussion tends rather to assume the form of either intracommunal debate or one group justifying its position toward another.

The communal framework also sharply limits freedom in altering political allegiances. Any change in political identification generally requires a change in one's social and personal relationships; conversely, any change in social relations tends to result in a change in political identification. The fortunate village youth who receives a modern education tends to move to the city, establish himself in a new sub-society, and become associated with a political group that may in no way reflect the political views of his original community. Even among the national politicians in the city, shifts in political ties are generally accompanied by changes in social and personal associations.

2. *Political parties in non-Western societies tend to take on a world view and represent a way of life.* The lack of a clearly differentiated political sphere means that political associations or groups cannot be clearly oriented to a distinct political arena but tend to be oriented to some aspect of the communal framework of politics. In reflecting the communal base of politics, political parties tend to represent total ways of life. Attempts to organize parties in terms of particular political principles or limited policy objectives generally result either in failure or in the adoption of a broad ethic which soon obscures the initial objective. Usually political parties represent some sub-society or simply the personality of a particularly influential individual.

Even secular parties devoted to achieving national sovereignty have tended to develop their own unique world views. Indeed, successful parties tend to become social movements. The indigenous basis for political parties is usually regional, ethnic, or religious groupings, all of which stress considerations not usually emphasized in Western secular politics. When a party is merely the personal projection of an individual leader it is usually not just his explicitly political views but all facets of his personality which are significant in determining the character of the movement.

In the past, the tendency for political parties to adopt world views was in some instances strongly encouraged by the desire of traditional authoritarian governments or colonial regimes to suppress all explicitly political associations, and such associations found it expedient to adopt a religious cloak to hide the character of their activities. In time, however, the religious aspect came to have genuine significance in determining the character of the group and maintaining its continuity. This was the case with most of the secret societies common to traditional Chinese society. The same development also took place in French Indo-China, where political activity took the form of organizing quasi-religious sects. Both the Cao Dai and the Hoa Hao began as political movements masking as religions, and, although they never lost their political character, they found a basis of integration in their religious aspects.

The history of the secular nationalist movements reflects a similar tendency for parties essentially to represent ways of life. Even after independence the tendency remains strong because such parties are inclined to feel that they have a mission to change all aspects of life within their society. Indeed, such parties often conceive of themselves as representing a prototype of what their entire country will become in time. Members of such movements frequently believe that their attitudes and views on all subjects will become the commonly shared attitudes and views of the entire population. Those committed to modernizing their societies can see few aspects of life which must not be altered, while those more attached to tradition have equally broad concerns.

3. *The political process in non-Western societies is characterized by a prevalence of cliques.* The lack of a distinct political sphere and the tendency for political parties to have a world view together provide a framework within which the most structured units of decision-making tend to be personal cliques. Although general considerations of social status determine the broad outlines of power and influence, the particular pattern of political relationships at any time is largely determined by decisions made at the personal level. This is the case because the social structure in non-Western societies is characterized by functionally diffuse relationships; individuals and groups do not have sharply defined and highly specific functions and thus do not represent specific interests that distinguish them from other groupings. There is no clearly structured setting that can provide a focus for the more refined pattern of day-to-day political activities. Hence, in arriving at their expectations about the probable behavior of others, those involved in the political process must rely heavily upon judgments about personality and the particular relations of the various actors to each other. The pattern of personal associations provides one of the firmest guides for understanding and action within the political process. Personal cliques are likely to become the key units of decision-making in the political process of most non-Western societies.

Western observers often see the phenomenon of cliques as being symptomatic of immoral and deviously motivated behavior. This may actually be the case. Considerations of motive, however, cannot explain either the prevalence of cliques in non-Western societies or their functions. It should also be noted that the fact that cliques are based on personal relations does not mean that there are no significant differences in their values and policy objectives. Since the members of a given clique are likely to have a common orientation toward politics, if their views were fully articulated they might appear as a distinct ideology that would be significantly different from those of the other factions.

In order to understand the workings of the political process in most non-Western countries it is necessary to analyze the character of interclique reactions. To ignore the importance of cliques would be comparable to ignoring the role of interest groups and elections in analyzing the behavior of American Congressmen.

4. *The character of political loyalty in non-Western societies gives to the leadership of political groups a high degree of freedom in determining matters of strategy and tactics.* The communal framework of politics and the tendency for political parties to have world views means that political loyalty is governed more by a sense of identification with the concrete group than by identification with the professed policy goals of the group. The expectation is that the leaders will seek to maximize all the interests of all the members of the group and not just seek to advance particular policies or values.

So long as the leaders appear to be working in the interests of the group as a whole, they usually do not have to be concerned that the loyalties of the members will be tested by current decisions. Under such conditions, it is possible for leadership to become firmly institutionalized within the group without the particular leaders having to make any strong commitments to a specific set of principles or to a given political strategy.

Problems relating to the loyalty of the membership can generally be handled more effectively by decisions about intra-group relations than by decisions about the goals or external policies of the group. So long as harmonious relations exist within the group, it is generally possible for the leaders to make drastic changes in strategy. Indeed, it is not uncommon for the membership to feel that matters relating to external policy should be left solely to the leadership, and it may not disturb them that such decisions reflect mainly the idiosyncracies of their leaders.

5. Opposition parties and aspiring elites tend to appear as revolutionary movements in non-Western politics. Since the current leadership in non-Western countries generally conceives of itself as seeking to effect changes in all aspects of life, and since all the political associations tend to have world views, any prospective change in national leadership is likely to seem to have revolutionary implications. The fact that the ruling party in most non-Western countries identifies itself with an effort to bring about total change in the society makes it difficult to limit the sphere of political controversy. Issues are not likely to remain as isolated and specific questions but tend to become associated with fundamental questions about the destiny of the society.

In addition, the broad and diffuse interests of the ruling elites make it easy for them to maintain that they represent the interest of the entire nation. Those seeking power are thus often placed in the position of appearing to be, at best, obstructionists of progress and, at worst, enemies of the country. Competition is not between parties that represent different functional specific interests or between groups that claim greater administrative skills; rather, the struggle takes on some of the qualities of a conflict between differing ways of life.

This situation is important in explaining the failure of responsible opposition parties to develop in most non-Western countries. For example, the Congress Party in India has been able to identify itself with the destiny of the entire country to such a degree that the opposition parties find it difficult to avoid appearing either as enemies of India's progress or as groups seeking precisely the same objectives as Congress. Since the frustration of opposition groups encourages them to turn to extremist measures, they may in fact come to be revolutionary movements.

6. The non-Western political process is characterized by a lack of integration among the participants, and this situation is a function of the lack of a unified communications system in the society. In most non-Western societies there is not a single general political process that is the focus of most political activities throughout the population; rather, there are several distinct and nearly unrelated political processes. The most conspicuous division is that between the dominant national politics of the more urban elements and the more traditional village level of politics. The conflicts that are central to the one may hardly appear in the other.

Those who participate, for example, in the political life of the village are not an integral part of the national politics, since they can act without regard to developments at the central level. Possibly even more significant is the fact that at the village level all the various village groups have their separate and autonomous political processes.

This situation is a reflection of, and is reinforced by, the communication system common to non-Western societies, where the media of mass communication generally reach only to elements of the urban population and to those who participate in the national political process. The vast majority of the people participate only in the traditional word-of-mouth communication system. Even when the media of mass communications do reach the village, through readers of newspapers or owners of radios, there is almost no "feedback" from the village level. The radio talks *to* the villagers but does not talk *with* them. The views of the vast majority of the population are not reflected in the mass media. Indeed, it is often the case that the Westerner has less difficulty than the majority of the indigenous population in understanding the intellectual and moral standards reflected in the media of mass communication, not only because these media are controlled by the more Westernized elements but also because the media may be consciously seeking to relate themselves more to the standards of the international systems of communication than to the local scene.

The lack of a unified communication system and the fact that the participants are not integrated into a common political process limit the types of political issues that can arise in non-Western societies. For example, although these are essentially agrarian societies in which industrial development is just beginning to take place, there has not yet appeared (in their politics) one of the issues basic to the history of Western politics: the clash between industry and agriculture, between town and countryside. Questions of agriculture usually arise in politics when the urbanized leaders advance plans for increasing production and developing village life. The values and concepts of the rural element are not effectively represented in the national political process largely because its fragmented character and the lack of a unified communications system leave the rural elements without a basis for mobilizing their combined strength and effectively advancing their demands on the government. It is possible that in time the rural masses, discovering that they have much in common, will find ways to mobilize their interests and so exert their full potential influence on the nation's political life. Such a development would drastically alter the national political character. In the meantime, however, the fragmented political process of the non-Western societies means that fundamentally agrarian countries will continue to have a form of national politics that is more urbanized than that commonly found in the industrial West. In many cases one city alone dominates the politics of an entire country.

7. *The non-Western political process is characterized by a high rate of recruitment of new elements to political roles.*[3] The spread of popular politics in traditional societies has meant a constant increase in the number of participants and the types of organizations involved in the political process. This development has been stimulated by the extraordinary rise in the urban population, which has greatly

[3] Kahin, Pauker, and Pye, *loc. cit.*, p. 1024.

increased the number of people who have some understanding about, and feeling for, politics at the national level. A basic feature of the acculturation process which creates the sub-society of the elite is the development of attitudes common to urban life. It is generally out of the rapid urban growth that there emerge the aspiring elites who demand to be heard. In almost all non-Western societies, there is a distinct strata of urban dwellers who are excluded from direct participation in national politics but whose existence affects the behavior of the current elite.

The more gradual reaching out of the mass media to the countryside has stimulated a broadening awareness that, although participation in the nation's political life is formally open to all, the rural elements actually have little access to the means of influence. In some places political parties, in seeking to reach the less urbanized elements, have opened up new channels for communicating with the powerful at the nation's center which may or may not be more effective than the old channels of the civil administration. In any case, the existence of multiple channels of contact with the national government tends to increase the number of people anxious to participate in national decision-making.

8. *The non-Western political process is characterized by sharp differences in the political orientation of the generations.* The process of social change in most non-Western societies results in a lack of continuity in the circumstances under which people are recruited to politics. Those who took part in the revolutionary movement against a colonial ruler are not necessarily regarded as indispensable leaders by the new generations; but their revolutionary role is still put forward as sufficient reason for their continued elite status. As a result, in some countries, as in Indonesia and Burma, groups that were not involved in the revolution feel that they are now being arbitrarily excluded from the inner circle of national politics. For these people, the current elite is claiming its status on the basis of ascriptive rather than achievement considerations.

This problem in non-Western societies is further complicated by demographic factors, for such societies are composed of rapidly growing populations that have a high birth rate. In Singapore, Malaya, and Burma, over half the population is under voting age, and the median age in most non-Western countries is in the low twenties. There is thus a constant pressure from the younger generation, whose demands for political influence conflict with the claims of current leaders who conceive of themselves as being still young and with many more years of active life ahead. In most of the newly independent countries, the initial tendency was for cabinet ministers and high officials to be in their thirties and forties, a condition which has colored the career expectations of the youth of succeeding generations, who now face frustration if they cannot achieve comparable status at the same age.

This telescoping of the generations has sharpened the clash of views so that intellectually there is an abnormal gap in political orientations, creating a potential for extreme reversal in policy, should the aspiring elites gain power. Ideas and symbols which are deeply felt by the current leaders, including those relating to the West, may have little meaning for a generation which has not experienced colonial rule.

9. *In non-Western societies there is little consensus as to the legitimate ends and means of political action.* The fundamental fact that non-Western societies are engrossed in a process of discontinuous social change precludes the possibility of a widely shared agreement as to the appropriate ends and means of political activities. In all the important non-Western countries, there are people who have assimilated Western culture to the point that their attitudes and concepts about politics differ little from those common in the West. At the other extreme there is the village peasant who has been little touched by Western influences. Living in different worlds, these individuals can hardly be expected to display a common approach toward political action.

The national leadership, recruited from people who have generally become highly urbanized, is in a position to set the standards for what may appear to be a widely shared consensus about politics. However, more often than not, this apparent national agreement is a reflection only of the distinct qualities of the elite sub-society. The mass of the population cannot fully appreciate the values and concepts which underlie the judgments of the elite and which guide its behavior.

The lack of a distinct political sphere increases the difficulties in achieving agreement about the legitimate scope and forms of political activities. The setting is not one in which political issues are relatively isolated and thus easily communicated and discussed. Instead, a knowledge of national politics requires an intimate acquaintance with the total social life of the elite. The fact that loyalty to the particular group rather than support of general principles is the key to most political behavior strengthens the tendency toward a distinct and individual rather than a shared orientation towards politics.

The situation is further complicated by the fact that, since most of the groupings within the political process represent total ways of life, few are concerned with limited and specific interests. The functionally diffuse character of most groups means that each tends to have its own approaches to political action in terms of both ends and means. Under these circumstances, the relationship of means to ends tends to be more organic than rational and functional. Indeed, in the gross behavior of the groups it is difficult to distinguish their primary goals from their operational measures. Consequently, the political actors in non-Western societies tend to demonstrate quite conspicuously the often-forgotten fact that people generally show greater imagination and ingenuity in discovering goals to match their means than in expanding their capabilities in order to reach distant goals.

Given the character of the groups, it is difficult to distinguish within the general political discourse of the society a distinction between discussions of desired objectives and analyses of appropriate means of political action.

10. *In non-Western societies the intensity and breadth of political discussion has little relationship to political decision-making.* Western observers are impressed with what they feel is a paradoxical situation in most non-Western countries: the masses seem to be apathetic toward political action and yet, considering the crude systems of communications, they are remarkably well informed about political events. Peasants and villagers often engage in lengthy discussions on matters related to the political

world that lies outside their immediate lives, but they rarely seem prepared to translate the information they receive into action that might influence the course of national politics.

The villagers are often responding in the traditional manner to national politics. In most traditional societies, an important function of the elite was to provide entertainment and material for discussion for the common people; but discussions in villages and teashops could center on the activities of an official without creating the expectation that discussion should lead to action. Thus the contemporary world of elite politics provides a drama for the common people, just as in many traditional cultures the popular forms of literature and drama stressed court life and the world of officialdom.

A second explanation for this pattern of behavior is that one of the important factors in determining social status and prestige within the village or local community is often a command of information about the wider world; knowledge of developments in the sphere of national and even international politics has a value in itself. But skill in discussing political matters again does not raise any expectations of actual participation in the world of politics.

Finally, many of the common people in non-Western societies find it desirable to keep informed about political developments in order to be able to adapt their lives to any major changes. Since their lives have often been drastically disrupted by political events, they have come to believe it prudent to seek advance warning of any developments which might again affect their lives; but it has not necessarily encouraged them to believe that their actions might influence such developments.

11. In the non-Western political process there is a high degree of substitutability of roles.[4] It seems that in non-Western societies most politically relevant roles are not clearly differentiated but have a functionally diffuse rather than a functionally specific character. For example, the civil bureaucracy is not usually limited to the role of a politically neutral instrument of public administration but may assume some of the functions of a political party or act as an interest group. Sometimes armies act as governments. Even within bureaucracies and governments, individuals may be formally called upon to perform several roles.

A shortage of competent personnel encourages such behavior either because one group may feel that the other is not performing its role in an effective manner or because the few skilled administrators are forced to take on concurrent asssignments. However, the more fundamental reason for this phenomenon is that in societies just emerging from traditional status, it is not generally expected that any particular group or organization will limit itself to performing a clearly specified function. Under these conditions there usually are not sharply defined divisions of labor in any sphere of life. All groups tend to have considerable freedom in trying to maximize their general influence.

12. In the non-Western political process there are relatively few explicitly organized interest groups with functionally specific roles. Although there are often large numbers of informal associations in non-Western countries, such groups tend

[4] See Almond, *loc. cit.*, p. 30.

to adopt diffuse orientations that cover all phases of life in much the same manner as the political parties and cliques. It is the rare association that represents a limited and functionally specific interest. Organizations which in name and formal structure are modeled after Western interest groups, such as trade unions and chambers of commerce, generally do not have a clearly defined focus.

In many cases groups, such as trade unions and peasant associations that in form would appear to represent a specific interest, are in fact agents of the government or of a dominant party or movement. Their function is primarily to mobilize the support of a segment of the population for the purposes of the dominant group, and not primarily to represent the interests of their constituency.

In situations where the associations are autonomous, the tendency is for them to act as protective associations and not as pressure groups. That is, their activities are concentrated on protecting their members from the consequences of governmental decisions and the political power of others. They do not seek to apply pressure openly on the government in order to influence positively the formation of public policy.

This role of the protective association is generally a well developed one in traditional societies and in countries under colonial rule. Under such authoritarian circumstances, since informal associations could have little hope of affecting the formal law-making process, they tended to focus on the law-enforcing process. Success in obtaining preferential treatment for their membership did not require that they mobilize general popular support. On the contrary, activities directed to broadly articulating their views were generally self-defeating. They were likely to be more successful if they worked quietly and informally to establish preferential relations with the policy-enforcing agents of the government. Under such conditions each association generally preferred to operate separately in order to gain special favors. The strategy of uniting in coalitions and alliances to present the appearance of making a popular demand on the government, as is common in an open democratic political process composed of pressure groups, would only weaken the position of all the informal associations in a traditional society, for it would represent a direct challenge to the existing governmental elite.

This approach to political activity common in traditional societies still lingers on in many non-Western societies. Informal associations tend to protect all the interests of their members in relations with the government. At the same time, many interests in the society are not explicitly organized. Although the process of social change is creating the basis for new interests, the formation of explicit interest groups rarely moves at the same pace. Often the new groups turn to the more traditional informal associations and only very gradually change their character. In other cases interest groups that fundamentally represent the newly developing aspects of the society perform according to the standards of the traditional groups.

From this brief discussion we may note that when interest groups act as protective associations, focusing on the law-enforcing process and seeking special treatment for their members, they are likely to avoid articulating publicly their goals and are likely to base their requests for special favors on particularistic rather than universalistic considerations. In appealing to policy-enforcement agents, prudence dictates the desirability of framing a request as an isolated demand; for any suggestion that a request constitutes a widespread demand, consistent with the general interest or the public good, would threaten the preserve of the law-makers, who were presumed to be unapproachable in most traditional societies.

We may sum up these observations by formulating a general hypothesis that would read: "Whenever the formally constituted law-makers are more distant from and more inaccessible to the general public than the law-enforcing agencies, the political process of the society will be characterized by a high degree of latency, and interests will be represented by informally organized groups seeking diffuse but particularistically defined goals which will not be broadly articulated nor claimed to be in the general interest." The corollary of this hypothesis would, of course, read: "Whenever the formally constituted law-makers are less distant from and more accessible to the general public than the law-enforcing agencies, the political process of the society will be open and manifest, and interests will be represented by explicitly organized groups seeking specific but universalistically defined goals which will be broadly articulated and claimed to be in the general interest."

13. In the non-Western political process the national leadership must appeal to an undifferentiated public. The lack of explicitly organized interest groups and the fact that not all participants are continuously represented in the political process deprive the national leadership of any easily available means for calculating the relative distribution of attitudes and values throughout the society. The national politician cannot easily determine the relative power of those in favor of a particular measure and those opposed; he cannot readily estimate the amount of effort needed to gain the support of the doubtful elements.

It is usually only within the circle of the elite or within the administrative structure that the national leaders can distinguish specific points of view and the relative backing that each commands. In turning to the population as a whole, the leaders find that they have few guides as to how the public may be divided over particular issues. Thus, in seeking popular support, the politician cannot direct his appeal to the interests of particular groups. Unable to identify or intelligently discriminate among the various interests latent in the public, the political leader is inclined to resort to broad generalized statements rather than to adopt specific positions on concrete issues. This situation also means that, whether the question is one of national or of merely local import, the leadership must appear to be striving to mobilize the entire population.

The inability to speak to a differentiated public encourages a strong propensity toward skillful and highly emotional forms of political articulation on the part of non-Western leaders. Forced to reach for the broadest possible appeals, the political leader tends at times to concentrate heavily on nationalistic sentiments and to present himself as a representative of the nation as a whole rather than of particular interests within the society. This is one of the reasons why some leaders of non-Western countries are often seen paradoxically both as extreme nationalists and as men out of touch with the masses.

14. *The unstructured character of the non-Western political process encourages leaders to adopt more clearly defined positions on international issues than on domestic issues.* Confronted with an undifferentiated public, leaders of non-Western countries often find the international political process more clearly structured than the domestic political scene. Consequently, they can make more refined calculations as to the advantages in taking a definite position in world politics than they can in domestic politics. This situation not only encourages the leaders of some non-Western countries to seek a role in world politics that is out of proportion to their nation's power, but it also allows such leaders to concentrate more on international than on domestic affairs. It should also be noted that in adopting a supra-national role, the current leaders of non-Western countries can heighten the impression that their domestic opposition is an enemy of the national interest.

15. *In non-Western societies the affective or expressive aspect of politics tends to override the problem-solving or public-policy aspect of politics.* Traditional societies have generally developed to a very high order the affective and expressive aspects of politics. Pomp and ceremony are usually basic features of traditional politics, and those who are members of the ruling elite in such societies are generally expected to lead more interesting and exciting lives than those not involved in politics. In contrast, traditional societies have not usually emphasized politics as a means for solving social problems. Questions of policy in such societies are largely limited to providing certain minimum social and economic functions and maintaining the way of life of the elite.

Although in transitional societies there is generally a greater awareness of the potentialities of politics as a means of rationally solving social problems, the expressive aspects of politics usually continue to occupy a central place in determining the character of political behavior. The peculiar Western assumption that issues of public policy are the most important aspect of politics, and practically the only legitimate concern of those with power, is not always applicable to non-Western politics. Indeed, in most non-Western societies the general assumption is not that those with power are committed to searching out and solving problems, but rather that they are the fortunate participants in the central drama of life. Politics is supposed to be exciting and emotionally satisfying.

In part the stress on the affective or expressive aspect of politics is related to the fact that, in most non-Western countries, questions of personal loyalties and identification are recognized as providing the basic issues of politics and the bond between leader and follower is generally an emotional one. In fact, in many non-Western societies, it is considered highly improper and even immoral for people to make loyalty contingent upon their leaders' ability to solve problems of public policy.

In the many non-Western societies in which the problem of national integration is of central importance, the national leaders often feel they must emphasize the symbols and sentiments of national unity since substantive problems of policy may divide the people. It should be noted that the governmental power base of many non-Western leaders encourages them to employ symbols and slogans customarily associated with administrative policy in their efforts to strengthen national unity. The Western observer may assume that statements employing such symbols represent policy intentions when in fact their function is to create national loyalty and to condition the public to think more in policy terms.

16. *Charismatic leaders tend to prevail in non-Western politics.*[5] Max Weber, in highlighting the characteristics of charismatic authority, specifically related the emergence of charismatic personalities to situations in which the hold of tradition has been weakened. By implication, he suggested that societies experiencing cultural change provide an ideal setting for such leaders since a society in which there is confusion over values is more susceptible to a leader who conveys a sense of mission and appears to be God-sent.

The problem of political communication further reinforces the position of the charismatic leader. Since the population does not share the leadership's modes of reason or standards of judgment, it is difficult to communicate subtle points of view. Communication of emotions is not confronted with such barriers, especially if it is related to considerations of human character and personality. All groups within the population can feel confident of their abilities to judge the worth of a man for what he is, even though they cannot understand his mode of reasoning.

So long as a society has difficulties in communication, the charismatic leader possesses great advantage over his opponents, even though they may have greater ability in rational planning. However, the very lack of precision in the image that a charismatic leader casts, especially in relation to operational policy, does make it possible for opposition to develop as long as it does not directly challenge the leader's charisma. Various groups with different programs can claim that they are in fact seeking the same objectives as those of the leader. For example, in both Indonesia and Burma, the Communists have been able to make headway by simply claiming that they are not directly opposed to the goals of Sukarno and U Nu.

Charisma is likely to wear thin. A critical question in most non-Western societies that now have charismatic leaders is whether such leadership will in the meantime become institutionalized in the form of rational-legal practices. This was the pattern in Turkey under Kemal Ataturk. Or will the passing of the charismatic leader be followed by confusion and chaos? The critical factor seems to be whether or not the leader encourages the development of functionally specific groups within the society that can genuinely represent particular interests.

17. *The non-Western political process operates largely without benefit of political "brokers."* In most non-Western societies there seems to be no institutionalized role for carrying out the tasks of, first, clarifying and delimiting the distribution of demands and interests within the population, and, next, engaging in the bargaining operation necessary to accommodate and maximize the satisfaction of those demands and interests in a fashion consistent with the requirements of public policy and administration. In other words, there are no political "brokers."

In the Western view, the political broker is a prerequisite for a smoothly operating system of representative government. It is through his activities that, on the one

[5]Kahin, Pauker, and Pye, *loc. cit.*, p. 1025.

hand, the problems of public policy and administration can be best explained to the masses in a way that is clearly related to their various specific interests and, on the other hand, that the diverse demands of the population can be articulated to the national leaders. This role in the West is performed by the influential members of the competing political parties and interest groups.

What is needed in most non-Western countries in order to have stable representative institutions are people who can perform the role that local party leaders did in introducing the various immigrant communities into American public life. Those party leaders, in their fashion, were able to provide channels through which the immigrant communities felt they could learn where their interests lay in national politics and through which the national leaders could discover the social concerns of the new citizens.

In most non-Western societies, the role of the political "broker" has been partially filled by those who perform a "mediator's" role, which consists largely of transmitting the views of the elite to the masses. Such "mediators" are people sufficiently acculturated to the elite society to understand its views but who still have contacts with the more traditional masses. In performing their role, they engage essentially in a public relations operation for the elite, and only to a marginal degree do they communicate to the elite the views of the public. They do not find it essential to identify and articulate the values of their public. Generally, since their influence depends upon their relations with the national leadership, they have not sought to develop an autonomous basis of power or to identify themselves with particular segments of the population as must the political "broker." As a consequence, they have not acted in a fashion that would stimulate the emergence of functionally specific interest groups.

CHAPTER 2

Is There a Non-Western Political Process?

ALFRED DIAMANT

Professor Pye constructs a generalized model of the non-Western political process in such a manner that it implies the existence of a Western process which is fundamentally different from its non-Western counterpart. For example, when he says that in non-Western societies the political and social spheres are not sharply differentiated, I take this to mean that in the West these spheres *are* so differentiated.

Reprinted from Alfred Diamant, "Is There a Non-Western Political Process? Comments on Lucian W. Pye's 'The Non-Western Political Process,'" JOURNAL OF POLITICS, *XI, No. 1 (February, 1959), 123–27, by permission of the publisher.*

If in the non-West new elements are recruited to political roles at a high rate, and there is a high degree of substitutability of roles, then in the West that rate or degree is low. This apposition of West and non-West could be carried out with regard to all seventeen points of Professor Pye's model of the non-Western political process.

I fear that Professor Pye's non-Western model will contribute to the increasing tendency in comparative politics to segregate sharply the study, research and teaching of Western and non-Western political systems. Those devoting their efforts to the "new" world, that is to Asia and Africa, tend to minimize the relevance of the study of Western politics to an understanding of non-Western systems. Because I believe that this drifting apart of Western and non-Western comparative politics ought to be stopped, I want to emphasize the continuities and similarities between the two political processes. Rustow, in the essay in *World Politics* quoted by Professor Pye, warns that there might be great danger in establishing a strict dichotomy between Western and non-Western systems and in treating non-Western politics *de novo*, as a field of inquiry to which the storehouse of data and concepts of Western politics could contribute little, if anything at all.[1] I should like to expand this point and propose that we view the two as poles of a spectrum or continuum between which there are few abrupt changes. Let me give a few illustrations of this lack of a clear dichotomy. For example, Professor Pye states that in non-Western systems "any change in political identification generally requires a change in one's social and personal relationships." (P. 470.) I would suggest that we look into the social stresses created by changing one's political affiliation from "Democratic" to "Republican" in South Carolina or Mississippi for an almost identical phenomenon. Professor Pye continues this particular argument by describing how the young man of the village moves into the city, acquires a Western education, and becomes "associated with a political group that may in no way reflect the political views of his original community." How different is this really from the change which takes place in political affiliation and general social outlook when the New York tenement dweller becomes a New Jersey suburbanite? Finally, professor Pye describes how administrators or politicians from the center must relate themselves to the social structure of the community in which they operate if they want to succeed in their tasks. (P. 469.) How different is this from the localism that characterizes American politics, which barely tolerates the outsider in elective politics and even manages to punish appointive administrative or judicial officers who violate the communal mores? Do we still recall how Charleston, South Carolina, treated Judge Waites Waring after he killed the white primary by handing down his decision in *Rice* v. *Elmore?*

In a recent critique of the literature on the modern corporation, Professor Latham suggested that Messrs. C. Wright Mills, A. A. Berle and John K. Galbraith cease behaving as if the problems created by the giant corporation and the power of private government were entirely new and unprecedented. He maintained that a study of some medieval and modern political theorists, chiefly Gierke and Maitland, would have provided them with the necessary tools to deal with these problems. In the same

[1] Dankwart A. Rustow, "New Horizons for Comparative Politics," *World Politics*, IX (July, 1957), 530–49.

manner I would suggest that the problem of rapid change in non-Western societies attempting to Westernize and industrialize is neither new, nor unprecedented. The theoretical tools for handling it are available in the literature on the "social question" produced in Europe, and in a lesser degree in this country, during the last hundred years. I refer here chiefly to the writings of laymen, as well as clerics, both Catholic and Protestant, who dealt with the political, social and economic impact of the French Revolution, the industrial revolution and capitalism on the traditional societies of Europe. There are at least four topics which were treated exhaustively in that literature and which recur in recent studies of the Westernization of underdeveloped areas: (1) the destruction of the traditional family and the atomization of the individual in the mass state; (2) the separation of work from residence and the impact of this change on the basic social institutions, a topic treated, for example, at great length in the UNESCO study, *Cultural Patterns and Technical Change*;[3] (3) the problem of leadership by an intelligentsia, or "detribalized" intelligentsia; and (4) the centralization which followed the creation of the nation-state and of industrialization, and which placed in the hands of the "state" matters formerly carried out by intermediate groups or private social institutions, a principal theme in corporative theory. The shattering effects of these changes on social institutions, groups and individuals in Europe can serve as a guide for assessing the effects of a similar process on non-European traditional societies and for predicting behavior of members of such societies in transition.

I am convinced that comparative studies based on the European experience would yield considerably more useful results if we abandoned the Western ideal-type which is based on the British two-party system and the strong consensus of British society. One could more successfully derive this ideal type from what Gabriel Almond has called the continental political system with its several sub-cultures.[4] Non-Western political systems would become more comprehensible and less remote if we would use this continental type which is based on a multi-racial (multi-national) society, lacking in strong consensus, and where conspiracy is always below the surface of politics. In fact, Professor Pye points to the existence of what he calls "sub-societies" as the cause of the *Weltanschauung* character of non-Western parties, a characteristic they share with the parties of continental politics.

I would like to select five of the seventeen characteristics of Professor Pye's non-Western model and show how much light can be thrown on them by the examination of certain aspects of European political systems. I propose to group three of them together. No. 2: "Political parties in non-Western societies tend to take on a world view and represent a way of life." No. 5: "Opposition parties and aspiring elites tend to appear as revolutionary movements in non-Western politics." No. 7: "In non-Western societies there is little consensus as to legitimate ends and

---

[2] Earl Latham, *Political Theories of Monopoly Power* (College Park, Maryland, 1957), p. 10.
[3] Margaret Mead (ed.), *Cultural Patterns and Technical Change* (New York, 1955).
[4] Gabriel A. Almond, "Comparative Political Systems," *Journal of Politics*, XVIII (August 1956), 391–409; see also my "The Group Basis of Austrian Politics," *Journal of Central European Affairs*, XVIII (July, 1958), 134–155.

means of political action." These three could be used without change for a description of Austrian politics between the two World Wars, and even during the last decades of the Hapsburg Empire when a system of political parties began to emerge. Austria has long been divided into three *Lager* (camps) which have tried to fashion a way of life for their followers and were not content simply to guide them in their political expression.[5] I might add that even in Great Britain, which gave birth to the prototypes of Western political parties, the Labor Party began as a social movement and has remained one, though to a lesser degree, to the present day. The Labor Party, the trade unions, the consumer cooperatives and a host of other institutions tried to encompass the entire life of the industrial worker and infuse it with a socialist spirit. The weakness of consensus and the revolutionary character of opposition movements described in Nos. 5 and 7 is, of course, the reason for the *Weltanschauung* character of parties and elites, not only in Austria but in many other European countries. After all, how far removed was Belgrade from non-Western politics on the day in 1929 when debate in the Skupština was terminated by pistol shots?

The fourth characteristic I have selected is No. 4, according to which political loyalty in the non-Western system is governed more by a sense of identification with a concrete group than by identification with the professed policy goals of the group. If this characteristic were lacking in the Western system, why would the Western petty bourgeoisie and small landholder class vote for parties and leaders whose principal interests are the protection of big business and agriculture, often at the expense of the "small man" in the city and on the farm? If there were no such identification with concrete groups in the West, why would the new middle class, "salariat," vote for so-called conservative parties?

Finally, I would argue that characteristic No. 6 is stated entirely too dogmatically: "The non-Western political process is characterized by a lack of integration of the participants, and this situation is a function of the lack of a unified communications system in the society." It is quite true that if we juxtaposed India and Great Britain we would discern several distinct political processes in the former, and only one highly integrated political system in the latter country. But what if we take the United States instead of Great Britain where the operation of a federal system and the existence of a wide range of economic, social and political diversities have created a number of distinct political processes? Huey Long, for example, had total control over one political sub-system, but never managed to get comparable power on the national level. The same was true of Eugene Talmadge.

I believe that the attempt to define some of the characteristics common to the political life of countries undergoing vast social change would profit a great deal from the "ideal-type" approach to the study of comparative politics suggested by Francis X. Sutton and applied to the study of comparative administration by Fred

---

[5]This analysis of Austrian politics in terms of three major *Lager* follows Adam Wandruszka, Österreichs politische Struktur. Die Entwicklung der parteien und der politischen Bewegungen," *Geschichte der Republik Österreich*, ed. Heinrich Bendikt (Vienna, 1954), pp. 289–485.

Riggs. In the paper in which he established the two ideal-types, "Agraria" and "Industria," Riggs suggested that any administrative system could be placed on a point along a spectrum ranging from the Industria type to the Agraria type pattern and could then be compared with any other administrative system.[6] This approach will prevent the division, against which Rustow warned, into water-tight compartments labelled "Western" and "non-Western" comparative politics because it will stress the existence of a continuum rather than of two sharply differentiated systems.

[6] Fred W. Riggs, "Agraria and Industria—Toward a Typology of Comparative Administration," *Toward the Comparative Study of Public Administration*, ed. William J. Siffin (Bloomington, Indiana, 1957), pp. 23–116.

CHAPTER 3

The General Characteristics of Totalitarian Dictatorship

CARL J. FRIEDRICH AND ZBIGNIEW BRZEZINSKI

Totalitarian regimes are autocracies. When they are said to be tyrannies, despotisms, or absolutisms, the basic general nature of such regimes is being denounced, for all these words have a strongly pejorative flavor. When they call themselves "democracies," qualifying it by the adjective "popular," they are not contradicting these indictments, except in trying to suggest that they are good or at least praiseworthy. An inspection of the meaning the totalitarians attach to the term "popular democracy" reveals that they mean by it a species of autocracy. The leaders of the people, identified with the leaders of the ruling party, have the last word. Once they have decided and been acclaimed by a party gathering, their decision is final. Whether it be a rule, a judgment, or a measure or any other act of government, they are the *autokrator*, the ruler accountable only to himself. Totalitarian dictatorship, in a sense, is the adaptation of autocracy to twentieth-century industrial society. (19)

Thus, as far as this characteristic absence of accountability is concerned, totalitarian dictatorship resembles earlier forms of autocracy. But it is our contention in this volume that totalitarian dictatorship is historically an innovation (cf. 133; 389; 52) and *sui generis*. It is also our conclusion from all the facts available to us that fascist and communist totalitarian dictatorships are basically alike, or at any rate more

Reprinted by permission of the publishers from Carl J. Friedrich and Zbigniew Brzezinski, TOTALITARIAN DICTATORSHIP AND AUTOCRACY *(Cambridge, Mass.: Harvard University Press, copyright, 1956, 1965, by the President and Fellows of Harvard College), Chap 2. The numbers in parentheses refer to the bibliographical notes at the end of the chapter.*

nearly like each other than like any other system of government, including earlier forms of autocracy. These two theses are closely linked and must be examined together. They are also linked to a third, that totalitarian dictatorship as it actually developed was not intended by those who created it—Mussolini talked of it, though he meant something different—but resulted from the political situations in which the anticonstitutionalist and antidemocratic revolutionary movements and their leaders found themselves. Before we explore these propositions, one very widespread theory of totalitarianism needs consideration.

It is a theory that centers on the regime's efforts to remold and transform the human beings under its control in the image of its ideology. As such, it might be called an ideological or anthropological theory of totalitarianism. The theory holds that the "essence" of totalitarianism is to be seen in such a regime's total control of the everyday life of its citizens, of its control, more particularly, of their thoughts and attitudes as well as their activities. "The particular criterion of totalitarian rule is the creeping rape [*sic*] of man by the perversion of his thoughts and his social life," a leading exponent of this view has written. "Totalitarian rule," he added, "is the claim transformed into political action that the world and social life are changeable without limit." (44a) As compared with this "essence," it is asserted that organization and method are criteria of secondary importance. There are a number of serious objections to this theory. The first is purely pragmatic. For while it may be the intent of the totalitarians to achieve total control, it is certainly doomed to disappointment; no such control is actually achieved, even within the ranks of their party membership or cadres, let alone over the population at large. The specific procedures generated by this desire for total control, this "passion for unanimity" as we call it later in our analysis, are highly significant, have evolved over time, and have varied greatly at different stages. They have perhaps been carried farthest by the Chinese Communists in their methods of thought control, but they were also different under Stalin and under Lenin, under Hitler and under Mussolini. Apart from this pragmatic objection, however, there also arises a comparative historical one. For such ideologically motivated concern for the whole man, such intent upon total control, has been characteristic of other regimes in the past, notably theocratic ones such as the Puritans' or the Moslems'. It has also found expression in some of the most elevated philosophical systems, especially that of Plato who certainly in *The Republic, The Statesman,* and *The Laws* advocates total control in the interest of good order in the political community. This in turn has led to the profound and unfortunate misunderstanding of Plato as a totalitarian (284; 111a; 353); he was an authoritarian, favoring the autocracy of the wise. The misunderstanding has further occasioned the misinterpretation of certain forms of tyrannical rule in classical antiquity as "totalitarian," on the ground that in Sparta, for instance, "the life and activity of the entire population are continuously subject to a close regimentation by the state." (114) Finally, it would be necessary to describe the order of the medieval monastery as totalitarian; for it was certainly characterized by such a scheme of total control. Indeed, much "primitive" government also appears then to be totalitarian (223) because of its close control of all participants. What is really the specific difference, the innovation of the totalitarian regimes, is the organization and methods developed and employed with the aid of modern technical devices in an effort to resuscitate such total control in the service of an ideologically motivated movement, dedicated to the total destruction and reconstruction of a mass society. It seems therefore highly desirable to use the term "totalism" to distinguish the much more general phenomenon just sketched, as has recently been proposed by a careful analyst of the methods of Chinese thought control. (217; 314)

Totalitarian dictatorship then emerges as a system of rule for realizing totalist intentions under modern political and technical conditions, as a novel type of autocracy. (301) The declared intention of creating a "new man," according to numerous reports, has had significant results where the regime has lasted long enough, as in Russia. In the view of one leading authority, "the most appealing traits of the Russians—their naturalness and candor—have suffered most." He considers this a "profound and apparently permanent transformation," and an "astonishing" one. (238a) In short, the effort at total control, while not achieving such control, has highly significant human effects.

The fascist and communist systems evolved in response to a series of grave crises—they are forms of crisis government. Even so, there is no reason to conclude that the existing totalitarian systems will disappear as a result of internal evolution, though there can be no doubt that they are undergoing continuous changes. The two totalitarian governments that have perished thus far have done so as the result of wars with outside powers, but this does not mean that the Soviet Union, Communist China, or any of the others necessarily will become involved in war. We do not presuppose that totalitarian societies are fixed and static entities but, on the contrary, that they have undergone and continue to undergo a steady evolution, presumably involving both growth and deterioration. (209f)

But what about the origins? If it is evident that the regimes came into being because a totalitarian movement achieved dominance over a society and its government, where did the movement come from? The answer to this question remains highly controversial. A great many explanations have been attempted in terms of the various ingredients of these ideologies. Not only Marx and Engels, where the case seems obvious, but Hegel, Luther, and a great many others have come in for their share of blame. Yet none of these thinkers was, of course, a totalitarian at all, and each would have rejected these regimes, if any presumption like that were to be tested in terms of his thought. They were humanists and religious men of intense spirituality of the kind the totalitarians explicitly reject. In short, all such "explanations," while interesting in illuminating particular elements of the totalitarian ideologies, are based on serious invalidating distortions of historical facts. (182; 126; 145.1; 280) If we leave aside such ideological explanations (and they are linked of course to the "ideological" theory of totalitarian dictatorship as criticized above), we find several other unsatisfactory genetic theories.

The debate about the causes or origins of totalitarianism has run all the way from a primitive bad-man theory (46a) to the "moral crisis of our time" kind of argument. A detailed inspection of the available evidence suggests that virtually every one of the factors which has been offered by itself as an explanation of the origin of totalitarian dictatorship has played its role. For example, in the case of Germany, Hitler's moral and personal defects, weaknesses in the German constitutional tradition, certain traits involved in the German "national character," the Versailles Treaty and its aftermath, the economic crisis and the "contradictions" of an aging capitalism, the "threat" of communism, the decline of Christianity and of such other spiritual moorings as the belief in the reason and the reasonableness of man—all have played a role in the total configuration of factors contributing to the over-all result. As in the case of other broad developments in history, only a multiple-factor analysis will yield an adequate account. But at the present time, we cannot fully explain the rise of totalitarian dictatorship. All we can do is to explain it partially by identifying some of the antecedent and concomitant conditions. To repeat: totalitarian dictatorship is a new phenomenon; there has never been anything quite like it before.

The discarding of ideological explanations—highly objectionable to all totalitarians, to be sure—opens up an understanding of and insight into the basic similarity of totalitarian regimes, whether communist or fascist. They are, in terms of organization and procedures—that is to say, in terms of structure, institutions, and processes of rule—*basically alike*. What does this mean? In the first place, it means that they are *not wholly alike*. Popular and journalistic interpretation has oscillated between two extremes; some have said that the communist and fascist dictatorships are wholly alike, others that they are not at all alike. The latter view was the prevailing one during the popular-front days in Europe as well as in liberal circles in the United States. It was even more popular during the Second World War, especially among Allied propagandists. Besides, it was and is the official communist and fascist party line. It is only natural that these regimes, conceiving of themselves as bitter enemies, dedicated to the task of liquidating each other, should take the view that they have nothing in common. This has happened before in history. When the Protestants and Catholics were fighting during the religious wars of the sixteenth and seventeenth centuries, they very commonly denied to one another the name of "Christians," and each argued about the other that it was not a "true church." Actually, and in the perspective of time, both were indeed Christian churches.

The other view, that communist and fascist dictatorships are wholly alike, was during the cold war demonstrably favored in the United States and in Western Europe to an increasing extent. Yet they are demonstrably not wholly alike. For example, they differ in their acknowledged purposes and intentions. Everyone knows that the communists say they seek the world revolution of the proletariat, while the fascists proclaimed their determination to establish the imperial predominance of a particular nation or race, either over the world or over a region. The communist and fascist dictatorships differ also in their historical antecedents: the fascist movements arose in reaction to the communist challenge and offered themselves to a frightened middle class as saviors from the communist danger. The communist movements, on the other hand, presented themselves as the liberators of an oppressed people from an existing autocratic regime, at least in Russia and China. Both claims are not without foundation, and one could perhaps coordinate them by treating the totalitarian movements as consequences of the First World War. "The rise [of totalitarianism] has occurred in the sequel to the first world war and those catastrophes, political and economic, which accompanied it and the feeling of crisis linked thereto." (31a) As we shall have occasion to show in the chapters to follow, there are many other differences which do not allow us to speak of the communist and fascist totalitarian dictatorships as wholly alike, but which suggest that they are sufficiently alike to class them together and to contrast them not only with constitutional systems, but also with former types of autocracy.

Before we turn to these common features, however, there is another difference that used to be emphasized by many who wanted "to do business with Hitler" or who admired Mussolini and therefore argued that, far from being wholly like the Communist dictatorship, the fascist regimes really had to be seen as merely authoritarian forms of constitutional systems. It is indeed true that more of the institutions of the antecedent liberal and constitutional society survived in the Italian Fascist than in the Russian or Chinese Communist society. But this is due in part to the fact that no liberal constitutional society preceded Soviet or Chinese Communism. The promising period of the Duma came to naught as a result of the war and the disintegration of tsarism, while the Kerensky interlude was far too brief and too superficial to become meaningful for the future. Similarly in China, the Kuomintang failed to develop a working constitutional order, though various councils were set up; they merely provided a facade for a military dictatorship disrupted by a great deal of anarchical localism, epitomized in the rule of associated warlords. In the Soviet satellites, on the other hand, numerous survivals of a nontotalitarian past continue to function. In Poland, Czechoslovakia, Hungary, and Yugoslavia we find such institutions as universities, churches, and schools. It is likely that, were a communist dictatorship to be established in Great Britain or France, the situation would be similar, and here even more such institutions of the liberal era would continue to operate, for a considerable initial period at least. Precisely this argument has been advanced by such British radicals as Sidney and Beatrice Webb. The tendency of isolated fragments of the preceding state of society to survive has been a significant source of misinterpretation of the fascist totalitarian society, especially in the case of Italy. In the twenties, Italian totalitarianism was very commonly misinterpreted as being "merely" an authoritarian form of middle-class rule, with the trains running on time and the beggars off the streets. (27) In the case of Germany, this sort of misinterpretation took a slightly different form. In the thirties, various writers tried to interpret German totalitarianism either as "the end phase of capitalism" or as "militarist imperialism." (263a) These interpretations stress the continuance of a "capitalist" economy whose leaders are represented as dominating the regime. The facts as we know them do not correspond to this view (see Part V). For one who sympathized with socialism or communism, it was very tempting to depict the totalitarian dictatorship of Hitler as nothing but a capitalist society and therefore totally at variance with the "new civilization" that was arising in the Soviet Union. These few remarks have suggested, it is hoped, why it may be wrong to consider the totalitarian dictatorships under discussion as either wholly alike or basically different. Why they are basically alike remains to be shown, and to this key argument we now turn.

The basic features or traits that we suggest as generally recognized to be common to totalitarian dictatorships are six in number. The "syndrome," or pattern of interrelated traits, of the totalitarian dictatorship consists of an ideology, a single party typically led by one man, a terroristic police, a communications monopoly, a weapons monopoly, and a centrally directed economy. Of these, the last two are also found in constitutional systems: Socialist Britain had a centrally directed economy, and all modern states possess a weapons monopoly. Whether these latter suggest a "trend" toward totalitarianism is a question that will be discussed in our last chapter. These six basic features, which we think constitute the distinctive pattern or model of totalitarian dictatorship, form a cluster of traits, intertwined and mutually supporting each other, as is usual in "organic" systems. They should therefore not be considered in isolation or be made the focal point of comparisons, such as "Caesar developed a terroristic secret police, therefore he was the first totalitarian dictator," or "the Catholic Church has practiced ideological thought control, therefore . . ."

The totalitarian dictatorships all possess the following:

1. An elaborate ideology, consisting of an official body of doctrine covering all vital aspects of man's existence to which everyone living in that society is supposed to adhere, at least passively; this ideology is characteristically focused and projected toward a perfect final state of mankind—that is to say, it contains a chiliastic claim, based upon a radical rejection of the existing society with conquest of the world for the new one.

2. A single mass party typically led by one man, the "dictator," and consisting of a relatively small percentage of the total population (up to 10 per cent) of men and women, a hard core of them passionately and unquestioningly dedicated to the ideology and prepared to assist in every way in promoting its general acceptance, such a party being hierarchically, oligarchically organized and typically either superior to, or completely intertwined with, the governmental bureaucracy.

3. A system of terror, whether physical or psychic, effected through party and secret-police control, supporting but also supervising the party for its leaders, and characteristically directed not only against demonstrable "enemies" of the regime, but against more or less arbitrarily selected classes of the population; the terror whether of the secret police or of party-directed social pressure systematically exploits modern science, and more especially scientific psychology.

4. A technologically conditioned, near-complete monopoly of control, in the hands of the party and of the government, of all means of effective mass communication, such as the press, radio, and motion pictures.

5. A similarly technologically conditioned, near-complete monopoly of the effective use of all weapons of armed combat.

6. A central control and direction of the entire economy through the bureaucratic coordination of formerly independent corporate entities, typically including most other associations and group activities.

The enumeration of these six traits or trait clusters is not meant to suggest that there might not be others, now insufficiently recognized. It has more particularly been suggested that the administrative control of justice and the courts is a distinctive trait (see Chapter 10); but actually the evolution of totalitarianism in recent years suggests that such administrative direction of judicial work may be greatly limited. We shall also discuss the problem of expansionism, which has been urged as a characteristic trait of totalitarianism. The traits here outlined have been generally acknowledged as the features of totalitarian dictatorship, to which the writings of students of the most varied backgrounds, including totalitarian writers, bear witness.

Within this broad pattern of similarities, there are many significant variations to which the analysis of this book will give detailed attention. To offer a few random illustrations: at present the party plays a much greater role in the Soviet Union than it did under Stalin; the ideology of the Soviet Union is more specifically committed to certain assumptions, because of its Marx-Engels bible, than that of Italian or German fascism, where ideology was formulated by the leader of the party himself; the corporate entities of the fascist economy remained in private hands, as far as property claims are concerned, whereas they become public property in the Soviet Union.

Let us now turn to our first point, namely, that totalitarian regimes are historically novel; that is to say, that no government like totalitarian dictatorship has ever before existed, even though it bears a resemblance to autocracies of the past. It may be interesting to consider briefly some data which show that the six traits we have just identified are to a large extent lacking in historically known autocratic regimes. Neither the oriental despotisms of the more remote past nor the absolute monarchies of modern Europe, neither the tyrannies of the ancient Greek cities nor the Roman empire, neither yet the tyrannies of the city-states of the Italian Renaissance and the Bonapartist military dictatorship nor the other functional dictatorships of this or the last century exhibit this design, this combination of features, though they may possess one or another of its characteristic traits. For example, efforts have often been made to organize some kind of secret police, but they have not even been horse-and-buggy affairs compared with the terror of the Gestapo or the OGPU (afterwards MVD, then KGB). Similarly, though there have been both military and propagandistic concentrations of power and control, the limits of technology have prevented the achievement of effective monopoly. Again, certainly neither the Roman emperor nor the absolute monarch of the eighteenth century sought or needed a party to support him or an ideology in the modern party sense, and the same is true of oriental despots. (389c) The tyrants of Greece and Italy may have had a party—that of the Medicis in Florence was called *lo stato*—but they had no ideology to speak of. And, of course, all of these autocratic regimes were far removed from the distinctive features that are rooted in modern technology.

In much of the foregoing, modern technology is mentioned as a significant condition for the invention of the totalitarian model. This aspect of totalitarianism is particularly striking in the field of weapons and communications, but it is involved also in secret-police terror, depending as it does upon technically advanced possibilities of supervision and control of the movement of persons. In addition, the centrally directed economy presupposes the reporting, cataloging, and calculating devices provided by modern technology. In short, four of the six traits are technologically conditioned. To envisage what this technological advance means in terms of political control, one has only to think of the weapons field. The Constitution of the United States guarantees to every citizen the right to bear arms (fourth amendment). In the days of the Minutemen, this was a very important right, and the freedom of the citizen was indeed symbolized by the gun over the hearth, as it is in Switzerland to this day. But who can "bear" such arms as a tank, a bomber, or a flamethrower, let alone an atom bomb? The citizen as an individual, and indeed in larger groups, is simply defenseless against the overwhelming technological superiority of those who can centralize in their hands the means with which to wield modern weapons and thereby physically to coerce the mass of the citizenry. Similar observations apply to the telephone and telegraph, the press, radio and television, and so forth. "Freedom" does not have the same potential it had a hundred and fifty years ago, resting as it then did upon individual effort. With few exceptions, the trend of technological advance implies the trend toward greater and greater size of organization. In the perspective of these four traits, therefore, totalitarian societies appear to be merely exaggerations, but nonetheless logical exaggerations, of the technological state of modern society.

Neither ideology nor party has as significant a relation to the state of technology. There is, of course, some connection, since the mass conversion continually attempted by totalitarian propaganda through effective use of the communication monopoly could not be carried through without it. It may here be observed that the Chinese Communists, lacking the means for mass communication, fell back upon the small group effort of word-of-mouth indoctrination, which incidentally offered a chance for substituting such groups for the family and transferring the filial tradition to them. (346a) Indeed, this process is seen by them as a key feature of their people's democracy.

Ideology and party are conditioned by modern democracy. Totalitarianism's own leaders see it as democracy's fulfillment, as the true democracy, replacing the plutocratic democracy of the bourgeoisie. From a more detached viewpoint, it appears to be an absolute, and hence autocratic, kind of democracy as contrasted with constitutional democracy. (346b) It can therefore grow out of the latter by perverting it. (30) Not only did Hitler, Mussolini, and Lenin* build typical parties within a constitutional, if not a democratic, context, but the connection is plain between the stress on ideology and the role that platforms and other types of ideological goal-formation play in democratic parties. To be sure, totalitarian parties developed a pronounced authoritarian pattern while organizing themselves into effective revolutionary instruments of action; but, at the same time, the leaders, beginning with Marx and Engels, saw themselves as constituting the vanguard of the democratic movement of their day, and Stalin always talked of the Soviet totalitarian society as the "perfect democracy"; Hitler and Mussolini (347) made similar

*Lenin's Bolshevik Party was quite different in actuality from the monolithic autocratic pattern that he outlined in *What Is To Be Done?* (205e)

statements. Both the world brotherhood of the proletariat and the folk community were conceived of as supplanting the class divisions of past societies by a complete harmony—the classless society of socialist tradition.

Not only the party but also its ideology harken back to the democratic context within which the totalitarian movements arose. Ideology generally, but more especially totalitarian ideology, involves a high degree of convictional certainty. As has been indicated, totalitarian ideology consists of an official doctrine that radically rejects the existing society in terms of a chiliastic proposal for a new one. It contains strongly utopian elements, some kind of notion of a paradise on earth. This utopian and chiliastic outlook of totalitarian ideologies gives them a pseudo-religious quality. In fact, they often elicit in their less critical followers a depth of conviction and a fervor of devotion usually found only among persons inspired by a transcendent faith. Whether these aspects of totalitarian ideologies bear some sort of relationship to the religions that they seek to replace is arguable. Marx denounced religion as the opium of the people. It would seem that this is rather an appropriate way of describing totalitarian ideologies. In place of the more or less sane platforms of regular political parties, critical of the existing state of affairs in a limited way, totalitarian ideologies are perversions of such programs. They substitute faith for reason, magic exhortation for knowledge and criticism. And yet it must be recognized that there are enough of these same elements in the operations of democratic parties to attest to the relation between them and their perverted descendants, the totalitarian movements. That is why these movements must be seen and analyzed in their relationship to the democracy they seek to supplant.

At this point, the problem of consensus deserves brief discussion. There has been a good deal of argument over the growth of consensus, especially in the Soviet Union, and in this connection psychoanalytic notions have been put forward. The ideology is said to have been "internalized," for example—that is to say, many people inside the party and out have become so accustomed to think, speak, and act in terms of the prevailing ideology that they are no longer aware of it. Whether one accepts such notions or not, there can be little doubt that a substantial measure of consensus has developed. Such consensus provides a basis for different procedures from what must be applied to a largely hostile population. These procedures were the core of Khrushchev's popularism, as it has been called, by which the lower cadres and members at large of the party were activated and the people's (mass) participation solicited. By such procedures, also employed on a large scale in Communist China, these communist regimes have come to resemble the fascist ones more closely; both in Italy and Germany the broad national consensus enabled the leadership to envisage the party cadres in a "capillary" function (see Chapter 4). As was pointed out in the last chapter, such consensus and the procedures it makes possible ought not to be confused with those of representative government. When Khrushchev and Mao talk about cooperation, one is reminded of the old definition aptly applied to a rather autocratic dean at a leading Eastern university: I operate and you coo. There is a good deal of consensual cooing in Soviet Russia and Communist China, there can be no doubt. That such cooing at times begins to resemble a growl, one suspects from some of the comments in Russian and Chinese sources. There is here, as in other totalitarian spheres, a certain amount of oscillation, of ups and downs that they themselves like to minimize in terms of "contradictions" that are becoming "nonantagonistic" and that are superseded in "dialectical reversals."

In summary, these regimes could have arisen only within the context of mass democracy and modern technology.

BIBLIOGRAPHICAL REFERENCES

19. Berman, Harold J., *Justice in the USSR—An Interpretation of the Soviet Law*, Cambridge, 1950; 2nd ed. 1963.
27. Borgese, G. A., *Goliath, The March of Fascism*, New York, 1937, pp. 271–344. For a bitter criticism of this tendency, see Emil Ludwig, *Mussolini*, Berlin, 1932, p. 231.
30. Bracher, Karl Dietrich, *Die Aufloesung der Weimarer Republik*, Stuttgart, 1955; 2nd ed. 1957.
31. Bracher, Karl Dietrich, and Wolfgang Sauer and Gerhard Schulz, *Die Nationalsozialistische Machtergreifung—Studien zur Errichtung des totalitaeren Machtsystems in Deutschland, 1933–34*, Cologne, 1960. (a) pp. 4–5, where Bracher accepts our view. See also comments by Schulz, pp. 371ff.
44. Buchheim, H., *Totalitäre Herrschaft—Wesen und Merkmale*, Munich, 1962. (a) pp. 14, 24.
46. Bullock, Alan, *Hitler—A Study in Tyranny*, London, 1952. (a) *passim*.
52. Catlin, G. E. G., *Systematic Politics*, Toronto, 1962.
111. Friedrich, C. J., *Transcendent Justice—The Religious Dimension of Constitutionalism*, Durham, 1964. (a) *passim*.
114. Fritz, Kurt von, "Totalitarismus und Demokratie im alten Griechenland und Rom," *Antike und Abendland*, 3:47–74 (1948).
126. Glum, Friedrich, *Philosophen im Spiegel und Zerrspiegel—Deutschlands Weg in den Nationalsozialismus*, Munich, 1954. For Fichte, see pp. 79–103; for Hegel, pp. 104–133.
133. Hallgarten, George W. F., *Why Dictators? The Causes and Forms of Tyrannical Rule since 600 B.C.*, New York, 1954. This study, in contrast to ours, explores what totalitarian dictatorship has in common with former "tyrannies."
145.1. Hersch, Jeanne, *Idéologies et réalité*, Paris, 1956.
182. Kolnay, Aurel, *The War Against the West*, New York, 1938.
205. Lenin, V. I., *Selected Works*, 12 vols., New York, 1943. (e) "What Is To Be Done?" II, 152.
209. Leonhard, Wolfgang, *The Kremlin since Stalin*, New York, 1962 (Ger. ed., 1959). (f) pp. 504ff. He here puts it in terms of modernization rather than liberalization.
217. Lifton, Jay, *Thought Reform and the Psychology of Totalism: A Study of Brainwashing in China*, New York, 1961.
223. Machiewicz, J., *The Katyn Wood Murders*, London, 1951.
238. Mehnert, Klaus, *Soviet Man in His World*, New York, 1962 (Ger. ed., Stuttgart, 1958). (a) p. 35.
263. Neumann, Franz, *Behemoth*, New York, 1942. (a) see also Maxine Y. Sweezy, *The Structure of the Nazi Economy*, Cambridge, Mass., 1941, and R. A. Brady,

The Spirit and Structure of Fascism, New York, 1937. Neumann's analysis is much the ablest of the three. The "imperialist" interpretation ties in with Thorstein Veblen's earlier analysis of German and Japanese militarism and imperialism.

280. Plessner, Hellmuth, *Die Verspätete Nation*, Stuttgart, 1959.
284. Popper, Karl R., *The Open Society and Its Enemies*, London, 1945, vol. 1, perhaps the most outspoken of the Platonic critics. This aspect of Plato's philosophy is the genuine link with the views of the totalitarians of our time. It has given rise to a heated controversy over whether Plato was or was not a totalitarian. In terms of our criteria, he clearly was not. See also ref. 353.
301. Rostow, W. W., *The Stages of Economic Growth: A Non-Communist manifesto*, Cambridge, Eng., 1960.
314. Schein, Edgar H., with Inge Schneier and Ithiel de Sola Pool, *Coercive Persuasion: A Sociological Analysis of the Brainwashing of American Civilians by the Chinese Communists*, New York, 1961.
346. Tang, Peter S. H., *Communist China Today*, vol. 1, Washington, D.C., 1957; 2nd ed. 1961. (a) pp. 87ff. (b) ch. 10; figs. on p. 462.
347. Tasca, A., *Nascita e avvento del Fascismo*, Rome, 1950. Mussolini's attitude toward democracy was ambivalent. Fascist theory was much more frankly elitist than Nazi ideology.
353. Thorson, Thomas L., *Plato, Totalitarian or Democrat?* New York, 1963. This author offers a convenient collection of writings, pro and con, on the theme of Plato's alleged totalitarianism, but the alternative suggested by the title is a false one.
389. Wittfogel, Karl, *Oriental Despotism*, New Haven, 1957. (c) He would identify what he calls hydraulic despotism with modern totalitarianism, but the argument is really about general "totalism," as discussed.

CHAPTER 4

Comparative Political Systems

GABRIEL A. ALMOND

What I propose to do in this brief paper is to suggest how the application of certain sociological and anthropological concepts may facilitate systematic comparison among the major types of political systems operative in the world today.

At the risk of saying the obvious, I am not suggesting to my colleagues in the field

Reprinted from Gabriel A. Almond, "Comparative Political Systems," JOURNAL OF POLITICS, *XVIII, No. 3 (August, 1956), 391–409, by permission of the publisher. Prepared for the Conference on the Comparative Method in the Study of Politics, held under the auspices of the Committee on Comparative Politics, Social Science Research Council at Princeton, June 2–4, 1955.*

of comparative government that social theory is a conceptual cure-all for the ailments of the discipline. There are many ways of laboring in the vineyard of the Lord, and I am quite prepared to concede that there are more musical forms of psalmody than sociological jargon. I suppose the test of the sociological approach that is discussed here is whether or not it enables us to solve certain persistent problems in the field more effectively than we now are able to solve them.

Our expectations of the field of comparative government have changed in at least two ways in the last decades. In the first place as American interests have broadened to include literally the whole world, our course offerings have expanded to include the many areas outside of Western Europe—Asia, the Middle East, Africa, and Latin America. Secondly, as our international interests have expanded and become more urgent, our requirements in knowledge have become more exacting. We can no longer view political crises in France with detached curiosity or view countries such as Indo-China and Indonesia as interesting political pathologies. We are led to extend our discipline and intensify it simultaneously.

It would simply be untrue to say that the discipline of comparative government has not begun to meet both of these challenges. As rapidly as it has been possible to train the personnel, new areas have been opened up to teaching and research; and there has been substantial encouragement to those who have been tempted to explore new aspects of the political process both here and abroad and to employ new methods in such research. It is precisely because of the eagerness and energy with which these challenges have been met that the field is now confronted with the problem of systematic cumulation and comparison. What appears to be required in view of the rapid expansion of the field are more comparative efforts in the tradition of Finer and Friedrich, if we are to gain the maximum in insight and knowledge from this large-scale research effort.

The problem to which this paper is a tentative and provisional answer is the following. With the proliferation of courses and special studies of specific "governments" and groupings of governments on an area or other bases, is it possible to set up and justify a preliminary classification into which most of the political systems which we study today can be assigned? The classifications which we now employ are particularistic (e.g., American Government, British Government, the Soviet Union, and the like); regional (e.g., Government and Politics of the Far East, Latin America, and the like); or political (e.g., the British Commonwealth, Colonial Government, and the like); or functional (e.g., the comprehensive comparative efforts limited to the European-American area, such as Finer and Friedrich, and the specific institutional comparisons such as comparative parties, and comparative administration).

Anyone concerned with this general problem of classification of political systems will find that all of the existing bases of classification leave something to be desired. Dealing with governments particularistically is no classification at all. A regional classification is based not on the properties of the political systems, but on their contiguity in space. The existing structural classifications, such as democracy-dictatorship, parliamentary-presidential systems, two-party and multi-party systems, often turn out to miss the point, particularly when they are used in the strikingly different political systems of the pre-industrial areas. There may be a certain use therefore in exploring the possibilities of other ways of classifying political systems. What is proposed here is just one of these ways, and because of the uneven state of our knowledge is necessarily crude and provisional.

In my own efforts to stand far off, so to speak, and make the grossest discriminations between types of empirical political systems operative in the world today, I have found a fourfold classification to be most useful: the Anglo-American (including some members of the Commonwealth), the Continental European (exclusive of the Scandinavian and Low Countries, which combine some of the features of the Continental European and the Anglo-American), the pre-industrial, or partially industrial, political systems outside the European-American area, and the totalitarian political systems. This classification will not include all the political systems in existence today, but it comes close to doing so. It will serve the purpose of today's discussion, which is not that of testing the inclusiveness of this classification but rather the usefulness of sociological concepts in bringing out the essential differences between these political systems.

The terms which I shall use in discriminating the essential properties of these classes have emerged out of the Weber-Parsons tradition in social theory.[1] I shall try to suggest why I find some of these concepts useful. First, a political system is a system of *action*. What this means is that the student of political systems is concerned with empirically observable behavior. He is concerned with norms or institutions in so far as they affect behavior. Emphasizing "action" merely means that the description of a political system can never be satisfied by a simple description of its legal or ethical norms. In other words, political institutions or persons performing political rôles are viewed in terms of what it is that they do, why they do it, and how what they do is related to and affects what others do. The term *system*[2] satisfies the need for an inclusive concept which covers all of the patterned actions relevant to the making of political decisions. Most political scientists use the term *political process* for these purposes. The difficulty with the term *process* is that it means any patterning of action through time. In contrast to *process*, the concept of *system* implies a *totality* of relevant units, an interdependence between the interactions of units, and a certain stability in the interaction of these units (perhaps best described as a changing equilibrium).

The unit of the political system is the rôle. The rôle, according to Parsons and Shils, ". . . is that organized sector of an actor's orientation which constitutes and defines his participation in an interactive process."[3] It involves a set of complementary expectations concerning his own actions and those of others with whom

---

[1] See in particular Max Weber, *The Theory of Social And Economic Organization*, trans. by A. M. Henderson and Talcott Parsons (New York: Oxford University Press, 1947), pp. 87 ff.

[2] See David Easton, *The Political System: An Inquiry into the State of Political Science* (New York: Alfred Knopf, 1953), pp. 90 ff.

[3] Talcott Parsons and Edward A. Shils, eds., *Toward a General Theory of Action* (Cambridge: Harvard University Press, 1951), p. 23.

he interacts. Thus a political system may be defined as a set of interacting rôles, or as a structure of rôles, if we understand by *structure* a patterning of interactions. The advantage of the concept of *rôle* as compared with such terms as *institutions, organizations,* or *groups,* is that it is a more inclusive and more open concept. It can include formal offices, informal offices, families, electorates, mobs, casual as well as persistent groupings, and the like, in so far as they enter into and affect the political system. The use of other concepts such as those indicated above involves ambiguity, forced definitions (such as groups), or residual categories. Like the concept of system it does not prejudice our choice of units but rather enables us to nominate them on the basis of empirical investigation.

While there appear to be certain advantages in these concepts of political system and rôle for our purposes, they confront the political scientist with a serious problem. While he intends the concept to have a general application, Parsons appears to have had before him in elaborating the concept the model of the primary group—family, friendship, and the like—and not complex social systems, the units of which are collectivities and not individual actors. In this sense the sociological concept of system and of rôle can only be a beginning of a conceptual model of the political system. The job of developing additional concepts necessary to handle macrocosmic social systems such as political systems—national and international—is still to be done.

My own conception of the distinguishing properties of the political system proceeds from Weber's definition—the legitimate monopoly of physical coercion over a given territory and population.[4] The political systems with which most political scientists concern themselves all are characterized by a specialized apparatus which possesses this legitimate monopoly, and the political system consists of those interacting rôles which affect its employment. There are, of course, simpler societies in which this function of maintenance of order through coercion is diffuse and unspecialized; it is combined with other functions in the family and other groupings. While these systems are also properly the subject matter of political science, there are few political scientists indeed with the specialized equipment necessary to study them.

It may be useful to add a few comments about this definition of politics and the political in order to avoid misunderstanding. To define politics as having this distinguishing property of monopolizing legitimate coercion in a given territory is not the same thing as saying that this is *all* that government does. It is the thing that government does and that other social systems ordinarily may not do legitimately. Other social systems may employ other forms of compulsion than physical coercion. Some indeed may legitimately employ physical coercion on a limited scale. But the employment of *ultimate, comprehensive,* and *legitimate* physical coercion is the monopoly of states, and the political system is uniquely concerned with the scope, the direction, and the conditions affecting the employment of this physical coercion. It is, of course, clear that political systems protect freedoms and provide welfare, as well

[4] From *Max Weber: Essays in Sociology*, trans. by H. H. Gerth and C. Wright Mills (New York: Oxford University Press, 1946), p. 78.

as impose order backed up by physical compulsion, but even their protection of freedom and their provision of welfare is characteristically backed up by the threat of physical compulsion. Hence it seems appropriate to define the political system as the patterned interaction of rôles affecting decisions backed up by the threat of physical compulsion.

The task of describing a political system consists in characterizing all the patterned interactions which take place within it. It takes us beyond the legal system into all the rôles which occur and involves our defining these rôles in action or behavioral terms. The concept of system implies that these rôles are interdependent and that a significant change in any one rôle affects changes in the others, and thereby changes the system as a whole. Thus the emergence of pressure groups in the present century produced certain changes in the party system and in the administrative and legislative processes. The rapid expansion of executive bureaucracy was one of the factors that triggered off the development of legislative bureaucracy and pressure group bureaucracy. Changes in the rôle of political communication have transformed the electoral process, the behavior of parties, the legislature, the executive. The concepts of system and of interdependence lead us to look for these changes when any specific rôle changes significantly. It suggests the usefulness of thinking at the level of the system and its interdependence rather than in terms of discrete phenomena or only limited bilateral relationships, or relationships occurring only within the formal-legal rôle structure.

The fourth concept is *orientation to political action*. Every political system is embedded in a set of meanings and purposes. We speak of "attitudes toward politics," "political values," "ideologies," "national character," "cultural ethos." The difficulty with all these terms is that their meanings are diffuse and ambiguous. The concepts of orientation to action and of the pattern variables are useful since they at least attempt logical distinctness and comprehensiveness. It is not essential for my purposes to go into the modes of orientation of action, or into the "pattern variables" in detail. Parsons and Shils tell us that any orientation to politics involves three components: the first is perception, or *cognition*; the second is preference, involvement, or affect (*cathexis*); the third is evaluation or choice through the application of standards or values to the cognitive and affective components. By *cognition* is meant the knowledge and discrimination of the objects, events, actions, issues, and the like. By *cathexis* is meant the investment of objects, issues, etc., with emotional significance, or affect. By *evaluation* is meant the manner in which individuals organize and select their perceptions, preferences, and values in the process of establishing a position *vis-à-vis political action*.[5]

Every political system is embedded in a particular pattern of orientations to political action. I have found it useful to refer to this as the *political culture*. There are two points to be made regarding the concept of political culture. First, it does not coincide with a given political system or society. Patterns of orientation to politics may, and usually do, extend beyond the boundaries of political systems. The second point is that the political culture is not the same thing as the general culture,

[5] Parsons and Shils, *op. cit.*, pp. 58 ff.

although it is related to it. Because political orientation involves cognition, intellection, and adaptation to external situations, as well as the standards and values of the general culture, it is a differentiated part of the culture and has a certain autonomy. Indeed, it is the failure to give proper weight to the cognitive and evaluative factors, and to the consequent autonomy of political culture, that has been responsible for the exaggerations and over-simplifications of the "national character" literature of recent years.

The usefulness of the concept of political culture and its meaning may perhaps be conveyed more effectively through illustration. I would argue that the United States, England, and several of the Commonwealth countries have a common political culture, but are separate and different kinds of political systems. And I would argue that the typical countries of continental Western Europe, while constituting individual political systems, include several different political cultures which extend beyond their borders. In other words, they are political systems with fragmented political cultures.

In an effort to overcome understandable resistances to the introduction of a new term, I should like to suggest why I find the concept of political culture more useful than the terms we now employ, such as *ideology* or *political party*. As I understand the term *ideology*, it means the systematic and explicit formulation of a general orientation to politics. We need this term to describe such political phenomena as these and should not reduce its specificity by broadening it to include not only the explicit doctrinal structure characteristically borne by a minority of *militants*, but also the vaguer and more implicit orientations which generally characterize political followings. The term *political party* also cannot serve our purpose, for we are here dealing with a formal organization which may or may not be a manifestation of a political culture. Indeed, we will be gravely misled if we try to force the concept of party to mean political culture. Thus the commonly used distinctions between one-party, two-party, and multi-party systems simply get nowhere in distinguishing the essential properties of the totalitarian, the Anglo-American, and the Continental European political systems. For the structure we call *party* in the totalitarian system is not a party at all; the two parties of the Anglo-American system are organized manifestations of a homogeneous political culture; and the multi-parties of Continental European political systems in some cases are and in some cases are not the organized manifestations of different political cultures.

But the actual test of the usefulness of this conceptual scheme can only come from a more detailed application of it in developing the special properties of the classes of political systems to which we earlier referred.

THE ANGLO-AMERICAN POLITICAL SYSTEMS

The Anglo-American political systems are characterized by a *homogeneous, secular* political culture. By a secular political culture I mean a multi-valued political culture, a rational-calculating, bargaining, and experimental political culture. It is a homogeneous culture in the sense that there is a sharing of political ends and means. The great majority of the actors in the political system accept as the ultimate goals of the political system some combination of the values of freedom, mass welfare, and security. There are groups which stress one value at the expense of the others; there are times when one value is stressed by all groups; but by and large the tendency is for all these values to be shared, and for no one of them to be completely repressed. To a Continental European this kind of political culture often looks sloppy. It has no logic, no clarity. This is probably correct in an intellectual sense, since this balancing of competing values occurs below the surface among most people and is not explicated in any very elegant way. Actually the logic is complex and is constantly referred to reality in an inductive process. It avoids the kind of logical simplism which characterizes much of the Continental European ideological polemic.

A secularized political system involves an individuation of and a measure of autonomy among the various rôles. Each one of the rôles sets itself up autonomously in political business, so to speak. There tends to be an arms-length bargaining relationship among the rôles. The political system is saturated with the atmosphere of the market. Groups of electors come to the political market with votes to sell in exchange for policies. Holders of offices in the formal-legal rôle structure tend to be viewed as agents and instrumentalities, or as brokers occupying points in the bargaining process. The secularized political process has some of the characteristics of a laboratory; that is, policies offered by candidates are viewed as hypotheses, and the consequences of legislation are rapidly communicated within the system and constitute a crude form of testing hypotheses. Finally, because the political culture tends to be homogeneous and pragmatic, it takes on some of the atmosphere of a game. A game is a good game when the outcome is in doubt and when the stakes are not too high. When the stakes are too high, the tone changes from excitement to anxiety. While "fun" is frequently an aspect of Anglo-American politics, it is rarely a manifestation of Continental European politics; and, unless one stretches the definition, it never occurs at all in totalitarian politics.

RÔLE STRUCTURE IN THE ANGLO-AMERICAN POLITICAL SYSTEMS

The rôle structure in this group of political systems is (1) highly differentiated, (2) manifest, organized, and bureaucratized, (3) characterized by a high degree of stability in the functions of the rôles, and (4) likely to have a diffusion of power and influence within the political system as a whole.

With regard to the first point, each one of the units—formal governmental agencies, political parties, pressure groups and other kinds of voluntary associations, the media of communication, and "publics" of various kinds—pursues specialized purposes and performs specialized functions in the system. As was already pointed out, each one of these entities is more or less autonomous—interdependent, but autonomous. Certainly there are striking differences in this respect as between the United States and the United Kingdom, but their similarity becomes clear in contrast to the other major types of systems which will be described below. Secondly, this rôle structure is manifest and on the surface. Most of the potential "interests" have been organized and possess bureaucracies. Thirdly, there is in contrast to some of the other systems a relatively high degree of stability of function in the various parts of the structure. Bureaucracies function as bureaucracies, armies as armies, parliaments as parliaments. The functions are not ordinarily substitutable as among these various institutions and organizations, in contrast to some of the other systems. This is another way of saying that the political division of labor is more complex, more explicit, and more stable. There are, of course, striking differences between the British and American versions in these respects. For the American system is at the same time more complex and less stable than the British. There are, for example, many more pressure groups and types of pressure groups in the United States for reasons of size, economic complexity, and ethnic and religious heterogeneity. Furthermore there is more substitutability of function in the American system, more policy-making by pressure groups and the media of communication, more intervention in policy-making through the transient impact of "public moods." But again if we are comparing the Anglo-American system with, for example, the pre-industrial or partially industrial systems, the British and American systems will stand out by virtue of their similarities on the score of complexity, manifestness, and stability of rôle structure.

Finally the Anglo-American type of political system is one in which there is a diffusion of power and influence. This is only partially expressed in the formal legal phraseology of a democratic suffrage and representative government. There is an effective as well as a legal diffusion of power, resulting from a system of mass communications, mass education, and representation by interest groups. Here again the British and American versions differ sharply in terms of formal governmental structure, the relations between parties and pressure groups, and the system of communication and education. The net result is a more centralized, predictable rôle structure in Britain than in the United States.

THE PRE-INDUSTRIAL POLITICAL SYSTEMS

The political systems which fall under this very general category are the least well-known of all four of the classes discussed here. But despite our relative ignorance in this area and our inability to elaborate the many sub-types which no doubt exist, a discussion of this kind of political system is analytically useful since it presents such a striking contrast to the homogeneous secular political culture, and the complex and relatively stable rôle structure of the Anglo-American political system.

The pre-industrial—or partially industrialized and Westernized—political systems may be best described as mixed political cultures and mixed political systems. Nowhere does the need for additional vocabulary become clearer than in the analysis of these systems; for here parliaments tend to be something other than parliaments, parties and pressure groups behave in unusual ways, bureaucracies and armies often dominate the political system, and there is an atmosphere of unpredictability and gunpowder surrounding the political system as a whole.

Some clarity is introduced into the understanding of these systems if one recognizes that they are embedded in mixed political cultures. What this means is that as a minimum we have two political cultures, the Western system with its parliament, its electoral system, its bureaucracy and the like, and the pre-Western system or systems. In countries such as India there are many traditional political cultures which intermingle with the Western system. What kind of amalgam emerges from this impingement of different political cultures will depend on at least five factors: (1) the type of traditional cultures which are involved; (2) the auspices under which Westernization has been introduced (e.g., Western colonial powers, or native élites); (3) the functions of the society which have been Westernized; (4) the tempo and tactics of the Westernization process; (5) the type of Western cultural products which have been introduced. As a consequence of this impingement of the Western and traditional political cultures, there is a third type of political culture which frequently emerges in this type of system; what in Max Weber's language may be called a charismatic political culture. It often happens as a consequence of the erosion of a traditional political culture that powerful forces are released—anxieties over the violation of sacred customs and relationships, feelings of rootlessness and directionlessness because of the rejection of habitual routines. The impact of the Western rational system on the traditional system or systems often creates a large potential for violence. One of the typical manifestations of this conflict of political cultures is the charismatic nationalism which occurs so frequently in these areas and which may be in part understood as being a movement toward accepting a new system of political norms, or a movement toward reaffirming the older traditional ones, often both in peculiar combinations. To overcome the resistance of habitual routines backed up by supernatural sanctions, the new form of legitimacy must represent a powerful affirmation capable of breaking up deeply ingrained habits and replacing earlier loyalties. Thus, at the minimum, we must have in these political systems the old or the traditional political culture, or cultures, the new or the Western-rational political culture, and transitional or resultant political phenomena of one kind or another. Needless to say, this typical mixture of political cultures presents the most serious problems of communication and coordination. We are dealing with a political system in which large groups have fundamentally different "cognitive maps" of politics and apply different norms to political action. Instability and unpredictability are not to be viewed as pathologies but as inescapable consequences of this type of mixture of political cultures.

RÔLE STRUCTURE IN THE PRE-INDUSTRIAL POLITICAL SYSTEMS

These characteristics of the pre-industrial political systems may be brought out more clearly and systematically in an analysis of the political rôle structure which is more or less characteristic.

There is first a relatively low degree of structural differentiation. Political interest often tends to be latent and when it emerges into politics often takes the form of spontaneous, violent action. Political parties are unstable; they fragment and consolidate, appear and disappear. There is ordinarily only a rudimentary specialized system of communication. Unless there is a bureaucracy left by a Western colonial power, the bureaucratic structure may be only partially developed.

Secondly, because of the absence of a stable and explicit rôle structure, there is likely to be a high degree of *substitutability* of rôles. Thus bureaucracies may take over the legislative function, and armies may and often do the same. A political party may pre-empt the policy-making function, or a mob may emerge and take the center of the policy-making stage for a brief interval. In other words, in contrast to the Anglo-American political systems, there is no stable division of political labor.

A third and most important aspect of these political systems is the mixing of political rôle structures. Thus there may be a parliament formally based on a set of legal norms and regulations; but operating within it may be a powerful family, a religious sect, a group of tribal chieftains, or some combination of these. These are elements of the traditional rôle structure operating according to their own traditional norms. The student of these political systems would be greatly misled if he followed Western norms and expectations in describing such a decision-making system. What would be corruption in a Western parliament would be normatively oriented conduct in a "mixed parliament" of the kind often found in the regions outside of the Western-European American area.

Thus such concepts as mixed political culture and mixed political rôle structures may prepare the field researcher more adequately than the accepted political science theory and terminology; for in going to Indonesia or Thailand he will not only have in mind the Western conception of political process and system and a conception of the appropriate rôles of legislatures, bureaucracies, parties, pressure groups, and public opinion, but will rather look for the particular pattern of amalgamation of these rôles with the traditional rôles. His intellectual apparatus would enable him to grapple more quickly and more adequately with political phenomena which he might otherwise overlook, or treat as pathologies.

TOTALITARIAN POLITICAL SYSTEMS

The totalitarian political culture gives the appearance of being homogeneous, but the homogeneity is synthetic. Since there are no voluntary associations, and political communication is controlled from the center, it is impossible to judge in any accurate way the extent to which there is a positive acceptance of the totalitarian order. One can only say that in view of the thorough-going penetration of the society by a centrally controlled system of organizations and communications, and the special way in which coercion or its threat is applied, the totalitarian system, in contrast to the others, tends to be non-consensual. This is not to say that it is completely non-consensual. A completely coercive political system is unthinkable. But if one were to place the totalitarian system on a continuum of consensual-non-consensual it would be located rather more at the non-consensual end of the continuum than the others described here. Unlike the other systems where some form of legitimacy—whether traditional, rational-legal, or charismatic—underlies the acquiescence of the individual in the political system, in the totalitarian order the characteristic orientation to authority tends to be some combination of conformity and apathy. This type of political system has become possible only in modern times, since it depends on the modern technology of communication, on modern types of organization, and on the modern technology of violence. Historic tyrannies have no doubt sought this kind of dominion but were limited in the effectiveness of their means.

Totalitarianism is tyranny with a rational bureaucracy, a monopoly of the modern technology of communication, and a monopoly of the modern technology of violence.

RÔLE STRUCTURE IN TOTALITARIAN POLITICAL SYSTEMS

I believe Franz Neumann in his *Behemoth*[6] was one of the first students of totalitarianism who rejected the *monocratic* model as being useful in understanding these systems. He spoke of the peculiar shapelessness of the Nazi régime, of the fact that there was no stable delegation of power among the bureaucracy, party, the army, the organizations of big business, and the like. He concluded, as you recall, that there was no state under the Nazis. I believe what he meant to say was that there was no *legitimate* state. Later students of totalitarianism such as Hannah Arendt,[7] Merle Fainsod,[8] Carl Friedrich,[9] Alex Inkeles,[10] and Barrington Moore, Jr.,[11] have been led to similar conclusions about totalitarianism in general, or about Soviet totalitarianism. Hannah Arendt has painted the most extreme picture, which, while an exaggeration, is useful analytically. She argues that the "isolation of atomized individuals provides not only the mass basis for totalitarian rule, but is carried through at the very top of the whole structure." The aim of this process of atomization is to destroy solidarity at any point in the system and to avoid all stable delegations of power which might reduce the freedom of manoeuver of those at the very center of the system. "As techniques of government, the totalitarian devices appear simple and ingeniously effective. They assure not only an absolute power monopoly, but unparalleled certainty that all commands will always be carried out; the multiplicity of the transmission belts, the confusion of the hierarchy, secure the dictator's complete independence of all his inferiors and make possible the swift and surprising changes in policy for which totalitarianism has become famous."[12]

There are thus at least two distinctive characteristics of the totalitarian rôle structure: (1) the predominance of the coercive rôles, and (2) the functional instability of the power rôles—bureaucracy, party, army, and secret police. The predominance of the coercive rôle structure is reflected in its penetration of all of the other rôle structures. Thus all forms of organization and communication become saturated with a coercive flavor. This predominance of coercion is reflected in the celebrated definition of the state as "bodies of armed men" in Lenin's *State and Revolution*. It is also reflected in the doctrine of the "potential enemy of the state," a

---

[6]Franz Neumann, *Behemoth: The Structure and Practice of National Socialism* (New York: Oxford University Press, 1942), pp. 459ff.

[7]Hannah Arendt, *The Origins of Totalitarianism* (New York: Harcourt, Brace and Company, 1951), p. 388.

[8]Merle Fainsod, *How Russia is Ruled* (Cambridge: Harvard University Press, 1953), pp. 354 ff.

[9]Carl J. Friedrich, ed., *Totalitarianism* (Cambridge: Harvard University Press, 1954), pp. 47ff.

[10]Alex Inkeles in *ibid.*, pp. 88 ff.

[11]Barrington Moore, Jr., *Terror And Progress USSR: Some Sources of Change and Stability in the Soviet Dictatorship* (Cambridge: Harvard University Press, 1954), pp. 154 ff.

[12]Arendt, *op. cit.*, p. 389.

conception under which almost any behavior may be arbitrarily defined as disloyal behavior. This eliminates the predictability of the impact of coercion and renders it an omnipresent force, however limited its application may be in a quantitative sense.

The functional instability among the power rôles has as its main purpose the prevention of any stable delegation of power, and the consequent diffusion of power and creation of other power centers. This pattern was apparently quite marked in the development of the Nazi régime and has been observable in the uneasy balance established in the Soviet Union between party, bureaucracy, army, and secret police. In the nature of the case there must be a stabler delegation of power among the economic allocative rôles, but even these rôles are penetrated by the coercive rôle structure and manipulated within limits. A third class of rôles is illustrated by the electoral process and the representative system, as well as the practice of "self-criticism" in the party. While there is a set of norms under which these activities are supposed to influence power and policy-making, they are rather to be understood as mobilizing devices, as devices intended to create a façade of consent.

THE CONTINENTAL EUROPEAN POLITICAL SYSTEMS

We refer here primarily to France, Germany, and Italy. The Scandinavian and Low Countries stand somewhere in between the Continental pattern and the Anglo-American. What is most marked about the Continental European systems is the fragmentation of political culture; but this fragmentation is rather different from that of the non-Western systems. For in the non-Western systems we are dealing with mixed political cultures involving the most striking contrasts. The Western political culture arising out of a very different development pattern is introduced bodily, so to speak, from the outside. In the Continental European systems we are dealing with a pattern of political culture characterized by an uneven pattern of development. There are significant survivals, "outcroppings," of older cultures and their political manifestations. But all of the cultural variations have common roots and share a common heritage.

In view of this developmental pattern it may be appropriate to speak of the Continental European systems as having political subcultures. There is indeed in all the examples of this type of system a surviving pre-industrial sub-culture (e.g., the Catholic *Ancien Régime* areas in France, Southern Italy, and the Islands, and parts of Bavaria). The historical background of all three of these systems is characterized by a failure on the part of the middle classes in the nineteenth century to carry through a thorough-going secularization of the political culture. Thus another political sub-culture in these political systems constitutes remnants of the older middle classes who are still primarily concerned with the secularization of the political system itself. A third group of political sub-cultures is associated with the modernized and industrialized parts of these societies. But because they emerged in an only partially secularized political culture, their potentialities for "political market" behavior were thwarted. As major political sub-cultures there are thus these three: (1) the pre-industrial, primarily Catholic components, (2) the older middle-class components, and (3) the industrial components proper. But the political culture is more complex than this. Since in the last century the political issues have involved the very survival of these sub-cultures, and the basic form of the political system itself, the political actors have not come to politics with specific bargainable differences but rather with conflicting and mutually exclusive designs for the political culture and political system. This has involved a further fragmentation at the level of ideology and political organizations. Thus the pre-industrial, primarily Catholic element has both an adaptive, semi-secular wing and an anti-secular wing. The middle classes are divided into conservative wings in uneasy alliance with clerical pre-republican elements, and left-wings in uneasy friendship with socialists. Finally, the industrial workers are divided according to the degree of their alienation from the political system as a whole. The organized political manifestations of this fragmented political culture take the form of "movements" or sects, rather than of political parties. This means that political affiliation is more of an act of faith than of agency.

Perhaps the most pronounced characteristic of the political rôle structure in these areas is what one might call a general alienation from the political market. The political culture pattern is not adapted to the political system. For while these countries have adopted parliaments and popular elections, they are not appropriately oriented to these institutions. The political actors come to the market not to exchange, compromise, and adapt, but to preach, exhort, convert, and transform the political system into something other than a bargaining agency. What bargaining and exchanging does occur tends to take the form of under-the-counter transactions. Thus demoralization (*"transformism"*) is an almost inescapable consequence of this combination of political culture and system. In contrast, the normatively consistent, morally confident actor in this type of political system is the *militant* who remains within the confines of his political sub-culture, continually reaffirms his special norms, and scolds his parliamentarians.

This suggests another essential characteristic of this type of rôle structure, which places it in contrast to the Anglo-American. There is not an individuation of the political rôles, but rather the rôles are embedded in the sub-cultures and tend to constitute separate sub-systems of rôles. Thus the Catholic sub-culture has the Church itself, the Catholic schools, propaganda organizations such as Catholic Action, Catholic trade unions, or worker organizations, a Catholic party or parties, and a Catholic press. The Communist sub-culture—the sub-culture of the political "alienates"—similarly has a complete and separate system of rôles. The socialist and "liberal" sub-cultures tend in the same direction but are less fully organized and less exclusive. Thus one would have to say that the center of gravity in these political systems is not in the formal legal rôle structure but in the political sub-cultures. Thus "immobilism" would appear to be a normal property of this kind of political system, and it is not so much an "immobilism" of formal-legal institutions as a consequence of the condition of the political culture. Needless to say, this portrayal of the Continental European political system has been exaggerated for purposes of contrast and comparison.

Two other general aspects of the rôle structure of these countries call for comment. First, there is a higher degree of substitutability of rôles than in the Anglo-American political systems and a lesser degree than in the non-Western systems. Thus parties may manipulate pressure groups in the sense of making their decisions for them (the Communist case); interest groups such as the Church and Catholic Action may manipulate parties and trade unions; and interest groups may operate directly in the legislative process, although this pattern occurs in the Anglo-American system as well. The "immobilism" of the formally political organs often leads to a predominance of the bureaucracy in policy-making.

A second general characteristic, which is a consequence of the immobilism of the political system as a whole, is the ever-present threat of what is often called the "Caesaristic" breakthrough. As in the non-Western area, although the situations and causes are different, these systems tend always to be threatened by, and sometimes to be swept away by, movements of charismatic nationalism which break through the boundaries of the political subcultures and overcome immobilism through coercive action and organization. In other words, these systems have a totalitarian potentiality in them. The fragmented political culture may be transformed into a synthetically homogeneous one and the stalemated rôle structure mobilized by the introduction of the coercive pattern already described.

In conclusion perhaps the point might be made that conceptual and terminological growth in the sciences is as inevitable as the growth of language itself. But just as all the slang and neologisms of the moment do not find a permanent place in the language, so also all of the conceptual jargon which the restless minds of scholars invent—sometimes to facilitate communication with their colleagues and sometimes to confound them—will not find its permanent place in the vocabulary of the disciplines. The ultimate criterion of admission or rejection is the facilitation of understanding, and this, fortunately enough, is not in the hands of the restless and inventive scholar, but in the hands of the future scholarly generations who will try them out for "fit." If I may be permitted to conclude with a minor note of blasphemy, it may be said of new concepts as it was said of the salvation of souls ". . . there shall be weeping and gnashing of teeth, for many are called but few are chosen."

CHAPTER 5

Typologies of Democratic Systems

AREND LIJPHART

Political theorists and scientists, from Plato and Aristotle on, have attempted to construct classificatory schemes of political systems to aid them in their explorations of politics, and such schemes have become particularly prevalent in comparative politics in recent years.[1] However, recent typologists have not followed the precedent set by their illustrious forebears in paying special attention to democracies. Aristotle distinguished five different kinds of democracies; but in the modern typologies, democracies usually either constitute one category, without any further elaboration within the single category, or are subsumed under a more comprehensive category.

In the two well-known typologies based on Talcott Parsons' pattern variables, a dichotomous distinction is made, in Francis X. Sutton's terms, between *intensive agricultural* and *modern industrial* societies, or, to use Fred W. Riggs' terms, between *agrarian* and *industrian* societies.[2] Modern democracies fit the latter of the two categories, in both cases. Riggs makes this explicit when he subdivides each category according to the distribution of power. The industrian type has two subtypes: *democracy* and *dictatorship*. In his later refinement of the typology, where

[1] See Plato, *Republic*, Books 8-9; Plato, *Statesman*; and Aristotle, *Politics*, Book 4, chaps. 4-5. A useful summary of recent typologies can be found in H. V. Wiseman, *Political Systems: Some Sociological Approaches* (N.Y.: Praeger, 1966), pp. 47-96. For a critical review of typologies in sociology, see John C. McKinney, *Constructive Typology and Social Theory* (N.Y.: Appleton-Century-Crofts, 1966).

[2] Francis X. Sutton, "Social Theory and Comparative Politics" (paper prepared for a conference of the Committee on Comparative Politics, Social Science Research Council, 1955), reprinted in Harry Eckstein and David E. Apter (eds.), *Comparative Politics: A Reader* (N.Y.: Free Press, 1963), pp. 67-81; Fred W. Riggs, "Agraria and Industria: Toward a Typology of Comparative Administration," in W. J. Siffin (ed.), *Toward the Comparative Study of Public Administration* (Bloomington: Indiana Univ. Press, 1959), pp. 23-116. A similar dichotomous classification is implicit in Lucian W. Pye, "The Non-Western Political Process," *J. Politics*, XX, 3 (Aug., 1958), 468-486.

Reprinted from Arend Lijphart, "Typologies of Democratic Systems," COMPARATIVE POLITICAL STUDIES, *I, No. 1 (April, 1968), 3-44. An earlier version of this paper was delivered at the Seventh World Congress of the International Political Science Association in Brussels, September 18-23, 1967. The author gratefully acknowledges the constructive criticisms of Jack Citrin, Ernst B. Haas, Val R. Lorwin, Rodney Stiefbold, and Patricia Taylor, and the generous financial support of the Institute of International Studies, University of California, Berkeley.*

a third basic type (the *prismatic*) appears between the polar types of *fused* and *refracted* systems, the refracted type (essentially the same as the earlier industrial type) is again classified further into *democratic* and *autocratic* subtypes.[3] David E. Apter constructs a typology based on the criteria of the degree of hierarchy and the type of values in the system. One of his four basic types is the *reconciliation* system which includes, but is not coextensive with, the democratic system: "Democracy is a reconciliation system of government . . . [but] reconciliation systems are not necessarily democratic."[4]

In S. N. Eisenstadt's typology of historical and modern political systems, the democratic system appears as one of the four modern systems.[5] Morris Janowitz distinguishes five categories, one of which consists of democratic competitive and semi-competitive systems.[6] Edward Shils first makes a distinction between democracies and oligarchies and then classifies the oligarchies into *modernizing, totalitarian, traditional*, and *traditionalistic* oligarchies, and the democracies into *political* democracies and *tutelary* ones. The tutelary democracy, according to his description, is not really a particular type of democracy but a political system being guided in the direction of democracy.[7] He might also have called it a *democratizing oligarchy*. Thus democracies again constitute a single class in his scheme. James S. Coleman adapts Shils' sixfold typology by adding three more types: two types of oligarchy and *terminal colonial democracy* which he describes as a special case of tutelary democracy.[8] This "democracy" is therefore also more democratic in intent than in actual practice.

Although it is generally true that recent typologies recognize democracies as a separate type but do not concern themselves with subtypes of democracies, there are two major exceptions. The first is Gabriel A. Almond's classification of political systems into four basic categories: *Anglo-American, Continental European, pre-industrial*, or *partially industrial*, and *totalitarian* systems.[9] Here two types of democracy are distinguished according to the criteria of political culture and role structure. The Anglo-American systems are characterized by a "homogeneous, secular political culture" and a "highly differentiated" role structure, while the Continental European systems are characterized by a "fragmentation of political culture"—that is, they have mutually separated "political subcultures"—and a role structure in which "the roles are embedded in the subcultures and tend to constitute

---

[3] Riggs, *op. cit.*, pp. 27–30; and *The Ecology of Public Administration* (Bombay: Asia Publishing House, 1961), pp. 93–97.

[4] David E. Apter, *The Politics of Modernization* (Chicago: Univ. of Chicago Press, 1965), pp. 22–38, 349, 400.

[5] S. N. Eisenstadt, *The Political Systems of Empires* (N.Y.: Free Press, 1963), pp. 10–12.

[6] Morris Janowitz, *The Military in the Political Development of New Nations: An Essay in Comparative Analysis* (Chicago: Univ. of Chicago Press, 1964), pp. 5–7.

[7] Edward Shils, "Political Development in the New States," *Comparative Studies in Society and History*, II, 4 (July, 1960), 382–406.

[8] James S. Coleman, "Conclusion: The Political Systems of the Developing Areas," in Gabriel A. Almond and James S. Coleman (eds.), *The Politics of the Developing Areas* (Princeton, N.J.: Princeton Univ. Press, 1960), pp. 561–576.

[9] Gabriel A. Almond, "Comparative Political Systems," *J. Politics*, XVIII, 3 (Aug., 1956), 391–409.

separate subsystems of roles."[10] Britain and the United States exemplify the first type, and Weimar Germany, France under the Third and Fourth Republics, and postwar Italy the second. It must be emphasized, however, that despite his geographically derived terminology, Almond does not use geographical location either as the fundamental or as an additional criterion for distinguishing between the Anglo-American and Continental European types of democracy. In fact, in the same article in which he proposes this typology, he specifically rejects any regional classification because it "is based not on the properties of the political systems, but on their contiguity in space," which is an irrelevant criterion.[11]

The distinction between these two types of democracy is maintained in the later and much more elaborate typology of political systems proposed by Almond in collaboration with G. Bingham Powell, Jr. The new typology is based on the same criteria of role structure and political culture, but encompasses no less than sixteen different categories.[12] Compared with the earlier fourfold classification, the increase is due in some cases to further subdivisions—for instance, the totalitarian systems are now classified as either *radical* (Soviet Union) or *conservative* (Nazi Germany)— and sometimes to the addition of new categories—for instance, two new types of democracy appear. Among the *mobilized modern systems*, there is one democratic type that is characterized by "low subsystem autonomy" (e.g., Mexico); it may also be referred to as the *one-party dominant* or *hegemonic party* system. The other new type Almond and Powell call *premobilized democratic* (e.g., Nigeria prior to the military coup of early 1966). With regard to both of these, one can raise the serious question of whether they are really democratic or only democratizing—like Shils' tutelary democracies. In any case, these types appear to be marginal. The other two democratic types in Almond and Powell's scheme are the Anglo-American and Continental European types in a new guise. They are both classified as mobilized modern systems, but distinguished from each other by the degree of *subsystem autonomy*—that is, the degree to which parties, interest groups, the communication media, etc., are independent of each other. Britain and the United States exemplify the systems with high subsystem autonomy, the pre-1958 France, postwar Italy, and Weimar Germany the systems with "limited" subsystem autonomy.[13]

The second exception is the traditional classification of political systems according to the number of parties operating in the system. It is primarily concerned with democracies, which are classified as either two-party or multiparty systems. Two further categories—the one-party system and the system without any significant political parties—can be added, of course, but in contemporary comparative politics literature the scheme is used mainly as a dichotomous classification of democratic

---

[10]*Ibid.*, pp. 398–399, 407 (italics omitted).

[11]*Ibid.*, p. 392. Arthur L. Kalleberg's statement that Almond's two categories of democracies "are based on criteria of geographic location and area" is clearly incorrect; see his "The Logic of Comparison: A Methodological Note on the Comparative Study of Political Systems," *World Politics*, XIX, 1 (Oct., 1966), 73.

[12]Gabriel A. Almond and G. Bingham Powell, Jr., *Comparative Politics: A Developmental Approach* (Boston: Little, Brown, 1966), p. 217.

[13]*Ibid.*, pp. 259–271, 285–298.

systems. It should also be emphasized that the scheme is commonly used not only to distinguish between *party systems* but between entire *political systems*. For instance, Sigmund Neumann argues that "these different party systems have far-reaching consequences for the voting process and even more so for governmental decision-making. . . . A classification along this line [according to the number of parties], therefore, proves to be quite suggestive and essential."[14] And Maurice Duverger concludes that "the distinction between single-party, two-party, and multiparty systems tends to become the fundamental mode of classifying contemporary regimes."[15]

This essay will be devoted to a critical examination and comparison of the above two typologies of democratic systems: *two-party versus multiparty* and *Anglo-American versus Continental European* systems. It will also propose certain modifications in order to improve their theoretical utility.

TYPOLOGIES AND EMPIRICAL THEORY

How is one to evaluate a particular classificatory scheme of political systems? In the first place, the logical requirement of any satisfactory classification is that its categories be jointly exhaustive and mutually exclusive; they must be constructed in such a way that any item in the domain under consideration can be assigned to exactly one—at least one and not more than one—category. Logically, classification is the step that precedes the formulation of general propositions. In the comparative analysis of political systems, this means that, as Kalleberg states, "comparison can only be made after classification has been completed."[16] In practice, this logical sequence cannot be adhered to rigorously. Robert Brown argues: "Problems and their answers are so closely linked to the categories and nomenclature adopted by the investigator that all these elements develop concurrently." And this is as it should be: classifications must be "developed with explanations in mind. . . . "[17] Overemphasizing the logical precedence of classification contains the distinct danger of not reaching beyond the classification stage to the establishment of empirical propositions.

Similarly, it is less important that the requirements of exhaustiveness and exclusiveness be met logically than that they be met empirically. Carl G. Hempel states that a classification in which "at least one of the conditions of exclusiveness and exhaustiveness is satisfied not simply as a logical consequence of the determining criteria but as a matter of empirical fact" is of greater significance for the generation of empirical theory than a classification which is merely logically correct. The reason

---

[14]Sigmund Neumann, "Toward a Comparative Study of Political Parties," in Sigmund Neumann (ed.), *Modern Political Parties: Approaches to Comparative Politics* (Chicago: Univ. of Chicago Press, 1956), pp. 402–403. The same claim is usually not made for the two other traditional dichotomous distinctions between democracies: presidential versus parliamentary, and unitary versus federal systems.
[15]Maurice Duverger, *Political Parties: Their Organization and Activity in the Modern State* (trans. Barbara and Robert North) (London: Methuen, 1959), p. 393.
[16]Kalleberg, *op. cit.* note 11, p. 75.
[17]Robert Brown, *Explanation in Social Science* (Chicago: Aldine, 1963), p. 171.

is that the former gets beyond the stage of classification: it "indicates an empirical law and thus confers some measure of systematic import upon the classificatory concepts involved."[18] For instance, the classification of democracies into two-party and multiparty systems is not logically exhaustive because it omits systems with fewer than two parties—unless, of course, the concept of democracy has been defined in terms of the existence of at least two parties. If one finds, however, that it is empirically exhaustive, one has an answer to the important question raised by Apter with regard to the developing countries: "Is more than one party essential to the workings of democratic government? Can representative institutions operate without several competitive parties?"[19]

Another example would be a classification based on two or more different and logically independent criteria. A classification of democracies according to their political culture (*homogeneous* or *heterogeneous*) and their role structure (*differentiated* or *undifferentiated*) logically leads to four categories. If one's typology has less than four classes and if it is exempirically exhaustive, it implies an empirical law: the existence of a relationship between the two attributes. It is this utility of classification which Paul F. Lazarsfeld and Allen H. Barton have in mind when they define "type" as not just a category defined according to a single criterion but as a "specific attribute *compound*."[20]

The necessary link between classification and generalization is also the principal thrust of the second requirement that a classificatory scheme be *natural* rather than *artificial*. The naturalness of a classification is a matter of degree; the more it aids in the discovery of empirical relationships, the more natural it is. A natural classification is, in Julian Huxley's words, "one which enables us to make the maximum number of prophecies and deductions."[21] Furthermore, a natural classification, as J. S. L. Gilmour points out, "can be used for a wide range of purposes," whereas an artificial one is "useful only for the limited purpose for which it was constructed."[22] In comparative politics, a natural typology of political systems should ideally perform two functions: (1) It should facilitate comparison among different types and aid in the discovery of significant characteristics that are logically independent of the criteria defining the types but empirically associated with the different types.[23] (2) It should also facilitate comparisons within each type, with the attributes held in common by all of the systems within the type serving as the "control" variables, or parameters.

[18]Carl G. Hempel, *Fundamentals of Concept Formation in Empirical Science*, International Encyclopedia of Unified Science, Vol. 2, No. 7 (Chicago: Univ. of Chicago Press, 1952), p. 51.
[19]David E. Apter, "Political Parties: Introduction," in Eckstein and Apter, *op. cit.* note 2, p. 329.
[20]Paul F. Lazarsfeld and Allen H. Barton, "Qualitative Measurement in the Social Sciences: Classification, Typologies, and Indices," in Daniel Lerner and Harold D. Lasswell (eds.), *The Policy Sciences* (Palo Alto: Stanford Univ. Press, 1951), p. 169 (emphasis added).
[21]Julian Huxley, "Towards the New Systematics," in Julian Huxley (ed.), *The New Systematics* (Oxford: Clarendon Press, 1940), p. 20.
[22]J. S. L. Gilmour, "Taxonomy and Philosophy," in Huxley, *ibid.*, p. 468.
[23]See Hempel, *op. cit.* note 18, pp. 52–54.

Because of the frequent neglect of the close link which ought to exist between typologies and the discovery of empirical laws, Robert Brown says, "it is often maintained, more in anger than in sorrow, that most classificatory schemes of sociology and political science, with their associated nomenclature, are of a completely useless variety."[24] Such a complaint cannot be made about the typologies of democratic systems. They are not merely classifications without theoretical implications. Instead it will be argued that these typologies require substantial modification, not because they are unrelated to empirical theory, but because the generalizations that are explicitly or implicitly based on them are not entirely satisfactory.

DEMOCRACIES: QUALITY AND STABILITY

The overriding concerns of scholars investigating democratic systems—and constructing typologies of democratic systems—have been the *quality* and the *stability* of democracy. The interest in the quality of democracy has taken different forms, but the paramount questions have been the rights of the individual and the establishment of governments both faithfully representative of and strictly accountable to the citizens. The term "stability" usually refers to a somewhat broader concept than the literal meaning of the word: the system's ability to survive intact, which depends on its capacity to deal effectively with the problems confronting it and to adjust flexibly to changing circumstances. Although both the quality and the stability (viability) of democracies have long been, and will probably continue to be, the ultimate foci of comparative research, there has been a decided shift in emphasis. In the nineteenth and early twentieth centuries, the question of quality received the greatest attention, whereas the question of stability has become the dominant concern of contemporary comparative studies of democratic systems. The change appears to have taken place in the interwar years, partly as a result of the unfortunate experience of the Weimar Republic, which started out optimistically with a qualitatively very democratic constitution but turned out to be insufficiently viable.

The vast majority of the scholars interested in democracy have been motivated, at least to some extent, by their favorable attitude toward it and their desire to improve and strengthen it. Both the quality and the stability of democracy are manifestly normative concerns. However, when these people—who are both democrats and scholars—attempt to establish relationships between types of democratic systems and democratic stability and/or quality, they are engaged in empirical analysis. It is from such efforts at explanation that the typologies of democratic systems derive their theoretical significance.

In spite of the shift in emphasis from quality to stability, there is a remarkable congruence between the theoretical propositions based on these typologies and also, as will be shown later, between the two typologies themselves. First, the two types of democracy that are distinguished from each other by the number of parties are usually related both to the quality and the stability of the system. Compared with multiparty systems, two-party systems are said to be both more "democratic" and

[24]Brown, *op. cit.* note 17, p. 169.

more stable. For instance, Neumann points out that a two-party system leads to "certain majorities" and therefore also to "unmistakable party responsibilities," whereas in multiparty systems political responsibility is less clearly focused: "The direct mandate of the people may be blurred by the parliamentary struggle for majorities until the voice of the electorate can be heard again." Not only the quality but also the stability of democracy is greater in a two-party system, according to Neumann. Unlike the two-party system, a multiparty system does not have a "unifying and centralizing order" and, consequently, "does not hold great promise of effective policy formation."[25]

Duverger stresses a different aspect of the quality of democracy—the degree to which the parties genuinely represent the electorate—but also argues that a two-party system is more conducive to good democracy than a multiparty system. Because "public opinion seems to manifest a deepseated tendency to divide into two rival major fractions," and because "almost always there is a duality of tendencies," a system with two parties "seems to correspond to the nature of things," and a multiparty system is unnatural. Moreover, a two-party system tends to be more stable, because it is more moderate than a multiparty system. In the former one finds a "decrease in the extent of political divisions" restricting the demagogy of parties, whereas in the latter there is an "aggravation of political divisions and an intensification of differences" coinciding with "a general 'extremization' of opinion."[26] It should be noted that Anthony Downs reaches similar conclusions deduced from his economic model of democracy: "Rational voting in a multiparty system is . . . more difficult . . . than in a two-party system," and "in a multiparty system governed by a coalition, the government takes less effective action to solve basic social problems . . . than in a two-party system."[27]

For Almond, as for other systems theorists, stability is the foremost question, and this is reflected in his typology of political systems. The *Anglo-American* type, with its homogeneous political culture and its autonomous parties, interest groups, and communication media, is associated with stability, and the *Continental European* type, with its fragmented culture and mutual dependence of parties and groups, with instability. The same relationship is stated implicitly in the "functional approach to comparative politics," proposed by Almond.[28] William T. Bluhm argues that it contains "a theory of the most efficient [i.e., stable] system" and comments that "the features of the most efficient order . . . look amazingly like those of the modern

---

[25] Neumann, *op. cit.* note 14, pp. 402–403. The proposition regarding the quality of a two-party system is logically deduced rather than empirically derived. If democratic accountability requires that a single group carries executive responsibility, it logically follows that only a pure two-party system, in a parliamentary setting with a cabinet depending on majority support from the legislature, can guarantee this. There are two further conditions that must also be met, however: the two parties must be cohesive entities, and they must form one-party rather than coalition cabinets.

[26] Duverger, *op. cit.* note 15, pp. 215, 387–388.

[27] Anthony Downs, *An Economic Theory of Democracy* (N.Y.: Harper, 1957), pp. 148, 297. See also F. A. Hermens, *Democracy or Anarchy? A Study of Proportional Representation* (Notre Dame: Review of Politics, 1941), esp. pp. 15–19.

[28] Gabriel A. Almond, "Introduction: A Functional Approach to Comparative Politics," in Almond and Coleman, *op. cit.* note 8, pp. 3–64.

parliamentary democracy, and especially its British embodiment"[29]—or, in other words, the Anglo-American type. To use Almond's own terms, the Continental European type is associated with *immobilism* and "the ever-present threat of what is often called the 'Caesaristic' breakthrough." This unstable type of government cannot easily sustain democracy, and may lead to dictatorial rule; it even has, Almond says, a "totalitarian potentiality" in it.[30] In his most recent work, the immobilism characteristic of the Continental European type of democracy is said to have "significant [and presumably unfavorable] consequences for its stability and survival." In contrast, the British system is described as *versatile,* meaning that it "can respond more flexibly to internal and external demands than many, perhaps most, other systems."[31]

ALMOND'S TYPOLOGY, SEPARATION OF POWERS, AND OVERLAPPING MEMBERSHIPS

Almond's scheme is primarily concerned with political stability, but it has close parallels with the *separation-of-powers* doctrine which has as its main concern the freedom of the individual—a basic qualitative aspect of democratic systems. In his presidential address to the American Political Science Association in 1966, Almond contrasts the separation-of-powers theory with systems theory, and describes the former as the "dominant paradigm" of political science in the eighteenth and nineteenth centuries, which is now replaced by the systems paradigm.[32] On the other hand, he also emphasizes the close connection between the two when he calls the authors of the *Federalist Papers* "systems theorists."[33] This connection between separation of powers and Almond's functional approach is particularly important in this context because one of the criteria distinguishing Almond's Anglo-American and Continental European types is *role structure*: the degree to which the "roles" are autonomous—or *separate.*

The main difference between the separation-of-powers doctrine and Almond's scheme is that Almond extends the idea of separation of powers from the three formal branches of government (the legislative, executive, and judicial) to the informal political substructures (parties, interest groups, and the media of communication), and that he places much more emphasis on the latter (the *input* structures) than on the former (the *output* structures). The other differences are mainly terminological. Almond translates *powers* into *functions,* and *separation* becomes *boundary maintenance.* And just as separation of powers contributes to the quality of democracy, according to the *Federalist Papers,* proper boundary maintenance between political functions contributes to the stability of the democratic system. In Britain, as an example of the Anglo-American type, there is "effective boundary maintenance . . . among the subsystems of the polity," but in France,

---
[29]William T. Bluhm, *Theories of the Political System: Classics of Political Thought and Modern Political Analysis* (Englewood Cliffs, N.J.: Prentice-Hall, 1965), p. 150. See also Constance E. van der Maesen and G. H. Scholten, "De functionele benadering van G. A. Almond bij het vergelijken van politieke stelsels," *Acta Politica,* I, 1-4 (1965–66), 220–226.
[30]Almond, "Comparative Political Systems," *op. cit.* note 9, p. 408.
[31]Almond and Powell, *op. cit.* note 12, pp. 106, 262.
[32]Gabriel A. Almond, "Political Theory and Political Science," *Am. Polit. Sci. Rev.,* LX, 4 (Dec., 1966), pp. 875–876.
[33]Almond and Powell, *op. cit.* note 12, p. 11.

representative of the Continental European type, one finds "poor boundary maintenance . . . among the various parts of the political system." French parties and interest groups "do not constitute *differentiated, autonomous* political subsystems. They *interpenetrate* one another," especially within the Catholic, Socialist, and Communist subcultures. Similarly, the Anglo-American and Continental European types are distinguished by the degree of autonomy of their media of communication. The United States, Britain, and the Old Commonwealth nations have "to the greatest extent autonomous and differentiated media of communication," whereas France and Italy "have a 'press' which tends to be dominated by interest groups and political parties."[34]

Just as the separation-of-powers doctrine is supplemented by the idea of checks and balances, the *boundary maintenance* doctrine is supplemented by the twin concepts of *multifunctionality* and *regulatory role*. Perfect boundary maintenance never occurs, according to Almond. The formal branches of government, parties, interest groups, etc., invariably perform more than just a single function: "All political structure, no matter how specialized, . . . is multifunctional." What is important, therefore, is not so much that, for instance, political parties are the only interest aggregators and perform no function other than interest aggregation, but that this function becomes their special responsibility. In modern specialized systems, of which the Anglo-American democracies are the prototype, there are certain structures "which have a functional distinctiveness, and which tend to perform what we may call a regulatory role in relation to that function within the political system as a whole."[35]

In addition to the convergence of the first criterion of Almond's typology—role structure—with the separation-of-powers doctrine, there is also a close connection between the second criterion—*political culture*—and the theory of *overlapping memberships*. Stated briefly, this theory holds that when individuals belong to a number of different organized or unorganized groups with diverse interests and outlooks, their attitudes will tend to be moderate as a result of these psychological cross-pressures. Moreover, leaders of organizations with heterogeneous memberships will be subject to the political cross-pressures of this situation, and will also tend to assume moderate, middle-of-the-road positions. Such moderation is essential to political stability. Conversely, when a society is riven with sharp cleavages and when memberships and loyalties do not overlap but are concentrated exclusively within each separate segment of society, the cross-pressures that are vital to political moderation and stability will be absent. As David B. Truman argues, if a complex society manages to avoid "revolution, degeneration, and decay [and] maintains its stability, . . . it may do so in large measure because of the fact of multiple memberships."[36]

---

[34] Almond, "A Functional Approach," *op. cit.* note 28, pp. 37–38, 46 (emphasis added).
[35] *Ibid.*, pp. 11, 18.
[36] David B. Truman, *The Governmental Process: Political Interests and Public Opinion* (N.Y.: Knopf, 1951), p. 168. A good summary of the theory may be found in Sidney Verba, "Organizational Membership and Democratic Consensus," *J. Politics*, XXVII, 3 (Aug., 1965), esp. pp. 467–473. For a thorough critique, see William C. Mitchell, "Interest Group

In political culture terminology, overlapping memberships are characteristic of a homogeneous political culture, whereas a fragmented culture has little or no overlapping between its distinct subcultures. In Almond's typology, the stable Anglo-American systems have a homogeneous culture and the unstable Continental European systems have deep subcultural cleavages. Their immobilism and instability, Almond says, are a "*consequence* of the condition of the political culture."[37] For instance, Almond and Powell describe the French system under the Fourth Republic as divided into "three main ideological families or subcultures," with the main parties, interest groups, and the media of communication "coordinated in ideological families." The results were that demands "piled up and were not converted into policy alternatives or enacted into law," and that there were long "periods of immobilism with brief periods of crisis-liquidation."[38] Sometimes Almond himself adopts the language of the overlapping memberships theory: in a nation like France, "the individual may be exposed to few of the kinds of 'cross-pressures' that moderate his rigid political attitudes."[39] And in *The Civic Culture*, Almond and Verba state that "membership patterns differ from one country to the next. In the European Catholic countries, for example, the pattern tends to be ideologically cumulative. Family, church, interest group, and party membership tend to coincide in their ideological and policy characteristics and to reinforce one another in their effects on opinion. In the United States and Britain, however, the overlapping pattern appears to be more common."[40]

CONVERGENCE OF THE TWO TYPOLOGIES
OF DEMOCRATIC SYSTEMS

The two typologies of democratic systems converge not only with a series of theoretical objectives and empirical propositions, but also with each other. In modern developed political systems with proper boundary maintenance (i.e., the Anglo-American type), political parties have the "distinctive and regulatory relation" to the interest aggregation function, and this function is in the "middle range of processing" and is supposed to transform the articulated interests "into a relatively small number of alternatives. . . ."[41] The two-party system would appear to be ideally suited for this, and multiparty systems would appear to be less efficient aggregators. Nevertheless, at first Almond rejects the idea that his Anglo-American type is congruent with the two-party system and his Continental European type with the multiparty system: "the commonly used distinctions between one-party, two-party, and multi-party systems simply get nowhere in distinguishing the essential properties of the

---

Theory and 'Overlapping Memberships': A Critique" (paper presented at the annual meeting of the American Political Science Association, 1963). See also Mancur Olson, Jr., *The Logic of Collective Action: Public Goods and the Theory of Groups* (Cambridge, Mass.: Harvard Univ. Press, 1965), pp. 117–131.

[37] Almond, "Comparative Political Systems," *op. cit.* note 9, p. 408 (emphasis added).
[38] Almond and Powell, *op. cit.* note 12, pp. 263–265.
[39] *Ibid.*, p. 122.
[40] Gabriel A. Almond and Sidney Verba, *The Civic Culture: Political Attitudes and Democracy in Five Nations* (Princeton, N.J.: Princeton Univ. Press, 1963), pp. 133–134.
[41] Almond, "A Functional Approach," *op. cit.* note 28, pp. 39, 40.

totalitarian, the Anglo-American, and the Continental European political systems."[42]

In his later writings, however, Almond implicitly does accept the congruence between his own typology (at least that part of the typology which deals with democratic systems) and the typology based on the number of parties: "some party systems aggregate interests much more effectively than others. The *number of parties* is a factor of importance. Two-party systems which are responsible to a broad electorate are usually forced toward aggregative policies." On the other hand, "the presence of a large number of fairly small parties makes it increasingly likely that each party will merely transmit the interests of a special subculture or clientele with a minimum of aggregation." Two-party systems are not only the best aggregators, but also contribute to effective boundary maintenance. It is desirable, according to Almond, for aggregation structures to be differentiated from both the decision-making and the interest articulation structures, and "the competitive two-party system perhaps most easily secures and maintains this differentiation. . . ."[43] Both effective aggregation and proper boundary maintenance are directly related to democratic stability, and both are characteristic of the Anglo-American type of democracy.

It is interesting to note that another systems theorist, David Easton, reaches the same conclusion about the relationship between the number of parties and democratic stability, and that his argument runs parallel to Almond's: "Where the parties seek to embrace the widest mixture of groups and individuals, as in two- or three-party systems, the political divisions are softened. In such instances the parties are able to draw together a variety of groups and thereby bring about some prior reconciliation of points of view."[44] Easton cautiously refers to "two- or three-" party systems rather than two-party systems, but he clearly endorses the proposition that the fewer the parties the more aggregative they are likely to be. What he calls *prior reconciliation* appears to be synonymous with *interest aggregation*.

DEVIANT CASES

The two typologies of democratic systems have so far passed the tests applied to them. They are both closely linked to empirical generalizations concerning democracy, and these combine into a relatively cohesive and plausible set of theoretical propositions. The crucial test, however, is not so much their internal and mutual consistency and the prima facie plausibility of the theory they represent, as the validity of this theory. And here the typologies run into trouble, especially with regard to the question of political stability. Are the typologies really exhaustive, and if so, to which category do the Scandinavian countries, the Low Countries, Switzerland, and Austria belong? All of these countries, with the exception of Austria, have multiparty systems, but they are nevertheless stable democracies. Moreover, the Benelux countries, Switzerland, and Austria have the kind of subcultural cleavage and interpenetration of parties, interest groups, and media of

[42] Almond, "Comparative Political Systems," *op. cit.* note 9, p. 397.
[43] Almond and Powell, *op. cit.* note 12, pp. 102–103, 107 (emphasis added).
[44] David Easton, *A Systems Analysis of Political Life* (N.Y.: Wiley, 1965), p. 257.

communication characteristic of the unstable Continental European type, but they also have considerable stability; in fact, Switzerland and the Netherlands are usually counted among the most stable of the world's democracies.

The presence of so many deviant cases—which also happen to be the smaller European democracies, largely forgotten in the post-World War II shift of scholarly attention from the major European nations to non-Western areas—throws doubt on the hypothesized relationship between the Continental European and multiparty type and democratic stability. As far as the number-of-parties typology is concerned, Leslie Lipson states his agreement with the proposition that the number of parties is related to the *quality* of democracy: "There is only one irrefutable argument in favor of the two-party system. It has the great merit of focusing responsibility." But he rejects the notion that "multipartism breeds instability and is the parent of governmental weakness," pointing to the examples of Switzerland and Scandinavia.[45] And indeed, in the universe of democracies, there are more stable than unstable multiparty cases. Dankwart A. Rustow refers to the party systems of Scandinavia as "working multiparty systems."[46] But to divide the multiparty type into the subtypes of *working* and *immobilist* systems begs the question. It fails to provide an explanation of why some multiparty systems somehow "work" and others do not.

Giovanni Sartori attempts to refine the typology and to establish a non-tautological link between the number of parties and democratic stability by dividing the multiparty systems into *moderate* and *extreme* multiparty systems. Counting only the major parties and omitting the small marginal ones, "three or four . . . is the normal number" of parties in moderate multiparty systems. For extreme multipartism, "a minimum of five parties is required," and Sartori adds in a footnote that "the very interactions of more than four parties help to explain a centrifugal pattern of development."[47] The line between four-party and five-party systems, is according to Sartori, the essential dividing line rather than the distinction between two-party and multiparty systems. Otherwise, he agrees with Duverger's and Neumann's assertions of a relationship between the number of parties and both the quality and the stability of democracy. Both two-party and moderate multiparty systems are *centripetal* whereas extreme multiparty systems are *centrifugal,* according to Sartori: "When the drive of a political system is centripetal one finds moderate politics, while immoderate or extremist politics reflects the prevalence of centrifugal drives." Extreme multiparty systems, of which postwar Italy, France especially under the Fourth Republic, Weimar Germany, and the Spanish Republic from 1931 to 1936 are given as examples, are not only "conducive to governmental deadlock and

[45]Leslie Lipson, *The Democratic Civilization* (N.Y.: Oxford Univ. Press, 1964), p. 352. Robert A. Dahl also challenges the "assertions about the moderating effects of a two-party system [which] are surely among the oldest and most widespread in the modern study of political parties . . ."; see Dahl (ed.), *Political Oppositions in Western Democracies* (New Haven: Yale Univ. Press, 1966), p. 372 n.

[46]Dankwart A. Rustow, "Scandinavia: Working Multiparty Systems," in Neumann, *op. cit.* note 14, pp. 169–193.

[47]Giovanni Sartori, "European Political Parties: The Case of Polarized Pluralism," in Joseph LaPalombara and Myron Weiner (eds.), *Political Parties and Political Development* (Princeton, N.J.: Princeton Univ. Press, 1966), pp. 153, 155.

paralysis," but are also poor in terms of democratic quality. In Italy, typical of the extreme multiparty type, Sartori finds "a Byzantine and undecipherable party system whose end product is overcomplication and confusion. At least this is how the polity must look to the ordinary voter. The average citizen is surely not in a position to disentangle the mess. . . ."[48]

Duverger makes a similar distinction between types of multiparty systems: *tripartism, quadri-partism,* and, when there are more than four parties, *polypartism*. *Polypartism* appears to be synonymous with Sartori's *extreme multipartism*. In both cases, the essential dividing line is between party systems with four principal parties and those with five. Duverger is much more cautious than Sartori, however, and does not attempt to establish a link between polypartism and instability. He calls the classification of multiparty systems into separate categories "precarious," and as examples of polypartism he gives both the unstable French system and the stable Dutch system.[49] It should be added that Sartori does not consistently draw the line between moderate and extreme multiparty systems at the same point. In a different work, he defines the moderate system as one in which "the relevant parties are three, four, or *at most five*. . . ."[50] Here the dividing line is between five-party and six-party systems rather than between the four- and five-party types. These refinements of the typology, based on the number of parties, will be examined further below.

How does Almond deal with the problem of the deviant cases? In the article in which he originally sets forth the Anglo-American and Continental European types with their distinctive qualities, the Scandinavian and Low Countries are specifically excluded from the Continental European political systems, and Austria and Switzerland are not mentioned at all. For the Scandinavian and Low Countries, a separate category is set up which is not elaborated in detail. Almond merely says that these political systems "combine some of the features of the Continental European and the Anglo-American," and that they "stand somewhere in between the Continental pattern and the Anglo-American."[51] In his later writings, he specifies more fully in what respects this third type of political system differs from the other two. Adopting the language of the number-of-parties typology, he distinguishes between the *crisis* or *immobilist* multiparty systems of France and Italy and the *working* multiparty systems of the Scandinavian and Low Countries. In the latter, at least some of the parties are "broadly aggregative," such as the Scandinavian Socialist parties and the Belgian Socialist and Catholic parties.[52] This criterion is not very satisfactory, because it does not apply clearly to the Dutch parties, whereas the Italian Christian

---

[48]*Ibid.*, pp. 139, 152, 161.

[49]Duverger, *op. cit.* note 15, pp. 234–239. See also Jean Blondel, "Party Systems and Patterns of Government in Western Democracies" (paper presented at the Seventh World Congress of the International Political Science Association, 1967).

[50]Giovanni Sartori, "Opposition and Control: Problems and Prospects," *Government and Opposition*, I, 2 (Feb., 1966), p. 152 (emphasis added). See also his *Parties and Party Systems* (N.Y.: Harper and Row, forthcoming), chap. 8.

[51]Almond, "Comparative Political Systems," *op. cit.*, note 9, pp. 392–393, 405.

[52]Almond, "A Functional Approach," *op. cit.* note 28, p. 42. See also his "A Comparative Study of Interest Groups and the Political Process," *Am. Polit. Sci. Rev.*, LII, 1 (March, 1958), pp. 275–277.

Democratic party does seem to fit it. Almond adds that, to the extent that interests are not fully aggregated within, for example, the Scandinavian parties, interest aggregation continues at the parliamentary and cabinet levels within party coalitions. But he warns that this becomes more difficult as the number of parties increases and their size decreases.[53] The last argument appears to move in the direction of Sartori's proposition concerning extreme multipartism.

The second criterion for distinguishing the working multiparty system from the immobilist system is much more satisfactory, at least with regard to the Scandinavian countries: their "political culture is more homogeneous and fusional of secular and traditional elements."[54] In fact, the Scandinavian countries do not significantly differ from the Anglo-American type in this respect. In *The Civic Culture*, Almond and Verba mention the Scandinavian countries together with England, the Old Commonwealth countries, and the United States, as having "homogeneous political cultures."[55] This is also in line with Rustow's assessment of the Scandinavian countries as a "highly homogeneous group of societies."[56]

But a homogeneous political culture is not at all characteristic either of the other working multiparty systems or of the two-party Austrian system. Especially the Catholic, Socialist, and Liberal *familles spirituelles* of Belgium and Luxembourg, the Catholic, Calvinist, Socialist, and Liberal *zuilen* of the Netherlands, and the Catholic, Socialist, and Liberal-National *Lager* of Austria, are subcultures quite similar to the subcultures characteristic of Almond's Continental European type. In fact, these countries have an even more thoroughly fragmented political culture than France, Italy, and Weimar Germany, with a solid network of interpenetrating groups and media of communication within each subculture and with even less flexibility and overlapping of membership between different subcultures. One would, therefore, expect even more immobilism and instability than in the Continental European systems, but one finds the opposite. Almond and Verba state that not only the Scandinavian countries, but also the Low Countries and Switzerland—and they might have added post-1945 Austria—"appear to have worked out their own version of a political culture and practice of accommodation and compromise."[57] This is undoubtedly true, but what exactly does this special version entail, and how can we explain it?

CONSOCIATIONAL DEMOCRACY

In order to arrive at an explanation of the stability in these fragmented democracies, the deterministic aspects of the overlapping memberships theory must be abandoned. The most extreme case of fragmentation is described succinctly by Verba as "a political system made up of two closed camps [subcultures] with no overlapping of membership. The only channels of communication between the two

[53] Almond and Powell, *op. cit.* note 12, p. 103.
[54] Almond, "A Functional Approach," *op. cit.* note 28, p. 42.
[55] Almond and Verba, *op. cit.* note 40, pp. 28–29.
[56] Rustow, *op. cit.* note 46, p. 191.
[57] Almond and Verba, *op. cit.* note 40, p. 8.

camps would be at the highest level—say when the leaders of the two camps meet in the governing chambers—and all conflict would have to be resolved at this highest level." In such a system, instability would be endemic: its "politics comes to resemble negotiations between rival states; and war or a breakdown of negotiations is always possible."[58] One could state this conclusion even more forcefully and argue that instability in such a system is not just *possible*, but even *probable*. On the other hand, it is *not inevitable*. The leaders of rival groups may take actions to counter the unstabilizing effects of fragmentation, if they are aware of the dangers involved and desire to avoid them, and thus turn the overlapping memberships proposition into a self-denying hypothesis. The knowledge of these dangerous tendencies may become an important element in the situation, and may substantially affect the outcome. As Dahl argues: "The possibility of violence and civil war always lurks as a special danger in countries with hostile subcultures; and this danger undoubtedly stimulates a search for alternative responses."[59] This search may become especially urgent when the leaders' perception of the likelihood of civil strife between the antagonistic subcultures is reinforced by the actual occurrence of such conflict in the past. This possibility is what Hans Daalder has in mind when he states that "not only the severity and incidence of conflicts" are important, "but also the attitudes political elites take toward the need to solve them by compromise rather than combat." And one factor predisposing the elites toward compromise may be "the traumatic memory of past conflicts," which, of course, may perpetuate their antagonism, but which may also cause them "to draw together."[60]

The clearest example of such a development is Austria. After the First World War, an attempt was made to set up a grand coalition bringing together the Catholic and Socialist *Lager*. But these subcultures were too antagonistic toward each other, and their leaders were not sufficiently impressed by the dangerous potentialities of non-cooperation, to render the attempt successful. The First Republic drifted into civil unrest with both groups literally turning themselves into armed camps. Civil war and the establishment of a dictatorship ensued. The Catholic and Socialist leaders learned a great deal from this experience, and decided to cooperate in a grand coalition after the Second World War. According to Frederick C. Engelmann, "critics and objective observers agree with Austria's leading politicians in the assessment that the coalition was a response to the civil-war tension of the First Republic. . . ."[61] And Otto Kirchheimer also specifically attributes Austria's

---

[58]Verba, *op. cit.* note 36, p. 470. See also Gerhard Lehmbruch, "A Non-Competitive Pattern of Conflict Management in Liberal Democracies: The Case of Switzerland, Austria, and Lebanon" (paper presented at the Seventh World Congress of the International Political Science Association, 1967), p. 6.

[59]Dahl, *op. cit.* note 45, p. 358.

[60]Hans Daalder, "Parties, Elites, and Political Developments in Western Europe," in LaPalombara and Weiner, *op. cit.* note 47, p. 69. Thus, although it is *probable* that fragmentation will lead to political instability, it is also, as Clause Ake states, *"possible* for a country to achieve a degree of political stability quite out of proportion to its social homogeneity." See his *A Theory of Political Integration* (Homewood, Ill.: Dorsey Press, 1967), p. 113 (emphasis added).

[61]Frederick C. Engelmann, "Haggling for the Equilibrium: The Renegotiation of the Austrian Coalition, 1959," *Am. Polit. Sci. Rev.*, LVI, 3 (Sept., 1962), p. 651.

"carefully prearranged system of collaboration" to its "historical record of political frustration and abiding supicion. . . ."[62] Under the grand coalition, which ruled Austria until early 1966, the Second Republic achieved considerable stability.

Val R. Lorwin analyzes the experience of pre-democratic Belgium at the time of the achievement of its independence in very similar terms. He states that the Catholic and Liberal leaders "had drawn the great lesson of mutual tolerance from the catastrophic experience of the Brabant Revolution of 1789, when the civil strife of their predecessors had so soon laid the country open to easy Habsburg reconquest. It was a remarkable and *self-conscious 'union of the oppositions'* which made the revolution of 1830, wrote the Constitution of 1831, and headed the government in its critical years."[63] Here again the potential instability caused by subcultural cleavage was deliberately counteracted by a grand coalition of the opposing forces.

Grand coalitions violate the idea that in parliamentary systems cabinets should have, and normally do have, majority support, but not the support of an overwhelming majority. A small coalition not only allows the existence of an effective democratic opposition, but it is also formed more easily because there are fewer different viewpoints and interests to reconcile. This common-sense notion is also in accord with William H. Riker's "size principle" based on game-theoretic assumptions. This principle states: "In n-person, zero-sum games, where side-payments [private agreements about the division of the payoff] are permitted, where players are rational, and where they have perfect information, only minimum winning coalitions occur." When applied to social situations similar to such games, this means that the "participants create coalitions just as large as they believe will ensure winning and no larger."[64]

The size principle is most useful in illuminating the nature of the grand coalition, because it stipulates the conditions under which a minimum winning coalition will occur and therefore also, by implication, the conditions for other kinds of coalitions such as the grand coalition. Riker states, for instance, that the size principle must be modified by the "information effect": Minimum winning coalitions can be expected only when the players have perfect information, and larger coalitions become necessary to the extent that information is imperfect. Even more important is the zero-sum condition: "Only the direct conflicts among participants are included, and common advantages are ignored." When common advantages do play a role, the zero-sum condition does not apply and neither does the size principle. This is not only logically true, but experimental evidence from small-group research also

---

[62]Otto Kirchheimer, "The Waning of Opposition in Parliamentary Regimes," *Social Research*, XXIV, 2 (Summer, 1957), p. 137. See also Gerhard Lehmbruch, *Proporzdemokratie: Politisches System und politische Kultur in der Schweiz und in Österreich* (Tübingen: Mohr, 1967), p. 25.

[63]Val R. Lorwin, "Constitutionalism and Controlled Violence in the Modern State: The Case of Belgium" (paper presented at the Annual Meeting of the American Historical Association, 1965), p. 4 (emphasis added). W. Arthur Lewis urges a similar solution for the heterogeneous countries of West Africa in his *Politics in West Africa* (London: Allen and Unwin, 1965), pp. 64–84.

[64]William H. Riker, *The Theory of Political Coalitions* (New Haven: Yale Univ. Press, 1962), pp. 32–33.

supports it. Riker found that the size principle was operative in games played by close friends who regarded the game as purely a game and had no difficulty in accepting its zero-sum condition, but that groups of less well acquainted persons did not tend to perceive the game as zero-sum—"considerations of maintaining the solidarity of the group and the loyalty of members to it" took precedence over it—and therefore tended to form larger than minimum winning coalitions. The zero-sum condition "implies a limit, namely that no outcome can disrupt the body. That is, no decision can be taken in such a way that losers would prefer to resign rather than acquiesce."[65]

In real political life, the zero-sum condition limits the application of the size principle to coalition building in two kinds of societies: (1) homogeneous societies with a high degree of consensus where common advantages are taken for granted, and (2) their polar opposites, societies marked by extreme internal antagonisms and hostilities. In other words, the size principal applies when the participants in the political process perceive politics either as a *game* or as all-out *war*. In intermediate situations, there is at least some pressure for enlarging the coalition and perhaps even creating a grand coalition.

The function of a grand coalition can also be clarified by placing it in the context of the competing principles of consensus and majority rule in normative democratic theory. On the one hand, broad agreement among all citizens seems more democratic than simple majority rule, but on the other hand, the only real alternative to majority rule is minority rule—or at least a minority veto. Most democratic constitutions try to resolve the dilemma by prescribing majority rule for the normal transaction of business when the stakes are presumably not too high, and extraordinary majorities or several majorities over a period of time for the most vital decisions, such as for constituent purposes. In practice, majority rule works well when opinions are distributed unimodally and with relatively little spread—in other words, when there is considerable consensus and the majority and minority are in fact not very far apart. When the people are "fundamentally at one," as Lord Balfour once said, they "can safely afford to bicker."[66] But in a political system with hostile subcultures, virtually all decisions are perceived as entailing high stakes, and strict majority rule may not only be regarded as, in Herbert J. Spiro's words, "undemocratically exclusive,"[67] but also places a strain on the unity and peace of the system. As Dahl observes, conflicts involving subcultures are "too explosive to be managed by ordinary parliamentary opposition, bargaining, campaigning, and winning elections."[68] This argument therefore also points to the deliberate resort to large majorities, such as in a grand coalition, as the appropriate response for maintaining democratic stability in a fragmented system.

Democracies with subcultural cleavages and with tendencies toward immobilism

[65]*Ibid.*, pp. 29, 51, 88–89, 103.
[66]Cited in Carl J. Friedrich, *Constitutional Government and Democracy: Theory and Practice in Europe and America* (rev. ed.; Waltham, Mass.: Blaisdell, 1950), p. 422.
[67]Herbert J. Spiro, *Government by Constitution: The Political Systems of Democracy* (N.Y.: Random House, 1959), p. 341.
[68]Dahl, *op. cit.* note 45, p. 358.

and instability which are deliberately turned into more stable systems by the leaders of the major subcultures may be called *consociational* democracies. This term is borrowed from Apter, who uses it in the context of African political systems for the type that is "willing to accommodate a variety of groups of divergent ideas in order to achieve a goal of unity. . ."; it is "essentially a system of compromise and accommodation."[69] In addition to Austria, the best examples of consociational democracy in Europe are Switzerland and the Benelux countries. In other parts of the world, Lebanon, Colombia, and Uruguay (until March, 1967) may also serve as examples.

The grand coalition cabinet, as in Austria, can be regarded as the prototypal device for accommodating the disagreements between hostile subcultures in a consociational democracy. There are a variety of other responses, however, that serve the same purpose. In fact, the essential characteristic of consociational democracy is not so much any particular institutional arrangement as overarching cooperation at the elite level with the deliberate aim of counteracting disintegrative tendencies in the system. Joining in a grand coalition or national unity cabinet represents such cooperation in its most comprehensive form, but it is only one among many possibilities. One can apply the term *grand coalition* to all of these, as long as it is clear that it is not used narrowly to refer to cabinets in parliamentary systems.

Even in Austria, the cabinet itself was not the most important organ of accommodation. The crucial decisions were made by the small extraconstitutional "coalition committee" (called the "working committee" in its last years), on which the top Catholic and Socialist leaders were equally represented. A similar extra-constitutional steering committee, superior to both cabinet and parliament, was the *Petka* in interwar Czechoslovakia, which consisted of the leaders of the five principal Czech parties.[70] The Swiss collegial executive and its Uruguayan counterpart, which was fashioned after the Swiss example and functioned from 1952 to 1967, are examples of the grand coalition principle applied to the formal constitutional executive. In the Lebanese and Colombian presidential systems, we find other variations on the basic consociational theme. The Colombian Liberal and Conservative parties agreed in 1958 to alternate in the presidency for a period of sixteen years, as part of a consociational design to avoid the civil wars and dictatorships which had plagued the country. In Lebanon, the National Pact of 1943 stipulated that the two top executive posts would be shared by the two major religious groups: the

[69]David E. Apter, *The Political Kingdom in Uganda: A Study in Bureaucratic Nationalism* (Princeton, N.J.: Princeton Univ. Press, 1961). Nigeria, where the attempt at consociational politics turned out to be unsuccessful in early 1966, is given as an example. Ake's concept of a *consensual* system and Lehmbruch's *proportional democracy (Proporzdemokratie)* are similar to consociational democracy. Every author is, of course, free to adopt his own terminology and definitions, but the term *consociational* seems preferable, because it avoids the too-broad connotation of *consensual* on the one hand, and the too-narrow meaning of *proportional* on the other hand. See Ake, *op. cit.* note 60, pp. 111–113; Lehmbruch, *Proporzdemokratie, op. cit.* note 62, pp. 7–9. Note also the use of the term *consociatio* by Johannes Althusius in his *Politica Methodice Digesta*.

[70]The *Petka* was not truly a grand coalition because it excluded the ethnic minorities. The Austrian grand coalition, of course, was also not as "grand" as possible because it never included the Liberals.

President of the Republic must be a Maronite and the President of the Council a Sunni.[71]

In Belgium and the Netherlands we find a somewhat looser pattern: broadly based coalitions (particularly in the Dutch case) but not grand coalitions, complemented by either permanent or *ad hoc* "grand" councils and committees with formally not much more than an advisory function, but with actually often decisive influence. The outstanding examples are the Dutch Social and Economic Council—one of the few examples of an effective and powerful economic parliament—and the temporary grand coalitions of party leaders that concluded the "school pacts" in the Netherlands in 1917 and in Belgium in 1958.

The emphasis on cooperation and the avoidance of competition may be extended to the electoral level. The Colombian *alternación* principle entails such a far-reaching agreement which prevents the voters from upsetting the carefully arranged civil peace, but also denies them a meaningful exercise of their democratic rights. The deviation from the democratic norm is less serious when this involves only a single election, such as the Dutch parliamentary election of 1917: all of the parties agreed that the voters should not be allowed to tamper with the peaceful settlement of the questions of universal suffrage and of state aid to denominational schools, and therefore did not contest the seats held by incumbents. A final interesting example is the small Swiss canton of Uri, where, according to Roger Girod, "for many years the Catholic-Conservatives, who have a large majority, reserve two seats to themselves in the federal Council of States, and in exchange, do not oppose the election of a Radical to the single seat of the canton in the National Council."[72]

PREREQUISITES FOR CONSOCIATIONAL DEMOCRACY

The question of whether a democracy with a fragmented political culture will be stable or unstable depends primarily on the character of the elite's response to the potential or actual instability of the system. In order to establish and maintain a consociational democracy, the leaders of the rival subcultures must have the following behavioral attributes:

1. *Ability to recognize the dangers inherent in a fragmented system.* The leaders must be fully aware of the system's unstable tendencies caused by subcultural cleavages. This awareness is particularly important at the crucial stage of the initial establishment of consociational practices; but stability can never be taken for granted, and this prerequisite quality—as well as the other prerequisites enumerated below—remains important in later critical periods. Once the precedent has been set, however, subsequent applications of consociationalism are facilitated, because they tend to become habitual and do not represent a deliberate departure from established competitive practices. As Gerhard Lehmbruch states, such norms "are retransmitted

---

[71] A similar sharing of presidential and vice-presidential offices was attempted in newly independent Cyprus, but consociation failed there.
[72] Roger Girod, "Geography of the Swiss Party System," in Erik Allardt and Yrjö Littunen (eds.), *Cleavages, Ideologies and Party Systems: Contributions to Comparative Political Sociology*, Transactions of the Westermarck Society, Vol. 10 (Helsinki: The Academic Bookstore, 1964), p. 146.

by the learning processes in the political socialization of elites and thus acquire a strong degree of persistence through time."[73] Furthermore, elite behavior and mass political culture are not mutually independent, and consociational solutions are likely in the long run to reduce gradually the intensity of the cleavages in a fragmented culture.

2. *Commitment to system maintenance.* As Lipson states, it is obvious that "democratic institutions cannot be applied to the relations between two groups which feel so antagonistic that they prefer a complete separation to any kind of tie."[74] A high degree of solidarity is not necessary, but the leaders must have a certain degree of willingness to make an effort to halt and reverse the disintegrative tendencies of the system. Without such a minimum of commitment, consociational democracy will not even be attempted.

3. *Ability to transcend subcultural cleavage at the elite level.* In order to translate the first two conditions into effective consociational action, two further conditions must be met. The leaders must be able to break through the barriers to mutual understanding caused by subcultural differences, and to establish effective contacts and communication across these cleavages. If the cleavages and mutual antagonisms are insurmountable not only at the mass level but also at the elite level, attempts at consociational democracy are likely to founder.

4. *Ability to forge appropriate solutions for the demands of the subcultures.* The leaders must be able to develop both institutional arrangements and rules of the game for the accommodation of their differences. This is probably the most important of the four conditions, and also the most difficult one to satisfy. Many solutions may have undesirable side effects. For instance, the most prevalent rule of the game in consociational democracies is *proportionality*. This rule can solve the thorny problem of allocating resources and appointments among the subcultures, but recruitment to the civil service based on membership in a certain subcultural group rather than individual talent may be at the expense of administrative efficiency. An even more serious danger is inherent in another frequent rule of the consociational game: the *mutual veto*. Such a veto is essential in order to induce all of the subcultural groups to participate in grand coalitions, but if it is not handled with caution and restraint, it is likely to produce the very stagnation and instability that consociational democracy is designed to avoid.

These four prerequisites must all be fulfilled if consociational democracy is to succeed. It is, therefore, not a simple achievement to turn the overlapping memberships proposition into a self-denying hypothesis. The chronic instability of French politics can be explained in these terms. Eric A. Nordlinger calls attention to the paradox of the simultaneous existence of ideological (subcultural) cleavages at the mass level and the lack of ideological concerns at the elite level. He rejects the argument that it is "a confluence of the multiparty system and the ideological inundation of French politics that is responsible for the political system's weaknesses . . .," because it conveniently overlooks "the way in which the game of

[73]Lehmbruch, "A Non-Competitive Pattern of Conflict Management," *op. cit.* note 58, p. 6.
[74]Lipson, *op. cit.* note 45, p. 122.

politics is actually played in France. Although ideologism pervades the parties' electoral and propaganda efforts, this public ideological posturing of French politicians does not prevent them from playing out their game of compromise in the Assembly and its *couloirs*."[75] Of the four conditions necessary to turn an ideologically fragmented system toward political stability, at least one is met: the traditional French party elites (with the exception of the left and right extremes) were in fact able to get together and transcend the subcultural cleavages. The first and second conditions are also met to some extent. But what has been conspicuously lacking is the leaders' ability to devote their pragmatism to effective and lasting solutions for the ills of French society.

Nathan Leites also argues that the parliamentary political game in the Fourth Republic, as played by the center parties supporting the Republic, was almost totally non-ideological. But this did not contribute much to the system's stability, because a cardinal rule of the game was the politicians' "well-developed capacity for avoiding their responsibility."[76] They played the political game in the same fashion as the game of *boules* was played in Laurence Wylie's village in the Vaucluse: both players and audience were more interested in the game itself than in its outcome, and most of the fun and excitement arose out of the players' arguments designed to maneuver themselves out of the responsibility for making decisions.[77] Nordlinger's paradox can thus be resolved: it is the combination of both the cleavages in French society and the leaders' inability to take effective countermeasures—they are pragmatic but not purposeful—that accounted for the instability of the French system.

The four prerequisites for consociational democracy listed above all have to do with elite attitudes and behavior. This is, of course, no accident, because the role of leadership is a crucial element in it. Stanley Hoffmann's complaint that "efforts at theory have produced a glut of typologies and models of political systems, often at a level of abstraction that squeezes out the role and impact of political leaders,"[78] clearly does not apply to the consociational type of democracy. The emphasis on the role of the elite has the theoretical advantage of aiding in the explanation of political stability in systems where one would have expected instability. The *explanatory* power of this type is, therefore, quite considerable, but its *predictive* power is for the same reason rather limited. Elite behavior seems to be more elusive and less susceptible to empirical generalization than mass phenomena. If a consociational mode of democracy has been in operation for some time, an analysis of its institutional mechanisms and the elite's operational code would yield some grounds for predicting its successful continuation. But to predict whether an unstable

[75] Eric A. Nordlinger, "Democratic Stability and Instability: The French Case," *World Politics*, XVIII, 1 (October 1965), p. 143.
[76] Nathan Leites, *On the Game of Politics in France* (Palo Alto: Stanford Univ. Press, 1959), p. 2; see also the chapter entitled "The Struggle Against Responsibility," pp. 35–75.
[77] Laurence Wylie, *Village in the Vaucluse: An Account of Life in a French Village* (rev. ed.; N.Y.: Harper and Row, 1964), pp. 250–259.
[78] Stanley Hoffman, "Heroic Leadership: The Case of Modern France" (paper presented at the Annual Meeting of the American Political Science Association, 1966), p. 1.

democracy can or will become more stable by adopting consociational practices is much more difficult, because this entails a deliberate *change* in elite behavior.

The predictive power of the consociational type can be enhanced, however, by identifying certain conditions of the social structure and of mass political culture that are conducive to consociational democracy, in the sense that they facilitate overarching inter-elite cooperation. On the basis of an examination of the most striking features of the five principal cases of consociational democracy—Austria, Belgium, Lebanon, the Netherlands, and Switzerland—a number of hypotheses can be formulated. The following list of conditions favorable to consociational democracy is tentative, and is meant to be illustrative rather than exhaustive:

1. *Distinct lines of cleavage between subcultures.* As indicated earlier, the cases of consociational democracy are characterized by an even more clearly fragmented political culture than the systems belonging to Almond's Continental European type. This appears to be paradoxical, because one would expect instability to increase as fragmentation increases. But at the same time, the probability that there will be a positive response to this situation, which will make the expectation self-denying, will also increase. In addition, clear boundaries between subcultures have the advantage of limiting mutual contacts and, consequently, of limiting the chances of ever-present potential antagonisms to erupt into actual hostility. Quincy Wright argues that "ideologies accepted by different groups within a society may be inconsistent without creating tension." The danger of great tention arises only when these groups "are in close contact. . . ."[79] Easton argues in a similar vein that efforts to homogenize a fragmented system may not be the best way of achieving a stable, integrated system: "Greater success may be attained through steps that conduce to the development of a deeper sense of mutual awareness and responsiveness among *encapsulated* cultural units."[80]

And Lorwin makes the following remark about Belgium which is equally applicable to the other consociational democracies: "If meaningful personal contacts with people of other subcultures are few, so are the occasions for personal hostility."[81] One important factor in the explanation of political stability in religiously and linguistically heterogeneous Switzerland is that many of the cantons, where much of the country's decentralized politics takes place, are quite homoge-

[79] Quincy Wright, "The Nature of Conflict," *Western Polit. Q.*, IV, 2 (June, 1951), 196.

[80] Easton, *op. cit.* note 44, p. 250 (emphasis added). This proposition appears to find increasing acceptance among social scientists. See, for instance, Verba's remark that in modern Africa "differing subcultures [are brought] into contact with each other and *hence* into conflict"—in "Some Dilemmas in Comparative Research," *World Politics*, XX, 1 (Oct., 1967), 126 (emphasis added)—and Walker Connor's rhetorical question: "If one is dealing not with minor variations of the same culture, but with two quite distinct and self-differentiating cultures, are not increased contacts between the two apt to increase antagonisms?"—in "Self-Determination: The New Phase," *loc. cit.* 49–50. See also G. H. Scholten, "Het vergelijken van federaties met behulp van systeem-analyse," *Acta Politica*, II, 1 (1966–67), 51–68.

[81] Val R. Lorwin, "Belgium: Religion, Class, and Language in National Politics," in Dahl, *op. cit.* note 45, p. 187.

neous. For instance, the eight cantons and half-cantons of the Sonderbund plus Ticino and Appenzell Inner Rhoden are not only overwhelmingly Catholic but also linguistically homogeneous with only two exceptions (Valais and Fribourg).[82]

A fragmented but consociational system may be schematically portrayed as follows:

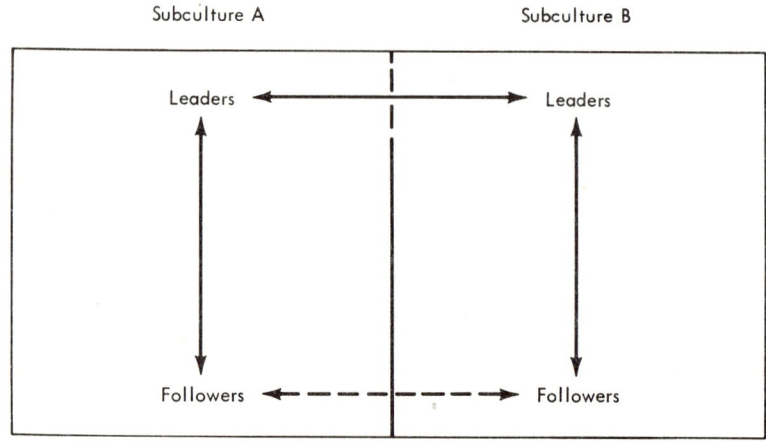

Close relations between the elites are essential, and their cooperative efforts are more likely to be successful if conflict at the mass level can be reduced—that is, if overlapping memberships are kept to a minimum. Furthermore, distinct lines of cleavage between the subcultures also contribute to consociational democracy to the extent that they indicate strong internal cohesion within the subcultures. A major challenge to A's elite is not only the establishment of fruitful cooperation with the leaders of subculture B, but at the same time the retention of the support and loyalty of their own followers. The elites must continually perform a difficult balancing act. There are two vital matters: not just, in Daalder's words, "the extent to which party leaders are more tolerant than their followers," but also the extent to which they "are yet *able to carry them along*..."[83]

With distinct lines of cleavage and with high internal cohesion, this support is more likely to be forthcoming than in a diffuse setting. To quote once more a remark by Lorwin on Belgium, but with wider applicability, the "bases of support [which the parties have] in the spiritual families have given them a certain security in the cooperation entailed by coalition government."[84] This point may be added as a

---

[82] Most of these cantons nevertheless have grand coalition executives; see Girod, *op. cit.* note 72, pp. 137–141. See also Lehmbruch, *Proporzdemokratie, op. cit.* note 62, pp. 33–34. In her doctoral dissertation, "Peasants Against Politics: Rural Organization in Finistere" (Harvard Univ., 1966), Suzanne D. Berger points out that there is no clear link between the degree of isolation and the degree of extremism: peasants prone to radicalism are not the most isolated ones, but those experiencing in some way the influence of the city (p. 12).

[83] Daalder, *op. cit.* note 60, p. 69 (emphasis added).

[84] Val R. Lorwin, "Conflict and Compromise in Belgian Politics" (paper presented at the Annual Meeting of the American Political Science Association, 1965), p. 3. However, deference and discipline within the subcultures have substantially decreased in recent years.

footnote to the explanation of Nordlinger's paradox of French politics. If one blames the French elite for their failure to play the political game responsibly and effectively, one also has to remember that the French subcultures (except the Communists) lack the cohesive social structure and tight organization of their Belgian counterparts, and consequently do not allow their leaders the necessary freedom of action. This is also related to French attitudes in general which, as Michel Crozier states, make it "impossible for an individual member of the group to become its leader."[85]

2. *A multiple balance of power among the subcultures.* If there are several subcultures in a fragmented system and if they are all minorities, there will be more incentive to cooperation and less temptation to attempt domination of rival groups than in a situation of a dual balance of power (or, of course, in the case of a clear hegemony of one group). A multiple balance of power has thus contributed to the success of consociational democracy in Switzerland, the Netherlands, and also in Lebanon. When there are only two major groups balancing each other, each group's tendency to dominate the other and its fear of being dominated complicate inter-elite cooperation Both Belgian linguistic communities (which do not coincide with the country's spiritual families) are afraid of being dominated by the other: the French-speaking Belgians fear the numerical superiority of the Flemish, and the Flemish fear and resent the economic and cultural dominance of the French-speaking element. When a division into two major subcultures is politically expressed as a two-party system, as in Austria, this dual pattern is less conducive to consociational democracy than a multiparty system. Almond and Powell argue that the two-party system of Austria, unlike the Italian multiparty system, has managed to aggregate interests rather effectively, and that this "has served to reduce . . . the relative strain on the Austrian system."[86] When one explains the stability of Austrian democracy in terms of the consociational type, one arrives at the opposite conclusion: Austria's stability was largely due to the cooperation of the rival elites in a grand coalition (until 1966), and the two-party system especially in the earlier years was a strain on this overarching cooperation rather than a support for it.[87]

3. *Popular attitudes favorable to government by grand coalition.* Whereas the first two factors conducive to consociational democracy were to some extent paradoxical, this third factor appears so obvious as to be almost tautological. It is a much too important factor, however, to be omitted. For instance, one of the problems confronting the Austrian grand coalition was the constant criticism of the "undemocratic" nature of grand coalition politics, according to the norm of especially British democracy. The Anglo-American stable two-party system is thus not only a particular empirical type of democracy, but also a *normative model* which may form an obstacle to alternative attempts to establish stable democracy. Similarly, the electoral

[85] Michel Crozier, *The Bureaucratic Phenomenon* (Chicago: Univ. of Chicago Press, 1964), p. 220.
[86] Almond and Powell, *op. cit.* note 12, p. 111.
[87] In Lehmbruch's words: "the bipolar structure of the coalition reinforced their [the parties'] antagonisms"—in "A Non-Competitive Pattern of Conflict Management," *op. cit.*—note 58, p. 8.

system for the Lebanese Chamber of Deputies is influenced by the overlapping memberships idea as a normative model. Proportional representation—the usual rule in consociational democracies—applies, but with an important modification: in each constituency, all voters have to choose among different proportionally constituted lists. The system, therefore, forces each candidate to have the support not only of his own community but also support from outside of his own group. The result is, as Pierre Rondot concludes, that the "typical champions of each community run the risk . . . of being passed over in favor of tamer individuals." Instead of its goal of achieving an integrated and efficient Chamber, the system only serves to make it largely irrelevant. The significant inter-elite contacts take place outside the Chamber: in the Council of Ministers and "noninstitutional substitutes," such as ad hoc or semi-permanent inter-community congresses.[88]

4. *External threats.* Consociational democracy is primarily a response to the national emergency created by internal divisions, but external threats may also contribute significantly to its establishment or reinforcement. In fact, a grave international emergency may cause by itself a temporary resort to consociational practices, as in homogeneous Britain and Sweden during the Second World War. Both countries had grand coalition cabinets, and inter-party cooperation in Britain was even extended to the electoral level by postponing parliamentary elections until 1945.

In the five principal consociational systems, the crucial steps toward this type of democracy were usually taken during times of international crisis or specific threats to the nation's existence. In Austria, Lebanon, and the Netherlands, the inception of consociational democracy can be traced plainly to a particular short span of time in their political histories, and it occurred without exception during, and also partly because of, an international emergency. Austria's government by grand coalition was primarily a response to the civil strife of the First Republic, but it was inaugurated, significantly, while Austria was occupied by the Allied powers after the Second World War. The Lebanese National Pact of 1943—the country's unwritten consociational constitution—was concluded during the Second World War; it followed an insurrection which united the religious communities against the external control of the French under the Mandate. And the comprehensive peaceful settlement of internal differences that paved the way for consociational democracy in the Netherlands was concluded in 1917 when the First World War was raging near its borders. In Belgium and Switzerland, consociational practices were adopted more gradually, but also partly under the influence of foreign threats. Belgian "unionism" (Catholic-Liberal grand coalitions) began during its struggle for independence but became infrequent when the nation's existence appeared to be secure. It was resumed again during the First World War, soon followed by the important step of admitting the Socialists to the consociational government. This final step of admitting the Socialists to the grand coalition was not taken until much later in Switzerland, but it also happened during a World War: in 1943.[89]

[88]Pierre Rondot, "The Political Institutions of Lebanese Democracy," in Leonard Binder (ed.), *Politics in Lebanon* (N.Y.: Wiley, 1966), pp. 133-134.

[89]In most cases, the changes brought about under the influence of external threats have proved quite durable. This contrasts with the finding of Karl W. Deutsch et al., that the

5. *Moderate nationalism.* In order to establish consociational democracy and to make it work, it is essential for the elites to have a commitment to the maintenance of the system Similarly, it is helpful to have some degree of such a commitment at the mass level. National attachment may offset somewhat the unstabilizing effects of deep social cleavages. One might hypothesize, therefore, that a very strong nationalism is particularly conducive to consociational democracy, but this hypothesis conflicts with the evidence of the five principal cases. In all five, there are some nationalistic sentiments but they are not strong at all.

In Belgium, one finds a Flemish nationalism, but not a strong Belgian nationalism. Lorwin states without exaggeration that national sentiment "is weaker . . . than in any other European nation."[90] A 1956 public opinion survey found that Austrians were deeply divided on the basic question of what nation they belonged to: 49% thought that there was an Austrian nation, but 46% considered the Austrians to be merely a part of the German nation, and 5% were undecided.[91] Swiss and Dutch nationalism are also weak, with local attachment being predominant in the former. And Shils reaches the following conclusion about Lebanon: "Lebanese democracy does not rest on a comprehensive countrywide attachment to Lebanon. People may know that they are Lebanese, but this is not as significant a fact for most of them as being Maronite, Orthodox Christian, Sunni, Shi'ite Muslims, or whatever else."[92]

What is the explanation? Two reasons may be tentatively advanced: First, one can argue that political stability exists in these countries *in spite of* their relatively weak nationalism. But, secondly, one can also make the case that their stability is achieved partly *because* their nationalism is moderate. A very strong nationalism is not necessarily a unifying force. Superpatriots tend to have an inflated image of their nation's worth and stature, and tend to attribute its weaknesses both to external and internal enemies. A strong nationalism may thus become a divisive force and a serious danger to an already fragmented society.

6. *A relatively low total load on the system.* The stability of any system can be considered in terms of the balance between its capabilities and the demands placed on it. Any system is more likely to be stable if it does not have to carry too heavy burdens. This is of particular significance in consociational systems. Here the management of subcultural cleavages is already a major burden requiring much of the leaders' energies and skills. This factor is very important in the explanation of the success of consociational democracy in Lebanon. The Lebanese version of this type of democracy is neither very developed nor very efficient. But its society has

---

effects of foreign military threats on integration were never very strong and not always positive, and that, even when positive, "their effects were transitory." See *Political Community and the North Atlantic Area: International Organization in the Light of Historical Experience* (Princeton, N.J.: Princeton Univ. Press, 1957), pp. 44–46. External threats may also contribute substantially to the internal political stability of more homogeneous countries, such as Finland and Israel. Leonard J. Fein argues that in Israel "the threat of annihilation . . . has made painfully manifest the dangers inherent in instability." See his *Politics in Israel* (Boston: Little, Brown, 1967), p. 31.

[90] Lorwin, "Belgium: Religion, Class and Language," *op. cit.* note 81, p. 176.

[91] Rodney Stiefbold et al., *Wahlen und Parteien in Österreich: Österreichisches Wahlhandbuch* (3 vols.; Vienna: Verlag für Jugend und Volk, 1966), Vol. 2, pp. 584–585.

[92] Edward Shils, "The Prospects for Lebanese Civility," in Binder, *op. cit.* note 88, pp. 3–4.

so far been relatively static, and the government's tasks have consequently been relatively simple and limited.

There is an additional characteristic which the five principal cases of consociational democracy have in common: they are all small countries. The largest of the five, the Netherlands, has a population of less than 13 million. The element of size seems to be indirectly related to the success of consociational practices. The last three of the factors listed above are more likely in small than in large countries. External threats are more likely to be considered serious and worthy of an unusual internal response; nationalism is less likely to be of an extreme variety; and the total load on the system is likely to be lighter, if only because of a small country's limited power on the international scene and its greater chance of avoiding difficult decisions in this realm.[93]

CENTRIFUGAL, CONSOCIATIONAL, AND CENTRIPETAL DEMOCRACIES

The consociational type of democracy is one category of a threefold typology of democratic systems. The three types are, in order of increasing stability:

1. *Centrifugal democracy.* This type is essentially the same as Almond's Continental European type, but the term *centrifugal* was chosen because it does not have a misleading geographical connotation and because it accurately indicates the characteristic feature of the system: a fragmented political culture (and poor boundary maintenance among the political structures within each subculture) leading to immobilism and instability. Italy, France under the Third and Fourth Republics, and Weimar Germany are the clearest examples of this type.[94]

2. *Consociational democracy.* The definition and several examples of this type were given above. Its political culture is similar to that of the centrifugal type, but it has achieved considerable stability, and is in that respect similar to the third type.

3. *Centripetal democracy.* This is the healthy type whose stability is based on a homogeneous political culture and is not threatened by normal inter-party competition. It includes not only the Anglo-American democracies and those of the Old Commonwealth, but also Ireland, the Scandinavian countries, Israel, and probably postwar West Germany as well. It can be further divided into the three subtypes of cohesive systems proposed by Harry Eckstein: (a) *consensus* systems, (b) *mechanically integrated* systems, and (c) *community* systems. The cohesion of these systems is a result of different causes. In consensus systems, "cohesion results from a low degree . . . of political division." Mechanically integrated systems do have political divisions but they are not mutually cumulative and, therefore, tend to moderate each

[93] See Paul Y. Hammond, "The Political Order and the Burden of External Relations," *World Politics*, XIX, 3 (April, 1967), 443–464. See also Lehmbruch, "A Non-Competitive Pattern of Conflict Management," *op. cit.* note 58, p. 4.

[94] Northern Ireland, with its distinct Protestant and Catholic subcultures, may also serve as an example, although its instability is not as marked as that of the other countries. It has its own government, but it is no more than an autonomous part of a larger sovereign entity. Moreover, what apparent stability it has is based on coercion rather than consensus. The Protestant subculture is numerically much larger than the Catholic subculture, and is consequently permanently in power. See Denis P. Barritt and Charles F. Carter, *The Northern Ireland Problem: A Study in Group Relations* (London: Oxford Univ. Press, 1962).

other, according to Truman's overlapping memberships theory. Finally, the community systems also have divisions, but these are in effect largely neutralized by "overarching sentiments of solidarity, whatever their source may be."[95] Eckstein's examples are, respectively, England, the United States, and Norway. The homogeneity of most of the centripetal democracies can be explained in terms of a combination of these cohesive tendencies. For instance, Truman's explanation of the stability of the American system does not rely solely on the "mechanical integration" of overlapping memberships. He also relies on what Eckstein calls "overarching sentiments of solidarity," or in his own words, "those interests or expectations that are so widely held in the society and are so reflected in the behavior of almost all citizens that they are, so to speak, taken for granted."[96]

This threefold typology differs in many respects from the typology proposed by Almond, but there are also certain similarities. In the first place, there is considerable overlap among the categories of these two typologies. The centrifugal type is by and large the same as Almond's Continental European or immobilist multiparty type. The centripetal democracies include the Anglo-American system but also Scandinavia. Of Almond's third category, the "in between" type of working multiparty system, the Scandinavian democracies belong to the centripetal type and the Low Countries—plus Switzerland and Austria—to the consociational type. Secondly, both typologies have their theoretical basis in the overlapping memberships theory. The typology proposed here differs from Almond's by adding the idea that this theory may become self-denying. It is in this sense that the category of consociational democracies is in between the centrifugal and centripetal types, not in the sense that they have an intermediate degree of overlap.

Thirdly, both typologies may be stated in terms of political culture, but the typology proposed here makes a distinction between elite and mass political culture. Actually, this third point is not so much an additional similarity as a restatement of the second similarity in different words. The crucial distinction between the two typologies concerns the attitudes and actions of the political elites in a potentially, but not inevitably, unstable system. Almond and Verba do call attention to the importance of *role cultures*—the political orientations of various elite groups: bureaucratic, military, party, interest group, etc.—to the operation of democracy. They argue that elite attitudes may be characterized by cultural heterogeneity, and thus "seriously affect the performance of political systems," if the elites are recruited from particular subcultures or if "the process of induction and socialization into these roles produces different values, skills, loyalties, and cognitive maps." But they admit that role culture may be *progressive* as well as *regressive*.[97] This is particularly true in consociational democracies, where the elites constitute an integrating force and are committed to democracy, even though they are recruited from different subcultures. And to the three sources of political attitudes mentioned by Almond and Verba—

[95] Harry Eckstein, *Division and Cohesion in Democracy: A Study of Norway* (Princeton, N.J.: Princeton Univ. Press, 1966), pp. 193-194. Eckstein's threefold typology of cohesive democracies does not include the consociational system.

[96] Truman, *op. cit.* note 36, p. 512.

[97] Almond and Verba, *op. cit.* note 40, pp. 29-30.

early socialization experiences, later socialization experiences during adolescence, and "post-socialization experiences as an adult"[98]—a fourth source must be added: the elites may be shrewd students of social history; they may not be only political practitioners but also capable, though amateur, theorists.

The proposed threefold typology does not correspond to the same extent with the number-of-parties typology in either its simple dichotomous form or in the refined version expounded by Sartori. As indicated earlier, Sartori distinguishes not only between two-party and multiparty systems, but also further between moderate multiparty systems (with three or four politically significant parties) which tend to be stable, and extreme multiparty systems (with five or more parties) which tend to be unstable. According to this criterion, however, both Norway and the Netherlands are extreme multiparty systems. These countries have five major durable parties, all of which have borne government responsibility. But they are also stable democracies, quite unlike Italy, Weimar Germany, and France under the Fourth Republic, which Sartori mentions as the typical examples of extreme multiparty systems. This problem can be avoided to some extent by fixing the line dividing the moderate from the extreme systems between five and six parties rather than between four and five. But this does not solve the problem of classifying the interwar Dutch system with its six separate parties. All six were sizable parties, and they all participated in cabinet coalitions at one time or another.

In the case of the Netherlands between the two World Wars, there can be no reasonable doubt that this was indeed a six-party system, but the question of which parties ought to be counted usually remains a major problem in classifying democracies according to the number of parties. Very small and insignificant parties should be disregarded, but when does a party become sufficiently large or important to be counted? Sartori explains in a footnote that "party numbers cannot be taken at their face value; they must be interpreted. A party system does not acquire different properties simply because some splinter party may happen to win a few seats, or because of the existence of very small marginal parties which play no significant role." He suggests that only those parties should be counted which have either "coalition potential" or "blackmail potential" (such as the French and Italian Communist parties).[99] However, the most objective and straightforward way of comparing the numbers and sizes of political parties in different systems is to examine the cumulative percentages of party strengths in descending order of party size: first the percentage of the total vote received by the largest party, then the combined percentage of the vote received by the two largest parties, etc. Such a comparison of the eight parties of Italy, which is Sartori's major contemporary example of an extreme multi-party system, with the same number of parties in several other multi-party systems yields interesting results. In the following table, the 1963 election results are used for Italy; for the other countries, the results of their parliamentary elections either in 1963 or as close as possible to this year are given:

[98] *Ibid.*, p. 326.
[99] Sartori, "European Political Parties," *op. cit.* note 47, p. 139 n.; *Parties and Party Systems, op. cit.* note 50, chap. 8.

	ITALY CHAMBER OF DEPUTIES (1963) %	SWITZERLAND NATIONAL COUNCIL (1963) %	NETHERLANDS SECOND CHAMBER (1963) %	DENMARK FOLKETING (1964) %	NORWAY STORTING (1965) %
Largest party	38.2	26.6	31.9	41.9	43.1
Two largest parties	63.5	50.6	59.9	62.7	64.2
Three largest parties	77.3	74.0	70.2	82.8	74.6
Four largest parties	84.3	85.4	78.9	88.6	84.5
Five largest parties	90.3	90.4	87.5	93.9	92.6
Six largest parties	95.5	92.6	90.5	96.4	98.6
Seven largest parties	97.2	94.8	93.3	97.7	100.0
Eight largest parties	98.8	96.6	95.6	98.9	

With only a few exceptions, the Swiss and Dutch percentages at each level are lower than the figures for Italy. If Italy is an extreme multiparty system, Switzerland and the Netherlands must be similarly classified.[100] Furthermore, the percentages for Denmark and Norway are generally, but not always, higher than those for Italy, but never by a very wide margin: the maximum difference is only about 5%. It is impossible, therefore, to make a clear distinction, based on the number and sizes of parties, between Italy and the other four countries. This means that there is no clear connection between the extreme multiparty type and democratic stability. This is indeed a fatal flaw, because it robs the typology of its major claim to theoretical significance.

The feature that does distinguish Italy from the other multiparty systems is the presence of a large Communist party, and Sartori uses the wide spread of the political spectrum as an additional criterion for defining extreme multiparty systems. This has nothing to do with the number of parties, however; extreme polarization may occur even in two-party systems, as in the Austrian First Republic. It is another way of saying that Italian political culture is fragmented by deep subcultural cleavages, which the elites have so far not been able or willing to bridge. This criterion thus resolves into one of the criteria of the classification of democracies into centripetal, consociational, and centrifugal systems proposed in this paper. Italy is a centrifugal democracy. The failure to achieve greater stability by adopting consociational practices must be blamed not only on the alleged anti-system tendencies of the Communists but also to some extent on the Christian Democratic leaders, particularly if one accepts Sidney G. Tarrow's judgment that "it is not clear that there is anything about the PCI [Communists] which would justify its classification as an anti-system party or a centrifugal force."[101]

[100]This also applies to Finland, which Sartori himself assigns to the extreme multiparty category, and to Israel, which Sartori calls "definitely atypical"—in "European Political Parties,' *op. cit.* note 47, p. 160n.
[101]Sidney G. Tarrow, "Political Dualism and Italian Communism," *Am. Polit. Sci. Rev.*, LXI, 1 (March, 1967), 40. See also Robert H. Evans, *Coexistence: Communism and Its Practice in Bologna, 1945–1965* (Notre Dame, Ind.: Univ. of Notre Dame Press, 1967).

The theoretical utility of schemes of classification depends on their purpose. If one is concerned with the quality of democracy, the number-of-parties typology may be retained, but if one's major theoretical concern is democratic stability, it ought to be abandoned. There is no strong empirical relationship between the number of parties in the system and its stability. Two-party systems may be centripetal (e.g., England), centrifugal (the Austrian First Republic), or consociational (the Austrian Second Republic until 1966). Moderate multiparty systems may be centripetal (Sweden) or consociational (Belgium). And extreme multiparty systems may also be centripetal (Norway), centrifugal (Italy), or consociational (the Netherlands). Consociational Lebanon cannot be classified according to the number of parties at all: there is an extreme multiplication of parties, but they have never occupied more than about a third of the seats in the Chamber.[102] Consequently, if the threefold typology proposed here is a natural classification for the purpose of analyzing the stability of democratic systems, the typologies based on the number of parties are artificial.

DEPOLITICIZED DEMOCRACY IN THE NEW EUROPE

The classification of democracies into the three types of centripetal, consociational, and centrifugal systems must be further elaborated in two respects. In the first place, the terms "classification" and "typology" have so far been used interchangeably. However, a typology may be regarded not as just any classificatory scheme but more specifically as a scheme for the classification of a series of interrelated models or ideal types. The three types of democratic systems may be regarded as such abstract models, of which no pure empirical instances exist. When a particular system is labeled as a consociational democracy, this means merely that its features correspond more closely to those of the consociational model than of the other models.

If, on the other hand, the typology is used as a scheme for classifying empirical cases, it is advisable to think of the criteria defining the types (categories) in terms of continua rather than dichotomies. There are no examples of either completely homogeneous or completely fragmented systems; all actual democracies fall somewhere in between these two extremes. Similarly, there are no examples of either pure grand coalition government or pure democratic competition without any consociational features; in practice, this is a matter of degree. The figure below presents the two continua on the horizontal and vertical axes. [See p. 77.]

If the two axes are so related that every point on the vertical axis represents the degree of consociation required to maintain stability in democracies characterized by a corresponding degree of cultural fragmentation on the horizontal axis, the diagonal line divides the stable from the unstable systems. The stable democracies are located above the line and the unstable ones below it. As fragmentation increases, the chances of achieving stability decrease but are not reduced to zero. A system's stability increases by moving upward or to the left, or, of course, in both directions at the same time. For instance, the Austrian First Republic should be located in the lower right-hand corner, well below the diagonal line. After the Second World War,

---

[102] Michael W. Suleiman, *Political Parties in Lebanon: The Challenge of a Fragmented Political Culture* (Ithaca, N.Y.: Cornell Univ. Press, 1967).

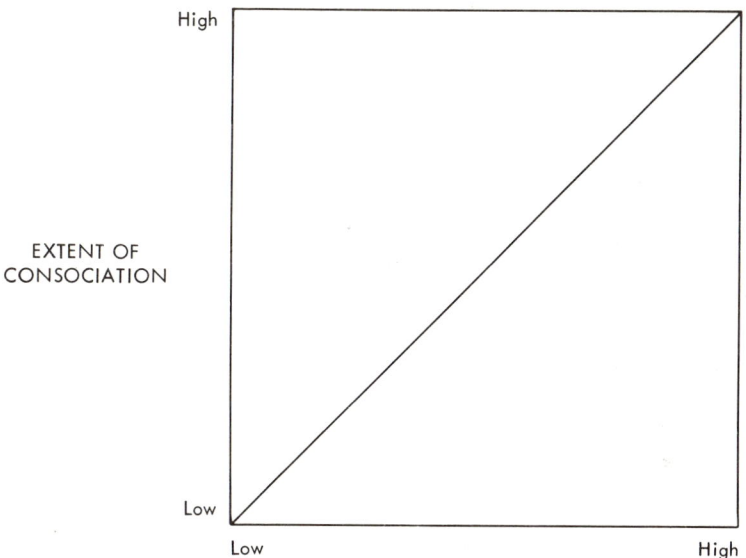

Austria moved into the upper right-hand corner above the diagonal line with the establishment of the Catholic-Socialist grand coalition, but also slowly toward the left as the hostility between subcultures gradually decreased. It moved down again in 1966 when the era of grand coalition cabinets came to an end, but because this did not mean the abandonment of all consociational practices and because fragmentation had significantly decreased, Austria will probably remain in the area of stability above the diagonal dividing line.

Such a lessening of ideological and religious tensions has occurred not only in Austria but almost everywhere in democratic Europe. Cultural fragmentation has been reduced and homogeneity increased in the years since the Second World War. Ernst B. Haas expresses this in terms of the overlapping memberships proposition: "If a citizen can bestow his support—or his indifference—to differing groups for purposes of education, welfare measures, religion, defense, recreation and ethnic identity, the logic of a pluralism based on cross-cutting cleavages will continue to mute ideology. Only if a citizen relies on his party or association for *all* of these aims will the logic of pluralism be defeated. . . . In the New Europe, however, this clustering of affections and expectations seems to be a thing of the past."[103] And as the democracies move toward the New Europe, this decrease of fragmentation and increase of overlapping loyalties will tend to bring greater political stability and, concomitantly, reduce the need for resorting to consociational practices.

[103] Ernst B. Haas, "Technocracy, Pluralism and the New Europe," in Stephen R. Graubard (ed.), *A New Europe?* (Boston: Houghton Mifflin, 1964), pp. 69–70. See also George Lichtheim, *The New Europe: Today—and Tomorrow* (2nd ed.; N.Y.: Praeger, 1964), esp. chap. 5.

It appears somewhat paradoxical, therefore, that the typical democracy of the New Europe is ruled in the grand coalition fashion. Especially in the economic realm, the trend is toward "democratic planning" which features, in Haas' words, "the continuous participation of *all major voluntary groups* in European society through elaborate systems of committees and councils."[104] Norway may serve as a specific example. Here, according to Stein Rokkan, the crucial economic decisions are taken in "yearly rounds of negotiations" among all interested groups: "the government authorities meet directly with the trade union leaders, the representatives of the farmers, the smallholders, and the fishermen, and the delegates of the Employers' Association" around the bargaining table.[105] The equivalent of this trend in American democracy is what Theodore Lowi has termed "interest-group liberalism." The essence of this philosophy, favored especially during the Kennedy Administration, is succinctly stated by Arthur Schlesinger, Jr.: "The leading interests in society are all represented in the interior processes of policy formation—which can be done only if members or advocates of these interests are included in key positions of government. . . ."[106] This "New America" idea is strikingly similar to the typical pattern of grand coalition politics in the New Europe.

The model democracy of the New Europe is characterized both by cultural homogeneity and by consociational patterns of government. In the figure presented above, it must be located in the upper left-hand corner. This means that a fourth type must be added to the typology of democratic systems. In order to avoid once again any possible unfortunate geographical connotations, this type will be referred to as *depoliticized* democracy rather than as the New Europe. The complete typology is as follows [on page 79].

The depoliticized type of democracy should have the greatest stability, followed in descending order by the centripetal, consociational, and centrifugal types. From the point of view of the quality of democracy, the depoliticized type does not rank so high. Most of the criticisms of the consociational type also apply to depoliticized democracy. Lowi condemns "interest-group liberalism" on a number of grounds, including its tendency to result in "an oligopolistic situation" and "the atrophy of institutions of popular control."[107] Just as the centripetal type as a normative model may adversely affect the operation of consociational democracy, it may also undermine the stability of depoliticized democracy. Dahl speculates that one of the sources of conflict in the democracies of the future may be "the new democratic Leviathan itself." There are already signs, he argues, "that many young people, intellectuals, and academics reject the democratic Leviathan . . . because, in their view, it is not democratic enough: this new Leviathan is too remote and bureaucratized, too addicted to bargaining and compromise, too much an instrument of political elites and technicians with whom they feel slight identification."[108]

[104] Haas, *op. cit.*, p. 68 (emphasis added).
[105] Stein Rokkan, "Norway: Numerical Democracy and Corporate Pluralism," in Dahl, *op. cit.* note 45, p. 107.
[106] Cited by Theodore Lowi, "The Public Philosophy: Interest-Group Liberalism," *Am. Polit. Sci. Rev.*, LXI, 1 (March, 1967), p. 15.
[107] *Ibid.*, pp. 18, 23.
[108] Dahl, *op. cit.* note 45, pp. 399–400.

TYPOLOGIES OF DEMOCRATIC SYSTEMS 79

POLITICAL CULTURE

	Homogeneous	Fragmented
Coalescent	Depoliticized democracy	Consociational democracy
Competitive	Centripetal democracy	Centrifugal democracy

ELITE BEHAVIOR

The experience of the consociational democracies should be scrutinized in order to throw light on this problem. In Kirchheimer's famous analysis of Austria's grand coalition politics, conducted in this spirit, he points out that for example the Austrian practice of *Bereichsopposition* ("opposition to what is happening under the agreed-upon jurisdiction of the other party"), as an alternative to the classic opposition, may contain useful lessons applicable to the waning of opposition in depoliticizing democracies as well.[109]

There are crucial differences between depoliticized and consociational democracy, however, and these imply definite limits to the extent to which the two types may be fruitfully compared. The abandonment of strictly competitive politics in consociational democracies is a deliberate response to the tensions of a fragmented society, whereas the adoption of grand coalition politics in depoliticized democracy is in response to the convergence of ideologies. In the latter, depoliticization occurs as a natural consequence of growing consensus, whereas it has to be imposed on an inherently highly political system in the former. An even more important, though closely related, difference is that grand coalition patterns in consociational democracies typically occur at the highest level: the party leaders are the pivotal political actors. In the depoliticized type of democracy, on the other hand, decision-making

---

[109] Kirchheimer, *op. cit.* note 62, pp. 127–156. See also Karl Dietrich Bracher, "Problems of Parliamentary Democracy in Europe," and Charles Frankel, "Bureaucracy and Democracy in the New Europe," in Graubard, *op. cit.* note 103, pp. 245–264, 538–559.

takes place in grand coalitions at lower levels: interest group representatives and bureaucrats are now the principal actors.

As the industrialized democracies of the West continue along the path of depoliticization, the questions of the quality and stability of democratic government will remain of crucial importance both to the scholar and to the citizen. The principal challenge will be to tame the "democratic Leviathan," and to keep it as democratic as possible—not only as an end in itself, but also because the stability and survival of democracy indirectly depend on it.

Part Two

THE BRITISH POLITICAL SYSTEM

INTRODUCTION

The five chapters in this part of the book deal comparatively with Great Britain. Winston Churchill's speech to the House of Commons on October 28, 1943, in Chapter 6, eloquently compares the characteristics of the British parliamentary system—especially the spirit and organization of the House of Commons—with those of other democracies. And, even though he indicates that he does not wish to make any "invidious criticisms of other nations," he proudly concludes that the British model serves democracy best.

Chapters 7 and 8 entail comparisons between Britain and the United States. Samuel H. Beer subjects the political parties and interest groups of the two countries to an intensive analysis, stressing the differences rather than the similarities. This has also been the tenor of traditional comparative treatments, which contrast the presidential and federal system of the United States with the parliamentary (or cabinet) and unitary system of Britain. (See, for instance, the lengthy debate on the parliamentary vs. the presidential system by Don K. Price and Harold J. Laski in the Public Administration Review *in 1943 and 1944.) On the other hand, some scholars have tended to emphasize the similarities of the two political systems. For instance, Gabriel A. Almond and Arend Lijphart argue that both systems belong to the same basic type (see Chapters 4 and 5 in Part I). In Chapter 8, Thomas J. Carbery presents a stimulating, and intentionally overstated, analysis of the disappearance of many of the striking differences between the two countries. British politics, he argues, is becoming more and more like American politics. His analysis, and his belief that this trend is likely to continue, challenge the reader to a consideration of the question how far this convergence can proceed, given the basic institutional differences between the two countries.*

Chapter 9 demonstrates how the comparative method can be used fruitfully in the discovery of general principles that govern politics. By comparing five countries which are similar in many respects—Britain and the four "older" democracies of the Commonwealth—Leslie Lipson is able to reach some general conclusions about the determinants of different party systems, thus making for a better understanding of the British system. Lipson's concluding sentence is worth repeating here: "The merit of a comparative method is that it serves, better than any other, to illuminate these relationships which, if observed within a single system, are more likely to remain obscure."

The final chapter in Part II is a comparison of party affiliation and opinions about international issues in Great Britain and France. Morris Davis and Sidney Verba adroitly use the results of many public opinion surveys in order to reach significant conclusions concerning both the differences and the similarities among party supporters in the two countries.

The reader is also referred to Chapter 20 (in Part IV on German politics), in which Giuseppe Di Palma compares disaffection, participation, and opposition in four countries, including Great Britan.

ABOUT THE AUTHORS

WINSTON S. CHURCHILL *was the prime minister of England during World War II and again from 1951 to 1955. He is generally considered to be one of the most brilliant politicians and orators of the twentieth century.*

SAMUEL H. BEER, *professor of government at Harvard University, is an expert on British politics. He authored* The Coordination of Financial and Economic Policy in Great Britain *and* British Politics in the Collectivist Age. *He is also the co-editor of* Patterns of Government.

THOMAS J. CARBERY *is a lecturer at the University of Strathclyde in Glasgow. His field of interest is comparative politics, and he is particularly interested in American-British comparisons.*

LESLIE LIPSON, *professor of political science at the University of California in Berkeley, is an expert on Great Britain and the Commonwealth, especially New Zealand, about which he wrote his book* The Politics of Equality. *He has also written* The American Governor, The Great Issues of Politics, *and* The Democratic Civilization.

MORRIS DAVIS, *associate professor of political science at the University of Illinois, has written extensively on European politics.*

SIDNEY VERBA, *professor of political science at the University of Chicago, has made significant contributions to both comparative politics and international relations. He is the author of* Small Groups and Political Behavior; *co-author, with Gabriel A. Almond, of* The Civic Culture; *and co-editor of* The International System *and* Political Culture and Political Development.

CHAPTER 6

Rebuilding the House of Commons

WINSTON S. CHURCHILL

I beg to move,

> "That a Select Committee be appointed to consider and report upon plans for the rebuilding of the House of Commons and upon such alterations as may be considered desirable while preserving all its essential features."

On the night of 10th May, 1941, with one of the last bombs of the last serious raid, our House of Commons was destroyed by the violence of the enemy, and we have now to consider whether we should build it up again, and how, and when. We shape our buildings and afterwards our buildings shape us. Having dwelt and served for more than 40 years in the late Chamber, and having derived very great pleasure and advantage therefrom, I, naturally, would like to see it restored in all essentials to its old form, convenience and dignity. I believe that will be the opinion of the great majority of its Members. It is certainly the opinion of His Majesty's Government and we propose to support this resolution to the best of our ability.

There are two main characteristics of the House of Commons which will command the approval of the support of reflective and experienced Members. They will, I have no doubt, sound odd to foreign ears. The first is that its shape should be oblong and not semi-circular. Here is a very potent factor in our political life. The semi-circular assembly, which appeals to political theorists, enables every individual or every group to move around the centre, adopting various shades of pink according as the weather changes. I am a convinced supporter of the party system in preference to the group system. I have seen many earnest and ardent Parliaments destroyed by the group system. The party system is much favoured by the oblong form of Chamber. It is easy for an individual to move through those insensible gradations from Left to Right but the act of crossing the Floor is one which requires serious consideration. I am well informed on this matter, for I have accomplished that difficult process, not only once but twice. Logic is a poor guide compared with custom. Logic which has created in so many countries semi-circular assemblies which have buildings which give to every Member, not only a seat to sit in but often a desk to write at, with a lid to bang, has proved fatal to Parliamentary Government as we know it here in its home and in the land of its birth.

Reprinted from PARLIAMENTARY DEBATES: HOUSE OF COMMONS, *October 28, 1943 (393 H.C. Deb. 5s., cols. 403–9).*

The second characteristic of a Chamber formed on the lines of the House of Commons is that it should not be big enough to contain all its Members at once without over-crowding and that there should be no question of every Member having a separate seat reserved for him. The reason for this has long been a puzzle to uninstructed outsiders and has frequently excited the curiosity and even the criticism of new Members. Yet it is not so difficult to understand if you look at it from a practical point of view. If the House is big enough to contain all its Members, nine-tenths of its Debates will be conducted in the depressing atmosphere of an almost empty or half-empty Chamber. The essence of good House of Commons speaking is the conversational style, the facility for quick, informal interruptions and interchanges. Harangues from a rostrum would be a bad substitute for the conversational style in which so much of our business is done. But the conversational style requires a fairly small space, and there should be on great occasions a sense of crowd and urgency. There should be a sense of the importance of much that is said and a sense that great matters are being decided, there and then, by the House.

We attach immense importance to the survival of Parliamentary democrcy. In this country this is one of our war aims. We wish to see our Parliament a strong, easy, flexible instrument of free Debate. For this purpose a small Chamber and a sense of intimacy are indispensable. It is notable that the Parliaments of the British Commonwealth have to a very large extent reproduced our Parliamentary institutions in their form as well as in their spirit, even to the Chair in which the Speakers of the different Assemblies sit. We do not seek to impose our ideas on others; we make no invidious criticisms of other nations. All the same we hold, none the less, tenaciously to them ourselves. The vitality and the authority of the House of Commons and its hold upon an electorate, based upon universal suffrage, depends to no small extent upon its episodes and great moments, even upon its scenes and rows, which, as everyone will agree, are better conducted at close quarters. Destroy that hold which Parliament has upon the public mind and has preserved through all these changing, turbulent times and the living organism of the House of Commons would be greatly impaired. You may have a machine, but the House of Commons is much more than a machine; it has earned and captured and held through long generations the imagination and respect of the British nation. It is not free from shortcomings; they mark all human institutions. Nevertheless, I submit to what is probably not an unfriendly audience on that subject that our House has proved itself capable of adapting itself to every change which the swift pace of modern life has brought upon us. It has a collective personality which enjoys the regard of the public and which imposes itself upon the conduct not only of individual Members but of parties. It has a code of its own which everyone knows, and it has means of its own of enforcing those manners and habits which have grown up and have been found to be an essential part of our Parliamentary life.

The House of Commons has lifted our affairs above the mechanical sphere into the human sphere. It thrives on criticism, it is perfectly impervious to newspaper abuse or taunts from any quarter, and it is capable of digesting almost anything or almost any body of gentlemen, whatever be the views with which they arrive. There is no situation to which it cannot address itself with vigour and ingenuity. It is the citadel of British liberty; it is the foundation of our laws; its traditions and its privileges are as lively to-day as when it broke the arbitrary power of the Crown and substituted that Constitutional Monarchy under which we have enjoyed so many blessings. In this war the House of Commons has proved itself to be a rock upon which an Administration, without losing the confidence of the House, has been able to confront the most terrible emergencies. The House has shown itself able to face the possibility of national destruction with classical composure. It can change Governments, and has changed them by heat of passion. It can sustain Governments in long, adverse, disappointing struggles through many dark, grey months and even years until the sun comes out again. I do not know how else this country can be governed other than by the House of Commons playing its part in all its broad freedom in British public life. We have learned—with these so recently confirmed facts around and before us—not to alter improvidently the physical structures which have enabled so remarkable an organism to carry on its work of banning dictatorships within this island and pursuing and beating into ruin all dictators who have molested us from outside.

* * *

His Majesty's Government are most anxious and are indeed resolved to ask the House to adhere firmly in principle to the structure and characteristics of the House of Commons we have known, and I do not doubt that that is the wish of the great majority of the Members in this the second longest Parliament of our history. If challenged, we must take issue upon that by the customary Parliamentary method of Debate followed by a Division. The question of Divisions again relates very directly to the structure of the House of Commons. We must look forward to periods when Divisions will be much more frequent than they are now. Many of us have seen 20 or 30 in a single Parliamentary Sitting, and in the Lobbies of the Chamber which Hitler shattered we had facilities and conveniences far exceeding those which we are able to enjoy in this lordly abode. I am, therefore, proposing in the name of His Majesty's Government that we decide to rebuild the House of Commons on its old foundations, which are intact, and in principle within its old dimensions, and that we utilise so far as possible its shattered walls. That is also the most cheap and expeditious method we could pursue to provide ourselves with a habitation.

I now come to some of the more practical issues which are involved. It is said that we should wait until the end of the war. . . . I must tell you, Mr. Speaker, that it would be a real danger if at the end of the war we find ourselves separated by a long period from the possibility of obtaining a restored and suitable House of Commons Chamber. We are building warships that will not be finished for many years ahead, and various works of construction are going forward for war purposes. But I am bound to say that I rank the House of Commons—the most powerful Assembly in the whole world—at least as important as a fortification or a battleship, even in time of war. Politics may be very fierce and violent in the after-war days. We may have all the changes in personnel following upon a General Election. We shall certainly have an immense press of Business and, very likely, of stormy controversy. We must have a good, well-tried and convenient place in which to do our work. The House owes it to itself, it owes it to the nation, to make sure that there is no gap, no awkward, injurious hiatus in the continuity of our Parliamentary life. I am to-day only expressing the views of the Government, but if the House sets up the Committee and in a few months' time the Committee give us their Report, we shall be able to take decisions together on the whole matter, and not be caught at a disadvantage in what must inevitably be a time of particular stress and crisis at the end of the war, from a Parliamentary point of view. Therefore, I ask that the Committee should be set up, and I feel sure that it will be able to make a good plan of action leaving the necessary latitude to the Government as to the time when this action can be taken and the speed at which it can be carried into effect having regard to the prime exigencies of the war.

We owe a great debt to the House of Lords for having placed at our disposal this spacious, splendid hall. We have already expressed in formal Resolution our thanks to them. We do not wish to outstay our welcome. We have been greatly convenienced by our sojourn on these red benches and under this gilded, ornamented, statue-bedecked roof. I express my gratitude and appreciation of what we have received and enjoyed, but

> "Mid pleasures and palaces though we may roam,
> Be it ever so humble, there's no place like home."

CHAPTER 7

Group Representation in Britain and the United States

SAMUEL H. BEER

We usually think of Great Britain as a country of strong parties and weak pressure groups; the United States as a country of weak parties and strong pressure groups. I wish to suggest some contrary views: that not only are British parties strong, but so also are British pressure groups; that in comparison both American parties and pressure groups are weak. The terms "strong" and "weak" cry out to be defined. The meanings I give them derive from a historical development—the rise of "collectivism"—that has similarly affected both parties and pressure groups.

What are the consequences for policy? Strong parties can more readily resist

Reprinted from Samuel H. Beer, "Group Representation in Britain and the United States," ANNALS OF THE AMERICAN ACADEMY OF POLITICAL AND SOCIAL SCIENCE, *CCCXIX (September, 1958), 131–40, by permission of the author and the publisher.*

pressure groups. They can also more readily yield them what they want. On the other hand, the dispersion of power may simply produce a self-defeating war of all against all in which even the special interests suffer. Centralized power at least creates the possibility of deliberate and orderly solutions.

THE COLLECTIVIST ECONOMY

The virtue of centralized power is worth examining if for no other reason than that the opposite doctrine holds so high a place in liberal democratic thought. Liberals and Radicals in both Britain and America have applied the doctrine of dispersed power to both the economy and the polity. In the Smithian model of the economy, for instance, the wealth of the nation and the satisfaction of consumers' wants will be maximized if the market is free. No unit, not even government, is to exercise "market power." Once power is removed, rational and voluntary exchange will result and along with it other desirable consequences in the allocation of resources and the satisfaction of the consumer.

Very similar is the Liberal-Radical model of the polity. Remove Burke's "established" aristocracy and all other agents of power that had historically guided the political process; reduce society to its individual, rational atoms: then, power removed, reason will reign. A free, competitive marketplace of ideas, automatic and self-regulating like the marketplace of the laissez-faire economy, will test the truth of opinions. Upon opinions so tested, popular government will base public policy.

In both the British and American economies in the nineteenth century, the market conditions required by the self-regulating model did actually exist in very great degree. And in both, to no inconsiderable extent, these conditions still exist. But in the past two generations or so, certain structural changes have taken place—reaching a further point of development in Britain than in the United States—that depart radically from this model. These developments, which we may call "collectivism," can be summarized under four headings. One is the tendency to a concentration of economic power among a few large buyers or sellers in a particular industry or complex of industries. Along with the increase in size of units has gone a change in internal structure that is referred to by terms such as bureaucracy and managerialism. Moreover, where such large units have grown up, they tend to deal with one another by a process of "bargaining"—or perhaps it is better to say, "collective bargaining." Finally, while bargaining tends to be confined to the relations of producers—whether business firms or trade unions—in their dealings with the mass of ultimate consumers, large units have learned to shape, even to create, the very "wants" that presumably they have come into existence to satisfy.

COLLECTIVIST PARTIES

In the polity as in the economy, there have been similar tendencies toward collectivism. By this I do not mean the increase in government intervention—the rise of the welfare state and the controlled economy. I mean rather that in the political structure have occurred certain changes analogous to those changes in economic structure summarized above. Starting from these contrasting models of the polity, the self-regulating and the collectivist, we may compare the distribution of power in Britain and the United States. It would appear that, as economic collectivism has proceeded farther in Britain than in the United States, so also has political collectivism.

We may look first at the relative number of units and their internal structure. Examined in the light of these criteria, both British parties and pressure groups present striking contrasts with the American models. While in both polities there are two major parties, the loose and sprawling parties of American politics make the British appear highly concentrated. In the American party, power is dispersed among many units—for example, personal followings or state and local machines—with the result that only occasionally and for limited purposes, such as nominating a Presidential candidate, does the national party act as a unit. In terms of density—that is, the per cent of eligibles organized as party members—American parties exceed British. But if we apply a measure of intensity, such as payment of dues, it is clear that British parties have mobilized the electorate far more highly than have American. In the British party, moreover, this membership is brought together for unified action by an elaborate and effective system of articulation, in particular active representative bodies extending from bottom to top and a bureaucratic staff running from top to bottom. There are still semiautonomous centers within the party that a perfected merger would obliterate. But to an American, a British party is a highly unified social body, remarkably well equipped for co-ordinated action: we think, for instance, of the fact that all candidates must be approved by a central-party agency and that they will all run on the same platform. No doubt, the most striking expression of this power of united action is the extent of party voting in the House of Commons. Judged even by Lowell's strict criteria, party voting has been on the increase for a hundred years and today reaches nearly one hundred per cent.[1]

Along with such concentration, and perhaps making it possible, goes a high measure of political homogeneity. (I do not mean social homogeneity, for, measured by nonpolitical criteria, the British are a very heterogeneous people.) This political homogeneity in the electorate as a whole is reflected in what students of voting behavior call the "nationalizing" of British politics. When political opinion moves, it moves in unison throughout the country: in a general election the "swing" from one party to the other is much the same in every constituency. In the United States, as Schattschneider and Paul David have shown, voting has also tended in this direction.[2] Sectionalism and the number of one-party states are on the decline. But—as 1956 illustrates—nothing like the uniformity of swing in British voting has been reached.

[1] Lowell counted as a party vote a division in which at least 90 per cent of one party voted in favor and at least 90 per cent of the other party voted against. A. L. Lowell, "The Influence of Party upon Legislation in England and America," *Annual Report of the American Historical Association for 1901*, Vol. 1 (Washington, 1902), pp. 319–542.

[2] E. E. Schattschneider, "The United States: The Functional Approach to Party Government," in Sigmund Neumann (Ed.), *Modern Political Parties* (Chicago: University of Chicago Press, 1956), pp. 194–215. Paul David, "Intensity of Inter-Party Competition and the Problem of Party Realignment"; a paper presented at the meeting of the American Political Science Association, September 5–7, 1957.

In spite of mass membership and representative bodies, however, the internal structure of the British party gives great power to central party leaders—far more, of course, than that possessed by American leaders. It is rather as if the Congressional caucus of post-Federalist days had been imposed upon the Jacksonian party system. In both British parties, as R. T. McKenzie has shown, the leaders of the parliamentary party, and especially the Leader, are dominant.[3] That is a loose description and needs must be, since the Leader's power is complex and certainly far from dictatorial. He must continually practice "the art of management," appeasing a dissident faction, finding a formula, keeping up party morale. Indeed, he is a "manager"—a modern-day manager committed to party principle, of course, but by his function compelled above all to think of the continuation of the organization.

COLLECTIVIST PRESSURE GROUPS

Turning from parties to pressure groups, we find that in Britain as in the United States, the center of the stage is occupied by organizations based on the great economic interest groups of modern society, especially the big three of business, labor, and agriculture. Given the nature of public policy, which affects these interests so often and so intimately, pressure groups claiming to speak for them are bound in turn to influence policy making more frequently and on the whole more effectively than pressure groups of other types.

In Britain as well as the United States, in addition to such "self-oriented" pressure groups, we must also deal with what S. E. Finer calls "promotional" groups.[4] Among the former we may classify such organization as the Federation of British Industries, the Trades Union Congress, the National Farmers Union, the British Medical Association, the National Union of Teachers, the British Legion, the National and Local Government Officers' Association. The "promotional" groups include the Howard League for Penal Reform, the National Council for Civil Liberties, the Peace Pledge Union, the Campaign for the Limitation of Secret Police Powers. As compared with the self-oriented groups, writes Finer, the latter "do not represent 'interests' in the same sense at all. They represent a cause, not a social or economic 'stake' in society."[5]

Such a broad distinction in the character of goals tends to have important consequences for structure and behavior. The promotional group, for instance, tends to be open to all like-minded persons, while the self-oriented group has, so to speak, a fixed clientele. By and large the self-oriented group can more readily extract money and work from its members on a continuing and regularized basis. It may also be less subject to splintering and more capable of continuous, unified action. At least in part for such reasons, the more powerful pressure groups of the British polity are self-oriented groups, based on a vocational interest, bureaucratic in structure, and continuing over a long period of time. While some form of group politics has long

---

[3] R. T. McKenzie, *British Political Parties* (New York: St. Martin's Press, 1955), *passim*.
[4] S. E. Finer, *Anonymous Empire: A Study of the Lobby in Great Britain* (London: Pall Mall Press, 1958), p. 3.
[5] *Ibid*.

flourished in the British as in other polities, this modern, collectivist type has emerged only in recent generations.[6] There is some sense in saying that one line of development in the history of British pressure groups has been from the promotional to the self-oriented, vocational type. Possibly a similar development has taken place in the United States, although here the third party has often played the role of the promotional group in Britain. We might also find that the promotional group remains a more important feature of the American polity than of the British.

Farm, Labor, and Business Organizations

Concentration and bureaucracy characterize British pressure groups as well as parties. Hardly without exception the big vocational pressure groups in Britain have a higher index of density and concentration. There, for instance, the National Farmers Union is the only significant organization of farmers and includes 90 per cent of its potential membership. In the United States, of course, only a fraction—no more than 30 per cent—of all farmers are organized and these are divided among three main groups and various minor ones. While absolute numbers are much smaller in Britain, we must remember that British agriculture is highly diversified as to crops, size of farms, and location. Yet through the NFU British farmers speak with one voice to a degree rarely achieved by farmers in the United States. No doubt this is true because to no small extent the organization is run from the top. In Bedford Square is a large and able bureaucracy and at its head stands one of the ablest managers in modern Britain, Sir James Turner—sometimes known as the "Sacred Bull of British Agriculture."

In the field of trade unions, just a little less than half the total working force has been organized, while in the United States the figure is around a quarter. To one peak organization, the TUC, nearly all unions are affiliated and it has been the undisputed spokesman for organized labor for generations. Its permanent secretary, even when Walter Citrine held the post, has never occupied the position of, say, a Gompers. The heads of the Big Three,[7] however, have as prominent a political role as our Reuther, Meany, and Lewis. The British labor leaders of this generation are more likely to have worked their way up the bureaucratic ladder by long and able management than to have emerged from heroic struggles for the right to organize or for better contracts. Contrary to popular impression and in strong contrast with American experience, the strike has almost ceased to be an instrument of labor-management relations in Britain since as far back as 1932.[8] If by bureaucracy, however, we mean full-time paid staff, then British unions generally are far less well endowed than American. The reluctance of the rank and file to pay dues sufficient to employ

[6] S. H. Beer, "The Representation of Interests in British Government: Historical Background," *American Political Science Review*, Vol. 51, No. 3 (September 1957), pp. 635–45; "Pressure Groups and Parties in Britain," *American Political Science Review*, Vol. 50, No. 1 (March 1956), p. 4

[7] The Transport and General Workers' Union; the National Union of General and Municipal Workers; the Amalgamated Engineering Union—which among them include 30 per cent of all unionists affiliated to the TUC.

[8] Hugh A. Clegg, "Strikes," *Political Quarterly*, Vol. 27, No. 1 (January–March 1956), pp. 31–35.

such staff—and to pay substantial salaries to any permanent official—seriously handicaps British unions.[9]

In the field of business, in Britain as in the United States the basic unit of political action is the trade association. Comparison is made a little easier if we consider only national manufacturing trade associations.[10] Of these there are 1,300 in Britain and some 950 in the United States. Density is high: a sample survey showed that 90 per cent of larger firms and 76 per cent of smaller firms in Britain belong to such associations. Concentration among manufacturing trade associations is considerably greater in Britain. The peak association is the Federation of British Industries (FBI) which represents, through its affiliated trade associations and directly through member firms, some 85 per cent of all manufacturing concerns employing ten or more workers.[11] In the United States, on the other hand, the National Association of Manufacturers has never represented more than 6 per cent of all manufacturing concerns.[12] If the same base as that used for the FBI were taken, however, there is reason to think that the NAM figure would be more like 20 per cent to 25 per cent. The contrast would still be striking.

BARGAINING IN THE POLITY

So much for the briefest sort of sketch of collectivism in the structure of the British polity. Let us turn to the modes of interaction of these massive unit actors, in particular the political party and the pressure group.

What we have called bargaining is a principal trait of the relationships of large producers in the collectivist economy. Its essence is that each of the negotiating units is highly dependent on the other as a seller or as a buyer. In a free market, on the other hand, each seller can turn to other buyers and each buyer to other sellers and none have significant market power. In bargaining, however, each unit has substantial market power; hence, the ultimate decision is made as a result of negotiations in which each gauges his offers in the light of expectations about the possible offers of the other.[13]

A similar kind of decision making occurs where a party enjoys large power over the authority of government, while a pressure group with which it deals enjoys similar power over something—such as votes—that the party wants. Such a situation is very different from one in which government authority is dispersed among many elected office-holders and voting power among an unorganized electorate. In the latter situation, there is a kind of bidding for votes on one side and for promises or policies on the other that has a limited, but real, analogy with the economic free

[9] John A. Mack, "Trade Union Leadership," *Political Quarterly*, Vol. 27, No. 1 (January–March 1956), p. 77.
[10] Data on British associations are from P.E.P., *Industrial Trade Associations; Activities and Organization* (London, 1957).
[11] S. E. Finer, "The Federation of British Industries," *Political Studies*, Vol. 4, No. 1 (February 1956), p. 62.
[12] R. W. Gable, "N.A.M.: Influential Lobby or Kiss of Death?", *Journal of Politics*, Vol. 15 (May 1953), p. 257.
[13] See Thomas C. Schelling, "An Essay on Bargaining," *American Economic Review*, Vol. 45, No. 3 (June 1956), pp. 281–83.

market. Where the centralized party in office confronts the massively organized pressure group, decisions are made quite differently. Indeed, some who have sat in on the Annual Price Review between the National Farmers Union in Britain and the Ministry of Agriculture have reported that the proceedings and the way in which a settlement is reached resemble nothing so much as collective bargaining. For both the farmers and the ministry there is a range of outcomes that would be better than no agreement at all. Each opponent pretty well knows what this range is. No wonder it has sometimes taken four months for a decision to be reached!

Consultation with interests is a feature of all modern Western democratic governments. Some years ago Leiserson, writing of representative advisory committees, traced their origin to "the delegation of discretionary rule-making powers under legislative standards to administrative agencies executing various types of social legislation."[14] Leiserson's statement, broadened somewhat, is a generalization valid for not only American, but also for Western European government: increasing government intervention for such purposes as social reform, economic stability, and national defense has led to the grant of rule-making power to administrative agencies and to increasing participation of interested groups in decision making at that level.

Different stages in this development, however, can be distinguished, depending upon how far the scope of policy has been expanded and the polity has become collectivist. The extent to which power has been mobilized and unified on each side—on the side of the party in power and on the side of the pressure group with which it deals—will determine whether bargaining predominates in the relationship. In the United States, we find administrative consultation on a vast scale both in Washington and in the state capitals. In Britain, a more collectivist polity, the situation is better described as "quasi-corporatism."

It is against the background of this power pattern that we must examine the emphasis that British pressure groups give to the various points in the process of decision making. The formal structure of authority—British parliamentary government as compared with the American separation of powers—will play its role. But we must recall that a hundred years ago Britain also had parliamentary government, yet pressure groups then gave far more attention to the legislature than they do now.

ADMINISTRATIVE CONSULTATION

In each polity we may distinguish four main phases of policy making: at elections, in the legislature, within the party, and at the administrative level. British pressure groups exert their major influence at the administrative level, taking this to include both ministerial and official contacts. Perhaps their second most important point of influence is within the party. In contrast American pressure groups, by and large, concentrate on the first two points: the electorate and the legislature.

There are, of course, many variations within these two broad patterns. A very important difference may result from the character of the power base of a group.

[14] Avery Leiserson, *Administrative Regulation: A Study in Representation of Interests* (Chicago: University of Chicago Press, 1942), p. 162.

There is a kind of power—and this is particularly important in Britain—that is created by the expansion of policy itself. "The greater the degree of detailed and technical control the government seeks to exert over industrial and commercial interests," E. P. Herring wrote, "the greater must be their degree of consent and active participation in the very process of regulation, if regulation is to be effective or successful."[15] This generalization, I should think, holds for most Western democracies and surely for Britain. There, certain types of control exercised in recent years—price control, materials allocation, tariffs, import control, and the encouragement of exports and productivity are only some of the more striking examples—simply could not be enforced effectively without the substantial cooperation of the groups concerned. The group's technical advice is often well-nigh indispensable. But co-operation—something more than grudging consent to "the law"—is a further necessity. Our farm programs with their referenda and farmer-elected committees recognize this necessity. But in Britain the far wider scope of regulation and planning—even after the various "bonfires of controls"—gives this power factor far greater weight.

A few examples: The farmers—meaning in effect the NFU—are represented on a set of local committees that have had important administrative duties under various agricultural programs, and the chance that the NFU might encourage these farmer representatives to withdraw from the committees has been a force in the annual price reviews. When the Conservatives in denationalizing part of the transport industry in 1946 dismantled the government haulage (that is, trucking) system, a standby scheme was organized by the industry itself. The Labour government's limitation of advertising expenditure was policed by the organized advertisers, and its important anti-inflationary effort to restrain both dividends and wage increases was carried out—and with remarkable success—on a voluntary basis by organized business and labor.

Neither the British nor the American system of consultation between government and pressure groups has been fully described.[15a] Some rough impressions, modestly intended, may be in order. In both countries the central device is the representative advisory committee. British examples range from high level bodies such as the Economic Planning Board, the National Joint Advisory Council of the Ministry of Labour, the National Production Advisory Council on Industry, on which the relevant peak organizations, the FBI, BEC and TUC, are represented, to the multitude of advisory committees of the main economic departments to which trade associations send representatives. The latter are connected with the system of "sponsoring" departments which grew up during and after World War II and which means today that every industry and every branch of it, no matter how small, has a sponsoring department or section of one, somewhere in the government machine. Apart from such committees, although often around them, a regular system of informal

[15] E. Pendleton Herring, *Public Administration and the Public Interest* (New York: McGraw-Hill Book Co., 1936), p. 192.
[15a] For a brief sketch of important aspects of American practice see *Consultation with Industry*. Historical Reports on Defense Production, Report No. 19. A History of the Office of Industry Advisory Committees of the N.P.A. (Washington, D. C.: U. S. Dept. of Commerce, 1953).

consultation has grown up. Private and public bureaucrats continually call one another on the telephone or meet for luncheon and discuss a problem on a first-name basis. Often several departments and several groups are concerned.

On the American side, the immense documentation on advisory committees in the federal government that was assembled by a subcommittee of the Government Operations Committee in 1957 has not yet been analyzed by political scientists.[16] But it is clear that from the time of the National Recovery Administration, the use of this device, from being relatively rare, has immensely increased. The number of advisory committees associated with government departments at the center—and in addition to many more at the local or regional level—runs into the hundreds. One major set established by statute are in the Department of Agriculture—for instance, the Commodity Stabilization Committees. Of the remainder, the vast majority it seems are associated principally with the defense effort—procurement, development, standards, stockpiling, and so on—and consist of industry advisory committees. In comparison with similar British industry advisory committees, the American appear to depend less on trade associations, the result at least in part of the Defense Production Act of 1950 that requires that nonmembers as well as members of trade associations be included. The peak associations—the NAM and United States Chamber of Commerce—also play a much less prominent role than their British counterparts not being represented as such, on even the Business Advisory Council. Certainly trade unions are not called in for advice so frequently or on so broad a front in the United States as in Britain. The TUC alone, for instance, is represented on some 60 committees touching all aspects of social and economic problems.

Of the broad character of the power relationship we can speak with confidence: the American executive possesses far less actual power than the British. Quite apart from the degree of delegated powers in this country, the political independence of Congress and the exercise of administrative oversight by Congressional Committees mean that the group interested in influencing policy must give great attention to the legislature. Some years ago Blaisdell found that pressure groups, while concerned with the administration, focused their attention principally upon Congress.[17] Broadly this must still be the case, although it would be interesting to know how far the defense effort may have shifted the balance.

PRESSURE ON PARTIES

At the Democratic National Convention in 1956 the number of trade-union officials sitting as delegates ran into the hundreds, while at the Republican convention there was no more than a scattering. Generally, however, in both national and state parties in the United States, the connection of pressure groups and parties is less

---

[16] *Advisory Committees* (Parts I–V), subcommittee of the House Committee on Government Operations, 84th Congress 2d Session (1956); Hearings before the same subcommittee on H.R. 3378, 85th Congress, 1st Session (1957); H.R. Report 576 on H.R. 7390, 85th Congress, 1st Session (1957).
[17] Donald C. Blaisdell, *Economic Power and Political Pressures*, Monograph 26, T.N.E.C., Investigation of Concentration of Economic Power, 76th Congress, 3rd Session (Washington, 1941), pp. 57 and 70.

close than in Britain. We do not have the formal affiliation of the trade union movement with one party. But the more important difference arises from the fact that American parties are so poorly unified that they do not provide an effective channel for influencing the use of government authority. In Britain, on the other hand, the party ranks second—although perhaps a poor second—to the administration as an object of pressure.

Where the power is, there the pressure will be applied. Where we see the pressure being applied, therefore, we shall probably find the seat of power. Judged by this rule, the central organs of the British party, especially the parliamentary party, are far more powerful than the party's representative assemblies. Pressure groups do not openly descend on a British party conference as they do on the platform hearings of an American party convention. Their representatives, however, may be present and spokesmen for various special interests—farmers, trade unionists, veterans, teachers, old-age pensioners, advertising men with a concern for commercial broadcasting—will take up a good deal of time at a party conference.

The important point of influence, however, is the parliamentary party—its regular, full meetings and its specialized committees—and to a lesser extent the party's central office. We are familiar with the way leaders of the Labour party while in power or in opposition will frequently consult with the trade unions on pending decisions. There is also an active alignment, if not formal affiliation, of organized business with the Conservatives. During the passage of the bill nationalizing transport in 1946-47, for instance, the Conservative opposition tabled several hundred amendments. Where had they come from? In practice the party's Parliamentary Secretariat—a body of party employees, not MPs—acted as intermediary between the transport committee of the parliamentary party and the various pressure groups, especially the General Council of British Shipping, the Dock and Harbors Association, and a joint committee of the Federation of British Industries, National Union of Manufacturers, and the Association of British Chambers of Commerce.[18]

Inseparable from these channels of influence is one of the, to an American, most curious phenomena of British politics. He is the "interested MP"—that is, the member who is connected with an outside interest group by direct personal involvement, such as occupation or ownership of property, or by membership or office holding in an outside organization speaking for an interest group. Today and for generations the House of Commons through the personal involvement of its members has represented a far wider range of interests than has the American Congress, notoriously inhabited by lawyers.

In Britain such personal involvement was a principal way in which interest groups of the nineteenth century made themselves heard in government. Of more importance in today's collectivist polity is the member connected with an outside organization. The MPs sponsored and subsidized by the trade unions are the best-known examples. But there is also a host of others: a joint Honorary Secretary

---

[18]Finer, *op. cit.* (note 3 *supra*), pp. 67–68. For other examples of pressure group activity in the House of Commons see J. D. Stewart, *British Pressure Groups: Their Role in Relation to the House of Commons* (Oxford: Oxford University Press, 1958).

of the Association of British Chambers of Commerce, the Chairman of the Howard League for Penal Reform, a Director of the Society of Motor Manufacturers and Traders, the President of the British Legion, the Secretary of the National Tyre Distributors Association—there seems to be hardly a member who fails to note some such connection in his biography in the *Times' House of Commons*. Perhaps some Congressmen also have similar connections. Amid their wide membership in churches, fraternal organizations, and "patriotic" groups as recorded in the *Congressional Directory,* however, they fail to mention them.

Perhaps, as S. E. Finer has suggested, the absence of such interested members from the Congress is one reason why American pressure groups must make up the deficiency by hiring lobbyists in such large numbers. For the interested MP is an active lobbyist within the legislature. His principal role is played within the parliamentary party, but his activity in the House itself is more observable. He may speak openly as the representative of a group, as the President of the British Legion often does in forwarding the Legion's campaign to increase disability pensions.[19] He is more likely to be effective at the amendment stage of a finance or other bill when, briefed by his association, he suggests changes, which perhaps at the same time are being urged on the Minister and civil servants by officers or staff of the pressure group.

INFLUENCING PUBLIC OPINION

Herring long ago observed how American pressure groups direct great attention to influencing public opinion: not only to win support for some immediate objective, but also to build up generally favorable attitudes. This he found to be a trait of the "new" lobby, and it is not irrelevant that this technique arose along with the development of modern mass-advertising methods and media. A major difference in Britain is that the big vocational pressure groups rarely mount such public campaigns. In the nineteenth century, this was not so. Beginning late in the century, however, this propagandist function seems more and more to have passed to political parties. Today and for many years now, the parties, in contrast with the pressure groups, have virtually monopolized communication with the voters as such—that is with the general public as distinguished from communication by a pressure group with its clientele.

This differentiation of function in political communication has gone very much farther in Britain than in the United States. A striking feature of nearly all the vocational pressure groups there is the extent to which they urge their demands simply and frankly as special interests. There is a significant contrast, I think, with American pressure groups which tend to base their claims on some large principle of social philosophy or national policy—as, for example, in the vast public-relations program of the NAM.

Yet the public campaign has sometimes been used by the big pressure groups of British politics and its use may be on the increase. Examples are the antinationali-

[19] J. H. Millett, "British Interest Group Tactics: A Case Study," *Political Science Quarterly* (March 1957).

zation campaign launched by the Road Haulage Association in 1946–47; Tate and Lyle's famous "Mr. Cube" campaign against the nationalization of sugar refining in 1949–50; and in general the growing use of Aims of Industry, a public-relations agency founded to defend and advocate free enterprise. Lesser efforts have been pressed by the National Union of Teachers and the British Legion. If this practice grows greatly, one might well expect it to weaken the position of the parties.

Such a development—which I do not expect—could have great consequences for the British polity. For without in any degree being cynical, one must acknowledge the large part played by British parties in creating the present political homogeneity of the British electorate—the national market for their brand-name goods. The British party battle is continuous and highly organized and so also is the stream of propaganda directed at the voter. Through it the party voter is strengthened, if not created, and the tight party majority in the legislature prepared. Even more important, the framework of public thinking about policy, the voter's sense of the alternatives, is fixed from above. Popular sovereignty in the polity has been qualified by the same means that have qualified consumers' sovereignty in the economy.

In this Americans are not likely to find much cause for self-congratulation. We will hardly say that we are more free of political propaganda. As in other aspects of the American power pattern, the difference is that the centers from which this weighty influence emanates are far more dispersed and unco-ordinated. Is this necessarily to our advantage? Some words of E. P. Herring's suggest an answer:

> A democracy inclines toward chaos rather than toward order. The representative principle, if logically followed, leads to infinite diversity rather than ultimate unity. . . Since the "voice of the people" is a pleasant fancy and not a present fact, the impulse for positive political action must be deliberately imposed at some strategic point, if democracy is to succeed as a form of government.[20]

[20]Herring, *op. cit.* (note 15 supra), p. 377.

CHAPTER 8

The Americanisation of British Politics

THOMAS J. CARBERY

Until the comparatively recent past, writers on the British Constitution have been somewhat insular, if not faintly xenophobic. The inference was that all was well with the British Constitution and that it had nothing to learn from outside example. This satisfaction with the state of affairs was given what was perhaps its most dramatic, certainly its best written expression by Professor Chrimes when he said—"This magic formula which better than any other reconciles the apparently irreconcilable fundamentals. . . ."

This was not to say that all was perfect. On the contrary it was conceded that it was not, but, as suggested, such reform as would ensue—and it is an accepted part of the Constitution that there will be reform—would be slow, subtle, conservative in nature if not Conservative inspired, and, above all, indigenous. Change would come naturally; it would come by a form of political spontaneous combustion but would not require any grafting from any alien body.

Now however the scene is changing. A phenomenon which someone with a happy turn of phrase could call something original and descriptive like "a wind of change" is making itself felt in British political circles. British politics have gone over, from a protectionist, to a free trade frame of mind. Thus, Professor Brian Chapman's "British Government Observed" which caused a stir in 1963 was of interest not only for its condemnation of the gifted amateur (and more particularly, the ungifted amateur) in British government, but provoked discussion on its preparedness to learn from foreign experience; mainly, though not exclusively, French experience.

Yet, in one sense, Professor Chapman and his critics—and here no one who has read Chapman's book should miss D. N. Chester's "review" of it in the Winter 1963 issue of Public Administration—have been looking in the wrong direction. The foreign experience which looks like having the greatest impact on British government and politics is not the French Conseil d'Etat, or their planning machinery, or their training of civil servants; nor is it the Scandinavian Ombudsman commendable though some of these may be. It is the United States and the brash, colourful,

Reprinted from Thomas J. Carbery, "The Americanisation of British Politics," JOURNAL OF POLITICS, *XXVII, No. 1 (February, 1965), 178–84, by permission of the publisher.*

American political scene which not only will influence, but already is influencing, British politics.

So far there seems little awareness of the extent of this process. To date the most noteworthy comments have come from R. T. McKenzie, himself a Transatlantic export to Britain. Writing in the Sunday Times in early December, Dr. McKenzie argued that "the 'Americanisation' of British Politics . . . has occurred with staggering rapidity." In essence McKenzie then concentrated on three aspects of the process which, he contended, have already occurred. These were our new protracted election campaign wherein it is now only a slight exaggeration to contend that the campaign for the next General Election begins as soon as the existing one is completed; the increasing attention paid to the top personalities of the parties thus endorsing Sir W. Ivor Jennings when he argues that a general election in this country is now primarily an election of a Prime Minister; and the growing lack of any real ideological distinction between the parties—an extension of Butskellism into what Dr. J. Dickson Mabon described as a situation wherein our two major parties are the Labour Party and the Conservasocialists. Doubtless Mr. Enoch Powell would reverse the trend but at the moment he rates as high on the Tory popularity charts as the Spectator's spectator, Mr. Iain Norman MacLeod, and, more important, the trend continues.

So much then, for those which have occurred or are occurring, but what of those which are liable to occur or at least being discussed?

Here the most likely contender is a system of stronger Parliamentary Committees. In the past the increasing use of, and more particularly, the continuing efficacy of, the Select Committee media in the Commons together with the continuing use of "specialist" M.P.'s, at the Committee stage of Bills, have given rise to speculation as to whether or not it would be to our advantage to use Committees even more extensively. Moreover, those concerned with Parliamentary Reform are increasingly concerned over the frustration experienced by many back-benchers. The result has been that many writers on Parliamentary Reform have advocated the introduction to the Westminster scene of specialist committees which would each review an aspect of governmental policy—Finance, Defence, Education, Health, Employment and so on. Such an extension of the Committee System would not necessarily be on the scale of the American Congressional Committees but would, nonetheless, be an admitted step in that direction.

But recent events and comments have suggested that it may be no bad thing were we to go even further and import that somewhat brash, exclusively American political institution—the Primary. This was advocated by the former Lord Altrincham in an article in the Observer of 10th March, 1963, and by the present writer in an academic journal of late 1963. Now it is advocated by Andrew Hill and Anthony Whichelow in their Penguin Special "What's wrong with Parliament?" To its advocates the primary has the great advantage that it faces up to the problem that in any single-member, simple majority electoral system a certain number of Seats are relatively safe. In such seats the representative is, to all intents and purposes, selected, not by the electorate on the great day, but by the dominant party in the area by its choice of a candidate. The asset of the Primary is that it extends the selection process in that basically all party voters (and not merely the party activists) participate in the selection process.

Another possible transatlantic intrusion could be politically appointed civil servants.

Since the great Civil Service reforms of the late 19th century the political impartiality of the Civil Service has been a sacred cow of British Government. The last time this whole question was examined was on the occasion of the Masterman Report and the ensuing infighting between the Civil Service Staff Associations and the Treasury, the outcome of which was the enshrining as sacrosanct of the principle of political neutrality of all "senior" civil servants down to and including Junior Executive Officers. The theory has it that the civil servants are dedicated men and women who will serve any government with unfailing fidelity. However as the Labour ministers of 1945-51 write their memoirs it is becoming increasingly clear that this theory is at least being questioned. Thus it is averred that the "strong" foreign policy of Ernest Bevin was not really Bevin's policy at all, but was a departmental policy sold to him by the Foreign Office. Because of such misgivings it has been suggested that a new government should be free to appoint a few temporary civil servants and appoint them on a political basis. Those appointed would work in close conjunction with the Ministers. Within each department they would be an 'alter ego', men of acceptable political opinions whom the Minister could occasionally consult, safe in the knowledge that at least they started from the same political base as himself. In this connection an article by R. H. S. Crossman in the Jubilee edition of the New Statesman is well worth reading, and the idea was given another whirl by Professor Chapman in his already mentioned book when he referred to the Minister's private 'cabinet.'

For all that, Harold Wilson seems to be unsympathetic towards this variation on the Kennedy Whiz Kids. Interviewed on the B.B.C. third programme by Norman Hunt, Wilson shied away from it. After Mr. Wilson had said that the idea of civil servants sabotaging the policy of Ministers was "nonsense." The interview contained the following:—

> HUNT: "Isn't there a strong case for any Minister who is taking over a new department bringing in as his own private little 'cabinet,' as it were, a few men whom he particularly trusts, who have the same political objectives as he has, who are dedicated to what he wants to achieve, to help him get his policies through that department?"
>
> WILSON: "I am rather hesitant about this in the departments, and I say this because of my own experience in the Cabinet Secretariat and also as a civil servant as well as a Minister. There is a danger that you get a sort of false division between his political 'cabinet,' if you want to use that phrase, and the civil servants. My own experience, having tried as a Minister to bring in one or two outside experts with the right political approach, was that I did far better when I relied in my private office on loyal civil servants who knew what I wanted and saw to it that the rest of the department knew what I wanted."

and later when talking about the Prime Minister's office:—

HUNT: "So you pick some top civil servants to strengthen your Cabinet offices—"
WILSON: "And some from outside"
HUNT: "—Who have the same political objectives as you have?"
WILSON: "No, I don't ask any civil servant to have any political objectives. I ask them to carry out the policy decisions of Ministers and of the Government"

yet later on the interview went:—

HUNT: "What sort of people do you envisage bringing in from outside into this Cabinet Secretariat?"
WILSON: "Here I think one can learn from the Kennedy experience. Of course this system is different, they changed the whole top civil service and we don't propose that at all: but he brought into the White House a number of people from universities, one or two top scientists, top administrators. I believe we are going to need a small number of those in the Cabinet Secretariat."

All of which left some people with the perhaps odd impression that Wilson thought the Whiz Kids approach was fair game in the Prime Minister's office, but not elsewhere. There is however another form of Americanisation on which Wilson seems to be completely sold. He likes the idea of projects in depth. Later, in the same interview with Hunt he said, "I would like to see our government doing what President Kennedy turned into a fine art, a project study with experts—not only Ministers—but top civil servants, planners within the department, and also, in some cases, people brought in from outside."

The next possible contender for an import certificate is the fixed term of office.

By the provisions of the Parliament Act 1911 the life of a Parliament in Britain is not more than five years. A Parliament can, however, in exceptional circumstances, extend its own life. Thus the M.P.'s elected in 1935 and due to stand down, or to seek re-election, in 1940, provided, with the co-operation of the House of Lords, for themselves to remain in the Commons, until 1945. But basically five years is the limit. It was of course, because of this situation that there arose so much speculation as to when in 1964, Sir Alec Douglas Home would call the election to replace the House elected in 1959. At the same time there is no obligation on a Prime Minister to allow the House to run a whole five years period. Attlee called an election in 1951 although there had been one the year before. Since 1951 they have tended to tumble along at four yearly intervals—1955-1959—rather than five. Indeed the present Parliament is the longest peace-time Parliament since the 1911 Act went on the Statute Book. In other words within the five year period the timing of an election in the United Kingdom is the prerogative of the Premier.

In the United States, on the other hand, both Congressional and Presidential elections take place acccording to a fixed time-table, e.g. the Presidential elections at four yearly intervals and that in November of each leap year.

It was the protracted indecision of Sir Alec re our "current" election which apparently gave rise to the argument that we too should work on fixed terms—preferably of a similar cycle to, though at different times from, that of the Americans. This it is argued would have two advantages. First, it would dispose of the uncertainty and secondly it would ensure that the two big Western Powers did not have, as they do this year, elections at much the same time.

The first occasion on which I heard this argument put forward was at a Conference on Government attended by academics and research workers from the parties, but surely there is some significance in the fact that not only did the Prime Minister deal with this matter at a dinner on the 10th April but that this was the only part of the speech which the B.B.C. chose to televise.

Two smaller points remain. The Americanisation of British politics was highlighted at the Conservative Conference in Blackpool in the autumn (we do not yet say the fall) of 1963. The dramatic news of the retiral of Harold MacMillan and the ensuing jockeying by the various contenders for applause and the other manifestations of approval, to say nothing of the great intrusion of disclaimer of peerage (and so indication of candidature) by the now Mr. Hogg, meant that the entire conference was reminiscent of a party convention in the United States.

Then again at that time both the party officials and the television and radio commentators were found to be using a considerable amount of American political phraseology: "dark-horse," "grass-roots," "band-wagon," "plurality" and similar remarks were bandied about. To me they did not seem the slightest bit out of context, did not seem unnatural. But the correspondence columns of the national press soon had a crop of letters arguing the trend was to the deplored, the inference being that the import of European institutions would be good, whereas that of American institutions and phraseology was not.

As one who unashamedly loves and revels in American politics, I cannot subscribe to such sentiments. In any event the Americanisation of British politics is only part of a bigger process—the Americanisation of all European life, the proof of which is to tune in to a French or German radio station. Lord Francis Williams and others may deplore it. I remember the first American voice I can recall. "Time marches on!" the man said.

CHAPTER 9

Party Systems in the United Kingdom and the Older Commonwealth: Causes, Resemblances and Variations

LESLIE LIPSON

1. STATEMENT OF THE PROBLEM

There are good reasons for the increasing attention which political scientists have recently given to the study of parties. In a democratic community the party system forms the focal point on which all political forces converge. Everything that is politically significant finds a place somewhere within the parties and in the relations between them. Conversely, everything that is done by the parties is relevant to politics. Thus, an extension of knowledge about the party system helps to unravel the politics by which a people is governed. Why is it that the parties have come to occupy so crucial a role in the functioning of modern democracies? This question is inevitably implied in analysing the subject of parties, but it is not always explicitly raised and answered. Since the argument of this paper does attempt an answer, it is fitting to start with a general picture of the place and functions of the party system on which the ensuing discussion will be based.

The parties owe their importance in a democracy to this fact: It is they that provide the bridge to connect the groupings of society with the institutions of the state. On the one side, society generates the clusters of interests which push their way into the political process. On the other side there stand the constitutional régime, the men and machinery of government, the institutional structures which compose the state. The state exists within society, and society permeates the state. But an intermediary is needed to provide a link between them—an intermediary which, to perform its role, must belong partly to both. Such are the political parties. The party system functions as a conductor of energy and as a transformer. It receives the impulses that flow from society, transforms them into the current of politics, and then distributes it among the institutions of which the state is constructed. Conversely, it is the party system which fastens the impress of the state upon the plastic associations wherein men combine.

Whatever lies in the middle will naturally be susceptible to influence from the two sides between which it is placed. This, of course, explains the difficulties that arise whenever one attempts to interpret the causes of a party system. Precisely because the parties are located in between the state and its society and belong in some measure to both, it has been possible for researchers to argue with considerable plausibility for either factor as the prime determinant of party politics. Hence, two kinds of answer have been given to the question: 'Why does this community have this kind of party system?' Some have thought that the institutions of government are the primary source for the formation of parties and their duration. Others believe that the party system derives in the main from the character of the society around it. The location of the parties in the middle has invited this divergence. But it remains the province of scholarship to continue the search for the factors which seem to incline the issue on one side or the other.

The purpose of this paper is to inquire into these problems in the case of five countries. These are the United Kingdom and the four ex-colonies which were the

---

Reprinted from Leslie Lipson, "Party Systems in the United Kingdom and the Older Commonwealth: Causes, Resemblances and Variations," POLITICAL STUDIES, *VII, No. 1 (February, 1959), 12–31, by permission of the author and the Clarendon Press, Oxford. This article is also incorporated in Leslie Lipson,* THE DEMOCRATIC CIVILIZATION *(New York: Oxford University Press, 1964). Originally presented as a paper at a panel of the Convention of the American Political Science Association in St. Louis, 1958. For the research on which it is based I [author] am greatly indebted to the Rockefeller Foundation for their support.*

first to attain full self-government within the Commonwealth—Australia, Canada, New Zealand, and South Africa. For comparative study of this topic, a more appropriate group than these five could scarcely be found. The party system which began to form in England after the Civil War of the seventeenth century was exported in the nineteenth to the overseas territories inhabited by emigrants of British stock. In Australia and New Zealand the indigenous populations were either too few in number or too retarded in culture to affect the imported system. But elsewhere the British tradition had to learn to tolerate that which it could not absorb: in Canada the descendants of France's *ancien régime*, and in South Africa an offshoot of seventeenth-century Holland and a large majority of tribally divided Africans. The United Kingdom, its influence fortified by ancient heritage and present might, provided the model from which the English-speaking colonists derived their political attitudes and governmental structures. Here then was a situation where an identical pattern of institutions and ideas was transplanted to geographical regions remote from the place of origin and scattered from one another. As it struck root and grew, the exported system retained some of its inherited features, but mingled these with the characteristics acquired in the new environment. Also, as the process of nation-building required amalgamation into larger units, the federal principle was borrowed from the United States and adapted to the circumstances of Canada and Australia. Thus, after the lapse of four or five generations, it was possible to observe in this group of countries both similarities and differences. It is these which provide the theme of this inquiry. The questions to be raised are these:

What have been the causes of the party systems in these five countries?
What are the reasons for the resemblances between the systems?
Where they differ, why have these variations occurred?

In order to answer those questions, this paper will attempt to describe the party systems which have emerged in the five countries under discussion, to assess the effect upon them of governmental institutions, to analyse how they were shaped by the social environment, and to sum up by suggesting how a comparative method may throw light on the nature and causes of the party system.

2. THE CHARACTER OF THE FIVE PARTY SYSTEMS

Since a causal interpretation must be preceded by a description of the subject under analysis, let us first survey the salient features of the five systems under review. In general, these can be summarized as examples either of a two-party system or of a fairly close approximation thereto. A two-party system can be defined as one which meets the following conditions:

1. At any given election the chance of coming to power is shared by two parties, and no more than two.

2. One of these is able to take office by itself, without requiring third-party support.

3. With the lapse of time, two parties alternate in office.

This list of criteria, besides what it affirms, contains certain implications. For example, it recognizes that, along with two major parties, the system may include some minor ones. Provided, however, that one of the major parties always has a clear majority and that no minor party succeeds in holding the balance, the system realistically considered still belongs to the two-party type. Also, allowance must be made for the time factor. As a people changes, so does its party system. Hence, if a particular party is based upon a certain interest and that interest declines, the party tends to decline with it; and another party, representing a rising interest, may take its place. During the transition, of course, a three-party system will develop. But the trend of a few elections will show whether this state of affairs is destined to be permanent or temporary. The occasional coexistence of three parties in a dynamic relationship does not seriously disturb the definition of a two-party system, if the latter be proven in time to be the norm to which the system always returns from its temporary deviations.

Judged by these criteria the United Kingdom nowadays has a two-party system and has had it for a long time. Its origins, as well as some phases of its evolution, remain a subject of disagreement among historians, and it is also arguable whether the term 'party' can properly be applied to some of the parliamentary clusters which were mobilized in the eighteenth century by the Crown or by influential factions. These are the difficulties which inevitably attend the problem of devising categories to fit the facts with accuracy. Yet, be that as it may, a fundamental truth of British politics holds good after all the attempts to explain it away. The seventeenth-century struggle between Royalists and Parliamentarians was followed by the eighteenth-century alignment of Tories against Whigs. This in turn evolved in the next century into a conflict of Conservatives and Liberals, which has been converted since 1931 into the battle between Conservatism and Labour. For three hundred years the main highway of British politics has channelled its traffic along two lanes. The highway has frequently been rerouted, reconstructed, and resurfaced, and it now bears a much heavier load of traffic. But it has successfully carried the British people through their evolution from oligarchy to democracy, from a largely agricultural economy to one that is heavily industrialized, and from the acquisition of Empire to the acceptance of Commonwealth. Small wonder that the two-party system has undergone transition and change. The greater wonder is that the system has been tough enough always to reappear, reassert itself, and endure.[1]

On a smaller scale, and in a society far younger and less complex than the British, the same story may be told of New Zealand, whose party system provides the closest parallel to that of the mother-country. This Dominion has now exercised responsible self-government for slightly more than a century. Its party system emerged through two formative decades (1856–76), in which the principal problems to be solved were those of immigration and public works, the Maori Wars, and central-provincial relations. The contest between groups then became polarized into a

[1] In order to avoid repeating here what I have said elsewhere in print, may I ask the reader to refer to my article 'The Two-Party System in British Politics', *American Political Science Review*, vol. xlvii, No. 2, June 1953, pp. 337–58.

Conservative-Liberal fight. As the working class came of age politically and universal suffrage was achieved,[2] the Liberal party broadened into Liberal-Labour. Later this hyphen lengthened into a split, and contemporaneously with Britain (from 1905 to 1931) New Zealand wrestled with the unstable equilibrium of three parties. The depression of the early nineteen-thirties restored the two-party system by impelling a coalition between the Conservatives (then called Reform) and what was left of the Liberals. Since that time to the present a normal alternation in office has continued between Labour and their opponents, who currently are named National. Minor parties have occasionally formed themselves during the last sixty or seventy years—usually in order to express and organize a special interest or some extreme viewpoint. But they have always evaporated quickly, like the morning dew.

Canada, being a more complicated country than New Zealand, has experienced more complications in maintaining a two-party system. Within a few decades after federation, political contests had already fallen into the classic pattern. Two principal parties had emerged, which took their names (Conservative, Liberal) from Britain, but in their inner dynamics rather resembled the Republicans and Democrats. Canada's size, along with the diversity of its groupings and their regional concentration, has always facilitated the breakaway of special interests or areas. As a consequence, third and fourth parties have been organized fairly frequently. Often these have been able to capture control of a province and there to maintain themselves in office throughout several elections, contemporary examples being Social Credit in Alberta and the CCF in Saskatchewan. At the Dominion level, however, Conservatives and Liberals have been alternating in and out of office for three-quarters of a century, and it has been a rare exception for a third party to hold the parliamentary balance.[3] Of recent decades the most remarkable feature of Canadian national politics was the long supremacy of the Liberals, which was accompanied by a decline of the Conservatives and some fragmentary representation for the Social Credit and the CCF. Had this paper been written two years ago it would have been possible to argue that Canada no longer exhibited a two-party system. At that time the four parties could have been evaluated in musical notation thus: the Liberals, a full note; the Conservatives, a half note; with a quarter note each for CCF and Social Credit. This would have counted up to two full notes. But, as every quantifier should know, an addition of fractions equalling two is not necessarily identical with a two-party system. Since the two federal elections of 1957 and 1958, however, the Liberal reign has terminated abruptly. What King passed on to Saint Laurent, the latter could not transmit to Pearson. The Conservatives, newly revivified and energetically led, are again ascendant; and, nationally at least, Social Credit was crushed and the CCF is crumbled. The swing of the pendulum—to mix the metaphor—was more like the swish of a scythe!

Australia became a Commonwealth thirty-three years later than Canada was created a Dominion. Hence, national party politics in Australia count less than six

---

[2]For men, by 1889; for women, by 1893.
[3]This did happen, however, after the election of 1921 when a revolt of the western prairies produced an agrarian movement called the Progressives.

decades. But a definite pattern has already emerged in this period. The facts are easy enough to state, although their meaning is in dispute. The first decade of this century was occupied with putting the federal system into motion and with a coalescing of groups. The fight between protectionists and free traders tended for a while to overshadow the other divisions and to twist the political interests into its particular mould. But once that issue had been resolved in favour of protection, the Australians developed the kind of party system with which they have now lived for five decades. On one side stands the Labour party. Its history has been a series of battles between a central discipline and the revolts against it. Arrayed against Labour are two parties. The chief of these, after frequent changes of name, has now resumed the Liberal label. The other is the Country party, which has maintained a separate existence ever since the early nineteen-twenties. How should the Australian system be classified? Plain though the answer seems, it is nevertheless in controversy. Some say that Australia has a three-party system, for the simple reason that the system does contain three parties! But others argue that this arithmetic has to be interpreted politically. The third member of the trio, the Country party, does not function in complete independence, nor does it fill the role of a balancer. It operates instead in a permanent coalition with the Liberals. Never does it combine, either electorally or in office, with Labour. Hence with a substance of truth one can assert that Australian politics are divided along the axis of Labour versus non-Labour. The latter side maintains two organizations, of which the weaker partner represents and increases the bargaining power of the rural elements. The Australian system, therefore, is a trio in form, a duet in function.

Younger yet than Australia is the misnamed 'Union' of South Africa. This is a geographical area which is occupied by a state that lacks a nation, a government that is one-fifth of a democracy, a minority living in perpetual fear of the majority. In a plural society where those of European descent are outnumbered by four to one and where government is the exclusive preserve of a minority which is itself acutely split, the party system functions amid a nightmarish network of restrictions and distortions. Since the minority has a monopoly of political participation, the parties 'represent' only the dominant fifth of the population. Yet all their calculations are powerfully influenced by the presence of the other four-fifths—who are disfranchised, discriminated against, and demoralized. Thus, in describing South Africa's party system one must allow for, but never forget, these basic limitations. Within their framework a complex of tangled groupings has tended to sort itself out into a rough alternation between two parties. One party is formed by the majority of the Afrikaners, and they are opposed by the minority of Afrikaners combining with the majority of those whose tongue is English. This division between groups, which runs primarily along linguistic and cultural boundaries, is widened by a difference of attitude towards the submerged majority. On the problem of race relations the fanatical extremists oppose the more moderate gradualists. Then, too, within the minority of dominant Europeans there are minorities which form splinter movements of their own—a Labour party to oppose the owners of capital, a Dominion or Federal party to resist the supremacy of Afrikanerdom over the English, a Liberal party to plead (valiantly but forlornly) the noble doctrine of human equality. Such groups have hitherto been unable to make much of a dent in the strength of the two big organizations—National and United.[4] At the present moment, since the three elections of 1948, 1953, and 1958 recorded rising percentages of the votes in their favour, the Nationalist fanatics unfortunately have everything their own way.

Such, then, is a short survey of the five systems. It supports the generalization that they conform principally to the two-party model. But the variations and deviations are quite as significant as the norm. What therefore remains to be explained are the reasons both for the norm and for the departures from it.

3. THE STRUCTURE OF GOVERNMENT AND ITS INFLUENCE ON THE PARTIES

According to the institutional explanation the primary factor which shapes the party system is the machinery of government. The stuff of politics is regarded as so much molten material, taking its form from the mould that the state imposes. Party warfare is conducted in order to capture control of the government; and consequently, on this reasoning, it is the nature of governmental institutions that imposes the organization and tactics to which the parties conform. Applying this logic, some scholars have considered two-party politics the result of such factors as the cabinet system, or the power of dissolution, or the electoral apportionment into single-member districts where a bare plurality wins. The present question, therefore, is this: In these five countries how uniform, how different, are the institutions which could have a bearing on the character of the party system? And is there a discernible connexion between these institutions and the parties?

At first glance one is struck by the amount of overall similarity in the institutions by which the five countries are governed. But at the same time certain differences are apparent and it is the explanation of the reasons for them which calls for further analysing. To begin with, there is complete uniformity in the use of the cabinet system. The fusion of parliamentary and executive leadership in the hands of the same persons is a distinctively British contribution to the art of constitution-making and is invariably followed throughout the Commonwealth. Rule by cabinet is the rule. To this there are no exceptions. Differences do exist, however, in the effects of the cabinet's power and in the techniques by which it is exercised. In certain of these countries, for example, the threat of dissolution has virtually atrophied and can no longer be considered essential—if indeed it ever was—either for maintaining discipline in the majority ranks or for perpetuating a two-party alignment. Seventy years elapsed in New Zealand (1881 to 1951) before a general election was called prior to the expiration of a parliament's full term. The reason for this is simply that New Zealand's elections have been triennial,[5] and it was therefore unnecessary to

---

[4]In the mid-twenties, however, Labour had sufficient votes to enter into coalition with Hertzog against Smuts, on a policy of monopolizing the best jobs for the white workers. Also, when Malan's Nationalists first came to office in 1948, they needed the parliamentary backing of the small Afrikaner party under Havenga which they later absorbed.

[5]Except for the occasions of World Wars I and II, and the depression of the early thirties, on which occasions Parliament extended its own duration.

shorten the life of a parliament. Notwithstanding this fact, both the two-party system and the discipline within the parties have continued in force.

In Australia, too, early dissolutions have become superfluous because the regular federal elections are held triennially, as in New Zealand. But in one important respect the Australian parliament differs from those of Great Britain, Canada, New Zealand, and South Africa. It retains its original bicameral character, in fact as well as in form. Following the pattern of the United States, the six States are represented in the Senate as equal units, and the terms of Senators are staggered. Consequently, a conflict between the two Houses can occur, if Labour controls one House while the non-Labour coalition controls the other. Because of this situation the Liberal Prime Minister, Mr. Menzies, took advantage in 1951 of the constitutional provision that authorizes a double dissolution. But this did not take place because of any defect in the working of the party system as such. It was the result, rather, of an institutional contradiction. Cabinet government, to be effective, requires not checks and balances, but a concentration of political authority in one place. Hence, whenever it functions in full force, the cabinet has either crippled or crushed bicameralism; and the ability of one chamber to eliminate its rival has been reinforced by the logic of the two-party system which necessitates a majority for one party over its opposition. Thus the power of dissolution, to the extent that it survives nowadays in Australia, is a consequence of the party system and not a cause of it.

The Australian case, however, contains other implications whose broad effects should be comparatively viewed. Australian bicameralism is the product of Australian federalism. One must therefore consider whether the institutional difference between a federal and a unitary constitution has any effect on the number of parties within the political system. Let us suppose that one were to rank the five countries on a scale according to their proximity to the poles of unitary or federal government. The order would be as follows:

UNITARY
New Zealand
United Kingdom
South Africa
Canada
Australia
FEDERAL

How does such a ranking correlate with the party systems of the same countries? In my judgement, fairly closely. The countries in which the two-party system is most firmly established are the very two whose institutions are most definitely centralized and unitary. On the other hand, the states which are federal or quasi-federal are precisely those in which the problem of the third, and even the fourth, party has continued to present itself. A person who favours the institutional argument will, therefore, be disposed to contend that federalism is an impediment to two-party monopoly. Just because an intermediate level of government exists, a party whose prospects for national power are slender may prevail in a state or province. Hence it would seem to follow that federalism tends to encourage a multiplicity of parties. If so, this indicates how the governmental structure helps to shape the party system.

On that point, briefly to anticipate the view for which I shall argue later, it would certainly appear that the party system is correlated with the degree of centralization or the reverse in the machinery of state. But to assert that one factor causes the other is an unproven inference. Rather, it may be held that both factors—the type of party system and the pattern of institutions—are themselves the joint product of another, more fundamental determinant that is as yet undisclosed. Should this be true, the correlation between the parties and either a unitary or a federal government would resemble the relation between siblings and not that between parent and child.

There remains one further institutional arrangement to consider, namely the electoral system. It is through the machinery of elections that the voters choose between alternative policies and candidates. By means of elections the parties place their representatives in political office. Hence it is self-evident that a connexion exists between parties and elections, that their relationship is direct and immediate, and that the influence of one factor upon the other is causal in character. What is in doubt, however, is the crucial issue: which is the cause, and which is the effect? Generally it has been assumed[6] that the electoral machinery is the cause which produces effects upon the party system. My own research, however, leads me to the opposite conclusion—that it is primarily the party system which moulds the electoral system. The parties, as it appears to me, determine which type they will employ among the available varieties of apportionment and voting. What they select is the type which best suits their own interest, and that interest is, of course, to preserve and perpetuate themselves. Naturally, once the electoral system has been adopted and goes into operation, it produces a reciprocal effect upon the parties which brought it into being. Normally this effect is to buttress and strengthen them. But, such being the nature of the political process, whenever the existing parties and electoral methods afford inadequate recognition to new interests and groupings which arise within society, the latter will succeed in making their own way. Thus, despite what is frequently considered the 'inevitable' tendency of the single-member district system to militate against the weakest party of three, the British Labour Party was able to survive and grow, and then to supplant the Liberals. And precisely the same has happened in New Zealand as in Great Britain.

A comparison of the electoral systems that are in use in these five countries discloses some significant facts. In Great Britain, Canada, New Zealand, and South Africa, the single-member district system prevails, the winner in the district being the candidate with a plurality. This method of apportionment, which evolved in the mother-country between 1832 and 1885, was exported to the maturing colonial territories during that period. Simultaneously the suffrage was extended to the whole adult male population, and the very parties which were responsible for the extension were opening their doors to new members. When two parties dominate the political scene, they are likely to favour the single-member district system just because its distortions, its lack of proportionalism, are an impediment to the growth of new parties. If, however, a third party does appear on the scene, one of two consequences follows. The third party may become potent enough to replace one of the existing

[6]For instance, by Professor Maurice Duverger in *Les Partis politiques*.

major parties and thereby restore the two-party system. If it can do this, as in Great Britain, the electoral system then continues without change because there is no reason to alter it. Or the third party may be permanently added to the other two, so that the party system itself changes. In this case it is both convenient and sensible to modify the electoral system subsequently. Indeed, this is what occurred in Australia, the one country of this group which forms the exception that proves the rule. There, the Country party was organized after World War I to increase the political power of the farmers on the anti-Labour side. The electoral system was then changed to conform with the altered character of the party system. Since it was expected that farmers would give their support first to the Country party, and next to the other non-Labour party, the electoral method which was adopted was that of the single transferable vote—a device precisely suited to the politics of a tri-party system wherein two of the members were normally expected to coalesce against the third. Clearly in this case, the party system formed itself first and then modified the electoral machinery to conform to its political requirements.

This point—namely, that the party system is supreme over the electoral system, and not the reverse—may be further illustrated by some evidence from New Zealand. There, in the first two decades of this century, the Labour group split off from the Liberals and organized their own party. The Liberals, who were dominant from 1890 to 1905, began to read the writing on the wall as Labour mobilized and as the old Conservative opposition regrouped itself under the name of Reform. In order to forestall the rise of a Labour party, the Liberals, while they still held a parliamentary majority, introduced a second ballot to provide for a run-off election in constituencies where no candidate had obtained a majority. This second ballot was in operation for two elections only—those of 1908 and 1911. It did not achieve the aims of its proponents. Despite its mechanics, the Labour party did grow and the Reform party increased its strength until in 1912 it was able to form a government. Then Reform used its parliamentary majority to abolish the second ballot. In short, it was the politics of the party system which proved triumphant over the electoral system, and not the mechanics of the latter which determined the shaping of the former.

To summarize this part of the discussion, one observes that the institutions of government are remarkably similar in the five countries. The most noteworthy exceptions are the electoral system in Australia and the use of a federal structure in that country and Canada. A relation does exist between these institutional differences and the variations in the party systems. It remains now to be seen whether the differences and similarities both in the party systems and the institutions of government are related to some further factor yet more pervasive and more penetrating. For that, I turn to the composition of the societies in question, since they are the sources of the politics manifested in both state and parties.

4. THE COMPOSITION OF SOCIETY AND THE POLITICS OF ITS PARTIES

To inquire into all aspects of a society which are politically relevant would go far beyond the scope of this paper. What can and must be done, however, is to look at one feature in the composition of society which has a direct and immediate bearing upon the character of the party system. Parties are groupings into which a people subdivides in order that policies may be debated and programmes executed. That groupings into parties are political in nature is self-evident. What needs clarifying is the relation between the parties and the other groupings of which a society is interlaced. Not all of these are always and necessarily political. But virtually any social grouping may acquire this significance, and many of them will from time to time become politically charged. The particular aspect of a society's structure which I am exploring here is its heterogeneity or homogeneity. The assumption is simply that a party system is certain to be influenced by the groupings in society, and is likely to reproduce either their complexity or simplicity. A people 'fundamentally at one,' to cite Lord Balfour's phrase,[7] would not be expected to sustain the same party politics as 'a house divided against itself.' Race, religion, language, occupation, class, property, and income are among the principal constituents of a social order. So, when the racial, religious, and linguistic groupings tend to coincide, political organizations will generally be more cohesive than if these groupings criss-cross with one another and if the same persons recombine differently in terms of economic status and occupation. To examine the validity of these assumptions, one must study the effect of the components of the social order upon the character of the party system.

The mother-country supplied to its colonies, as they matured, not only a framework of government but also a model for party politics. In point of time, Britain's two-party system stems from beginnings which long antedate the grant of responsible self-government to territories overseas. But that same party system did develop further because of Britain's own domestic evolution and was being more completely organized simultaneously with its transplantation to other continents. In a dual sense, therefore, the British model is both prior to, and contemporary with, the systems that sprang up in the Dominions.

The story of the British party system is so long and so complicated that the need for compression in a paper such as this involves the risks of oversimplification and apparent dogmatism. Nevertheless, there are certain salient features about which some reasonable generalizations can be offered. As long as the monarch possessed real power, any genuine alternation of parties was impossible. The Civil War of 1641–9 marked a division of society into two sides which could not compose their differences through the existing framework of government. The supremacy of Parliament was a necessary institutional accompaniment of a party system, since Parliament provided the forum where parties could engage in regular public debate. Gradually the controversies of the Restoration period crystallized into a Whig-Tory dichotomy, imparting a modicum of coherence to the eighteenth-century gambits of aristocratic interest, influence, and intrigue. Such a dualism corresponded with the social divisions of that era. Granted that politics were a function of the upper classes, i.e. of the minority, opinions tended to polarize around two issues—one economic, the other religious. In a society that was largely agricultural and rural, but was

[7]From his introductory essay to *The English Constitution* by Walter Bagehot, World's Classics edition (Oxford, 1928), p. xxiv.

acquiring colonies and increasing its external commerce, two interests were opposed. On one side were the majority of landowners. On the other, the minority of landowners were allied with the mercantile interest. In the religious sphere, the ecclesiastical settlement which had been imposed after the Restoration and reaffirmed after the flight of James II provided the boundary line for a second split. The dominant Church of England displayed its intolerance not only to Jews and Catholics, but also to Nonconformist Protestants. Hence, two parties existed in the religious field—those who were for the established Church and those who were not. Most important of all, however, was the fact that the alignment of groups on the religious issue tended to correspond with the economic bifurcation. The country squires, for the most part, were rural-minded, royalist, Anglican, and Tory. Their opponents were more friendly to commerce, more tolerant of dissent, more loyal to Parliament, and Whig.

With the groundwork thus laid for a two-party system, nineteenth-century Britain was required to adjust its politics to an economic transformation and its social aftermath. The economic change was widespread industrialization; the social change, urbanism; the political result, democracy. The effect upon the parties of industrialization, although delayed for a generation by the Napoleonic Wars, was to help the descendants of the Whigs, then called Liberals. Demanding a revolution in economic policy, the Liberals supported the reform of governmental institutions as its precondition. Thus in 1832 they won for the urban middle class a measure of representation in Parliament and therewith a political ascendancy which lasted for four decades. Conversely the Conservatives, identified with the landed interest, were split and crushed. Later in the century, however, by a strange irony, the same forces which had once aided the Liberals turned to their disadvantage and redounded to the benefit of their opponents. When the welfare of the working class became the overriding issue, some of the Liberals were unwilling to incur the higher taxation and the extension of governmental functions which social services involved. Thus commenced the succession of breaks which culminated in the disintegration of the once-mighty Liberals. But while the need for further change destroyed the party which had elevated reform into a philosophy, the Conservatives discovered that changes, after their adoption, acquire a conservative tinge and that the landed aristocrat could tax a factory owner for the welfare of the factory worker. With the enfranchisement of the industrial and agricultural labourers in 1867 and 1884, the underprivileged were finally in a position to use liberty to obtain equality; and, since their strength lay in the solidarity of numbers, they organized their own party. Caught in between, the Liberals cracked into further splinters, which flew off—left and right—to coalesce with Conservatism or Labour, helping thereby to liberalize the former and constitutionalize the latter.

Contemporary Britain, for all its size and inner diversities, has a compactness and solidarity which underlie the two-party system. Its economy is overwhelmingly industrial and commercial; its way of life, predominantly urban; its religions, strongly Protestant; its race, homogeneous;[8] and its political union, an English

---

[8]Though, because of West Indian immigration, an incipient colour question now exists on a small scale.

majority far outnumbering the Scots, Welsh, and Ulstermen. Under these circumstances the party system reflects, on the whole, the surviving elements of horizontal stratification in the British economy and social order. The Labour Party has mobilized the bulk of the wage-earners along with a section of the middle class and the intellegentsia. The upper class and well-to-do, the farmers, the majority of the middle class, and a minority of the wage-earners vote Conservative. Finally, those who have an allergy to mammoth organization, and who echo individualistically 'a plague on both your houses', will continue to inscribe a protest vote for the Liberals—especially at by-elections, where the control of Parliament is seldom affected.

I interpret the British party system, then, as the outcome of an ancient social order which has now attained a fundamental harmony; which therefore divides easily into two sides whose opposition is conducted within mutually agreed limits; which has evolved along with, and has helped to cause, the framework of institutions that suit its purpose. It is this system which has provided an example for the English-speaking inhabitants of Australia, Canada, New Zealand, and South Africa. But the societies which have developed in these countries are not identical with the British, nor are their peoples derived solely from the British Isles. In terms of homogeneity and its opposite, it is fairly evident that New Zealand and Australia contain societies which are more homogeneous than that of Britain, whereas the Canadian and South African societies are much more heterogeneous. How, then, are these social variations related to the party systems?

The simplest in structure of all these societies is the one in New Zealand, the country whose party system is the closest akin to the British. The people of New Zealand are homogeneous to a rare degree. They have the population of Philadelphia spread over an area as big as Oregon. The Europeans, who outnumber the Maoris by ten to one, are drawn, to the extent of 98 per cent, from the British Isles. Of these the great majority were English or Scots and Protestant, and are descended from emigrants of the working class and lower-middle class. Although farming provides nine-tenths of the country's export wealth, farmers ceased long ago to be in a majority and have also lost the advantage of special apportionment under the electoral law. The demand for higher living standards, equally spread, has led to a moderate urban development. But, instead of one dominant metropolis New Zealand has four main centres, spaced apart from north to south. For political purposes the people divide along two axes: the city folk versus the farmers, and those who have less against those who have more. The farmers form a bloc which seldom[9] splits. But, being insufficient to win an election, they need to coalesce with some urban interests in order to produce a majority. The result is their alliance with importers, businessmen, and higher-income professional people. Thereby the National Party is cemented together. On the other side the Labour Party enjoys the full allegiance of the industrial workers. With these it unites in the cities many of the lower-paid white-collar employees of government and of private firms. For the National Party

[9]Occasionally, however, and in times of falling prices overseas, the dairy farmers may break from the sheep farmers, and, as in 1935, some will vote Labour.

the problem is to hold together its rural and urban wings; for Labour, its industrial and white-collar supporters. A loosening of either tie is usually enough to produce an electoral swing and thus to change the party in power.

What New Zealand politics would be like if its social structure were projected on to a larger screen can readily be seen in the case of its neighbour, the Commonwealth of Australia. In many of their basic characteristics the two countries are identical. But the sub-groups which exist in Australia are proportionately larger than their New Zealand counterparts. This fact has intensified their sense of separateness and aggravated their frictions, with consequences which are immediately observable in the party system. In terms of race, language, and national origins the Australians are remarkably homogeneous. In religion, however, because rather more Irish emigrated to Australia than to New Zealand, the number of Catholics amounts to 21 per cent, as against 14 per cent, in the smaller Dominion. This is politically relevant because the bulk of the Catholics (for the same reasons that formed them into Democrats in the U.S.A.) early joined the Labour Party, where their concentration makes them a powerful force. Also owing to their geographical distances from one another, the regional sentiments are more pronounced in Australia than New Zealand and explain why one country adopted, and the other abandoned, a federal constitution. The antagonisms between classes, and between city and country, are also more sharply felt in Australia. In practically every state over two-fifths of the population are to be found in the capital city; and in the rural areas the pastoral industry had to adapt to conditions of climate which required huge 'runs' for the sheep to graze and led to cleavages between the owners and their labourers. Hence the Australian party system accurately reflects the same two lines of social division as in New Zealand—but with two important differences. The city-country conflict has fostered two parties on the anti-Labour side. The Country Party represents many of the farmers, while a minority of them join with the urban business interests in the Liberal Party. Labour has the backing of the industrial workers and many of those in the lower-paid clerical jobs. Maintaining unity on the Labour side, however, has proven difficult. The party was split acutely in World War I on the issue of conscription, in the depression of the early nineteen-thirties over the choice between radical or orthodox finance, and since 1947 between its Catholic right wing and a left wing which veered close to the Communists. When Labour are divided, the other side are sure to hold power. When Labour are unified, they can take advantage of the city-country feud and oust their opponents from office. Thus, the system described above as a duet in function and a trio in form corresponds appropriately to the patterns of the Australian society.

The country which comes next in order of social complexity is Canada. It has produced a distinctive brand of politics and a party system which supplies the nearest parallel to that of the United States. Canada consists of two cultures which have coexisted for nearly two centuries but have not fused. They are divided by language, religion, and political memories, and the capacity of the minority to survive is reinforced by its geographical concentration and its 'peculiar institution,' the Church. The art of governing Canada is exceedingly difficult. So far only three men—Macdonald, Laurier, and King—have fully mastered its secret. For the politics of the Dominion require the maintenance of a delicate equilibrium among forces which are for the most part centrifugal. Externally, Canadians are drawn to influential poles both east and south. The English-speaking are pulled by trade and sentiment to Britain; the French-speaking to France—but of the *ancien régime*, not of the contemporary Republic. Southward, the natural alignment of the North American continent has attracted Canadians to the United States for the mutual benefits of trade, defence, and propinquity. Internally, a country whose north is sparsely settled and whose people are stretched east-west in a narrow band from Newfoundland to Vancouver Island is divided into the five well-marked regions of the Maritimes, the St. Lawrence and Great Lakes, the Laurentian Shield, the Prairies, and the Pacific slope of the Rockies. The function of the federal government at the centre is to hold the balance between the United States and Western Europe. Under these conditions the party which keeps in power is the one which can discover the highest common factor of the contending interests. Thus, statesmanship in Canada becomes synonymous with compromise.

The two main parties of Canada are best described as federal unions within a federal union. The strategy of their opposition is dictated by two recurring factors: the unity of Quebec, and the tendency for people to vote one way provincially and another way nationally. For many decades Quebec was the 'solid South' of Canada. Each of the major parties owes its long period of ascendancy to support from that province. But the retention of power at Ottawa has been consistent with continued opposition from the provincial capitals. What was so significant during the long Liberal hegemony was that effective party conflict virtually ceased at the federal level and also within most of the provinces, but was still conducted between the two levels. Only a few years ago, many provinces for all practical purposes had produced one-party régimes, since for many elections in a row there had been no political change. Equally significant was the fact that the Liberals, though lording it at Ottawa, did not control the majority of provincial governments. Thus, Duplessis' Union Nationale was ascendant in Quebec; the Conservatives ruled Ontario; the CCF was dominant in Saskatchewan, as was Social Credit in Alberta. Yet some of the very districts which chose these parties provincially were voting for the Liberals nationally. What this must mean is that in Canada the effective line of political division is drawn between the nation and its component regions. Indeed, even when the party that controls the province is nominally the same as the one prevailing at Ottawa, conflict can still occur between province and Dominion—a striking example being the fight in the nineteen-thirties between Prime Minister MacKenzie King and Premier Hepburn of Ontario.

It is these fundamental tendencies that are reflected in the history of Canadian politics. Because of the regional diversities in the Canadian economy and the geographical distribution of languages and religions Canadians express their preferences on one set of issues locally, on another set nationally. Hence, as many as four or five parties may dominate the several provinces. At the centre, however, that party prevails which cements a winning coalition by compromises and conciliation.

Thus, following the same line of argument, one can see why the federal system, as such, is not the cause of the Canadian party-complex. Both the party system and federalism, in my judgement, are the consequences of the same social factors. The federal structure is well suited to the diversities of Canada's regions and cultures; and the parties, major and minor, which flow from these very diversities support and use that structure for their political ends.

Last to be considered is the most heterogeneous of the societies discussed in this study. Viewed cumulatively, South Africa exhibits all of the diversities of Canada—plus the tremendous complication of the size of the non-European majority. Also, in two respects the position of the English-speaking South Africans differs from that of English-speaking Canadians. In the former country they are the minority group within the European minority, whereas in Canada they still outnumber the fertile French. Also, in South Africa the memories of armed conflict between English-speaking and Afrikaners reach back to as recent a date as the beginning of this century; and the bitterness of military defeat has increased the determination of extreme Afrikanerdom for a political revenge. Because non-Europeans are not permitted to participate in elections (save to the limited extent that the Coloured may vote in the Cape), the pivot for South African parties lies at the point where the Afrikaner group splits into fanatics and moderates. When there were enough of the moderates, their alliance with the English-speaking minority could produce victories for the United Party—the party of which Botha and Smut were leaders. But the Afrikaner moderates are always threatened by their own extremists. In this way Hertzog rebelled against Botha, as Malan did later against Hertzog. Since 1948 the programme of apartheid has served as the rallying cry for the fanatics, who have thus far outmanoeuvred their opponents by appeals to prejudice, fear, and white supremacy.

South African society is subdivided both vertically and horizontally. The horizontal layers are racial, with the whites occupying the narrow apex at the top of the pyramid. Vertically, they are themselves split by language, religion, economic status, and the bitterness of the past. Since all lines of cleavage tend to coincide, each reinforces the other. In this way there has arisen among the dominant whites a system of two major parties flanked by a number of splinters. But the nub of this party system is that it does not include and represent the majority. Instead, it serves merely to organize the fear for the majority which the minority entertain. Since South Africa comprises not one society but several which have been superimposed in a hierarchy, the party system reflects the restrictions by which the country is crippled. Clearly, if the Liberal viewpoint were ever to prevail, one would expect to see many more parties in the arena.

Equally clearly one can observe that the institutions of government adopted in South Africa do not accord with the natural requirements of its plural society. The country was ill fitted for a constitution which enshrines the British principle of parliamentary supremacy and establishes no legal checks to restrain the majority in the parliament. A federal structure, and certain internal checks and balances, would have been the appropriate means of safeguarding minority rights—and also, in this case, the rights of the Bantu majority. But the parliamentary form and, with it, a structure that leans more to the unitary than the federal were preferred by Smuts because the minority's fear of the majority made them seek protection by concentrating the coercive powers of government at one focal point. South Africa, therefore, offers an example of both a party system and an institutional framework which result from the attempt of the few to preserve a position of privilege in a divided society.

5. COMPARISONS AND CONCLUSIONS

What conclusions can be drawn from the comparisons undertaken in this survey? In the process of transplanting the British way of governing the amount that has been retained is perhaps more striking than the amount that was modified. Modifications there were bound to be since structures and processes do not emerge the same when their social accompaniments are altered. The British who emigrated to the Colonies were not an exact statistical cross-section of the original society, since the upper-class element was far smaller and less influential. Moreover, in Canada and South Africa the British were condemned to live with other cultures which they had forcibly brought under their jurisdiction. In the homogeneous societies party politics are founded, almost exclusively, on economic divisions. Hence it is in the United Kingdom, Australia, and New Zealand that a Labour Party has become strong, and not in South Africa or Canada. In the societies where language, religion, and race create their own affiliations, party politics are influenced as strongly by these groupings as by the economic. In the plural societies the attitudes of sub-groups towards the structure of government and the principles of politics will vary with their numerical strength. The party system of a people fundamentally at one can readily absorb the small electoral swings which turn a minority into a majority and result in a change of Ministry. But the exercise of majority power is quite differently viewed in such a country as Canada and South Africa. French-speaking Canadians are tenaciously jealous of minority rights and normally uphold the provincial position against the Dominion. Likewise in South Africa, and notably in Natal, the English-speaking minority has become sensitive to the need for safeguards—especially nowadays when Afrikaner extremism is rampant.

It is these cross-currents within society which inject the motive force into the party system and, through it, determine how the institutions of the state will function. The same factors, moreover, supply a different content to such familiar labels and concepts as Liberal and Conservative. Conservatism in any system derives its ethos from the nature of that which it seeks to conserve. Thus an inhabitant of Quebec who wishes to keep his language, Church, and social order intact stands on the right wing of Canadian politics, although in many national elections he has found it his interest to vote for a party which calls itself Liberal. Also on the right wing, but defending a different *status quo*, are the financial interests of Toronto. In such a country as New Zealand the centre of political gravity has moved so far over on the left that the conservatism of the National Party would be damned as outright communism by many a conservative in other lands. On the other hand, in South Africa the politics of fear have given such an impetus to racial reaction that the vaunted moderation of the United Party appears wishy-washy and ineffectual on any comparative scale of liberalism.

To sum up, then, it is these groupings in society and their psychological consequence in human attitudes which provide the substance for a party system. These also will mould the structure of government and modify its use. That structure can itself impinge upon the party system. But the dominant parties will always alter the structure, by law or custom, when their interests demand it. The merit of a comparative method is that it serves, better than any other, to illuminate these relationships which, if observed within a single system, are more likely to remain obscure.

CHAPTER 10

Party Affiliation and International Opinions in Britain and France, 1947–1956

MORRIS DAVIS AND SIDNEY VERBA

There is no lack of reports of public opinion surveys made in Britain and France since World War II in which results have been tabulated by party affiliation. Each report, however, has in the main confined itself to the answers given by one set of respondents during one interviewing period. When polling agencies have explicitly collated material from more than one survey, this has usually been only for responses given to the same question (or, failing that, to similar questions) at various periods of time, their intention being to indicate trends in opinion. It has not been the aim of the commercial pollsters to ascertain the general and lasting effects of party affiliation on political attitudes. Among academic scholars of public opinion, too, there have been relatively few attempts to combine in one study data from many surveys; and even in these exceptions, the combinations, it is not unfair to say, have been not so much systematic as illustrative.

In this article, on the other hand, an attempt is made, by applying consistently certain simple measures to a large collection of poll results, to describe systematically some of the effects of party affiliation on opinions in Britain and France during the period 1947-1956. In order to keep the study within the compass of an article, we

Reprinted from Morris Davis and Sidney Verba, "Party Affiliation and International Opinions in Britain and France, 1947–1956," PUBLIC OPINION QUARTERLY, *XXIV, No. 4 (Winter, 1960), 590–604, by permission of the authors and the publisher. Material summarized in this article was collected while the authors were on the research staff of the Center of International Studies, Princeton University. We [authors] are indebted to Mrs. Annalisa Kelley for processing many of the results used here, and to Ithiel de Sola Pool, Morris Rosenberg, and Gabriel A. Almond for their helpful criticisms.*

have limited ourselves here to a single broad topic, namely international affairs, and in addition have focused primarily on two questions: (1) the extent to which party affiliation tends to correlate with attitudes favorable or unfavorable to American foreign policy objectives in various key issue areas, and (2) the extent to which party affiliation tends to generate or polarize attitudes on international problems rather than merely to mirror attitudes widespread throughout the entire country. While it cannot be claimed that these questions begin to exhaust the effects of party affiliation on opinion, they are sufficiently important to warrant detailed and careful examination.

THE NATURE OF THE RAW MATERIAL USED

The data used in this article consist of the results of all available questions on international issues asked by the French and British Gallup organizations[1] between 1947 and 1956 (and also of a few questions from declassified United States Department of State studies) and reported by party affiliation, provided that the questions either are dichotomous or can easily be dichotomized. All "Don't Know" answers have been eliminated. For each question one of the two alternatives has been characterized as the "positive" response (that is, in favor of American foreign policy, as defined more specifically below), and each question has been classified according to the ratio of positive to negative responses. Ten equal-interval ratio categories have been designated, whose limits are as follows:

"1"	0.00 – 0.19
"2"	0.20 – 0.39
"3"	0.40 – 0.59
"4"	0.60 – 0.79
"5"	0.80 – 0.99
"6"	1.00 – 1.24
"7"	1.25 – 1.66
"8"	1.67 – 2.49
"9"	2.50 – 4.99
"10"	5.00 – ∞

In addition, the subject matter of the questions has been classified under the following headings:

   A. *U.S. or U.S.S.R.* These are questions in which one or both of these countries is mentioned explicitly. A positive answer approves of the U.S. and/or disapproves of the U.S.S.R.
   B. *European alliances.* These ask about various schemes for European integration and other military or nonmilitary alliances. A positive answer expresses a favorable attitude toward such a union or alliance.
   C. *Internationalism.* Here fall questions on the UN's progress, on the usefulness or desirability of international conferences, and on cooperation within the international community. A positive answer favors internationalism (for example, the UN has accomplished much, conferences are good, nations ought to cooperate).

[1] For Britain we have relied on the rather complete files devoted to that country at the American Institute of Public Opinion in Princeton, N.J.; for France we have relied mainly on the findings printed in the journal *Sondages*.

D. *Armament.* These include such problems as defense spending, the development of new weapons, conscription, and the like. A positive answer favors armament of Western countries.
E. *Far East.* Here fall questions asked about the Korean War, Red China, Formosa, etc. A positive answer is one in accord with what has sometimes been labeled the "China Lobby" position; it would oppose recognizing Red China, support Chiang Kai-shek, favor carrying the Korean War to Manchuria, oppose Truman's dismissal of MacArthur, and so on.
F. *Colonialism.* These are questions on the nation's colonies or on its relations with countries that were once in a colonial position. A positive answer favors the European nation rather than the colony or underdeveloped country.
G. *Axis powers.* These include questions on Nazi war criminals, the Ruhr, and the like. A positive answer is anti-Axis, anti-German, or anti-Italian. (This category exists for France only.)

In addition, these seven categories have been combined into one omnibus issue labeled (with some trepidation) "American foreign policy." Answers that prefer the U.S. to the U.S.S.R. and favor European alliances, internationalism, armament of the West, the "China Lobby" position on the Far East, a strong colonial policy, and a firm policy against the Axis can together be considered as affirming the American foreign policy position. Except for the questions on colonialism and some on the Far East, the label is fairly accurate. In any case, the omnibus issue we call "American foreign policy" is introduced here as a useful device to give some order to the many disparate questions we use, and not as a description of the actual policy positon of the United States.

Table 1 summarizes the number of questions available to us from Britain and France (1947-1956) under each of these headings. In order to gauge the extent to which party affiliation correlates with a favorable or unfavorable attitude toward American foreign policy, an *index of issue cohesion* has been employed, defined as,

$$\tfrac{1}{5} [5("10"-"1") + 4("9"-"2") + 3("8"-"3") + 2("7"-"4") + ("6"-"5")],$$

where "1" is the percentage of questionnaire answers, on a particular subject matter and for a given party, that has a positive to negative ratio between 0.00 and 0.19, etc. This index, ranging from +100 to −100, gives a relatively accurate indication

Table 1. NUMBER OF QUESTIONS FROM BRITAIN AND FRANCE USED IN THIS STUDY, BY SELECTED ISSUES

QUESTIONS ON	BRITAIN	FRANCE*	
A. U.S. or U.S.S.R.	15	21	(16)
B. European alliances	13	16	(13)
C. Internationalism	13	3	(2)
D. Armament	27	8	(7)
E. Far East	20	5	(5)
F. Colonialism	12	14	(3)
G. Axis powers	0	8	(2)
	100	75	(48)

*Figures in parentheses are for the Gaullists; see below.

of the consistency and degree of unanimity with which party followers express opinions favorable to the United States and its international policies.

To suggest the extent to which party affiliation tends to differentiate foreign policy attitudes rather than simply reflect nationwide attitudes, two measures are used for each question and cross-tabulated: an *index of interparty agreement* and an *index of intraparty agreement*.

The formula for interparty agreement is

$$(|R_1 - \overline{R}| + |R_2 - \overline{R}| + \ldots + |R_n - \overline{R}|)/N$$

where N equals the number of parties considered, R equals the ratio category into which the responses of the adherents of a particular party (1 to n) fall, $\overline{R}$ equals the unweighted mean of the ratio categories of all the parties, and the vertical lines enclose expressions in which only the absolute difference (i.e. without sign) is considered.[2] This index measures the mean deviation from the unweighted mean of the ratio categories. If on a question all parties[3] fall in the same ratio category, the index would, of course, be zero; if they do not, the mean deviation can range as high as between 4.0 and 4.5, depending on the number of parties (if two parties, 4.5; if three, 4.0; if four, 4.5; if five, 4.3; if six, 4.5). In our data, in no case is the index number higher than 3.0.

As for the index of intraparty agreement, it measures the extent to which on each question all the parties on the average take within themselves a strong or weak position, and is defined as $\sum A_{1-n}/N$, in which N equals the number of parties, and A indicates the extent of agreement within a particular party and has the following values: for category "1" or "10," 5; for "2" or "9," 4; for "3" or "8," 3; for "4" or "7," 2; for "5" or "6," 1. This index may range from 1 to 5. A score of 5 would mean, for example, that all parties take an internally unified position on a question; it does not, of course, specify whether the parties strongly agree *among* themselves or whether they strongly disagree.

For both these indices we have tabulated the percentage of question responses that fall within certain index numbers, and cross-tabulated the results. Such a cross-tabulation indicates precisely the relationship between the extent of interparty agreement and the extent of intraparty agreement, and thereby permits an accurate assessment of the degree to which party affiliation polarizes opinions.

Finally, a brief explanation is needed of the categories used here to designate party adherents. Persons are considered as such because in response to some question like, "Which party do you usually support?" or, "Which party did you vote for in the last election?" or, "Which party do you intend to vote for?" they responded by naming a particular party. For Britain only three classifications are used here: Lab. (Laborites), Lib. (Liberals), and Cons. (Conservatives). Omitted from considera-

---

[2] It would be preferable to use the ratio category for the nation as a whole rather than $\overline{R}$ in this formula, but the figures for the nation as a whole are unfortunately often omitted in the French source material.

[3] "Party" when used in this article always means party-in-the-electorate. To say, for example, that the Communist Party in France takes a strong position on an issue is, in the context of this study, to say that, among those respondents who identify themselves with the Communist Party in France, a large percentage take a certain position.

tion are such headings as "Other," "Don't Know," "Communists," as well as such occasional semiparty headings as "Labor Changers."

The French material, as one might well expect, demonstrates greater variety than does the British. There are more parties and sometimes the names change. Parties have grown up and others have died or merged. In order to simplify procedures, "Don't Know," "Others," and "Others among whom are Poujadists" are omitted. The following classifications alone have been used: Com. (Communists), Soc. (Socialists), Rad. (radicals, including such parties as the R.G.R. and U.D.S.R.), M.R.P. (the major Catholic party in France), Mod. (moderates, including here the P.R.L., Independents, Peasants, and similar groups), and Gaul. (the Gaullists, whether as the R.P.F. or such later groups as the U.R.A.S.). Only the Gaullists, as Table 1 indicated, are missing from any great number of questions in France.

LIMITATIONS ON MATERIAL AND METHOD

Before proceeding to the substance of this article, it might be well to indicate briefly some of the limitations of the materials and methods used here.

1. Since the questions come from a number of surveys, they reflect numerous populations of respondents, occurred at different times, were asked by various interviewers under various conditions, etc., all of which differences can easily be imagined to affect the results.
2. Since the questions asked of the British were different from (and almost never coordinated with) the questions asked of the French, any comparisons between the two countries are affected not only by the difficulties listed under 1 above but also by differences in question wording and meaning.
3. Because our basic unit is not (as is almost always the case in scholarly analyses of public opinion data) the individual respondent but rather a cell in a table of printed results, we have been unable to perform any individual correlations either between party affiliation and any other sorts of ecological data or between one attitude and another.
4. Having considered only dichotomous or dichotomizable questions, we have doubtless introduced biases into the results, even though it is probable that any assignment of weights to open or multiple-choice questions would also result in serious biases.

These limitations, among others, indicate emphatically that comparisons must be made with caution and conclusions not necessarily treated as conclusive. Certainly, any individual index number must be treated *cum grano salis*. On the other hand, the uses to which the data are put here are, we believe, of some interest and importance; the data used are the best available for these purposes; and the results, if not definitive, are at least suggestive of hypotheses for further research.

THE EVIDENCE

Correlation of party affiliation in Britain and France with an attitude favorable or unfavorable to American foreign policy. Table 2 lists, under categories defined earlier, the various indices of issue cohesion—the measure used in this article to illuminate these attitudes—for the major British and French political parties. Looking at the British data first, we see that in two categories of international issue the British political parties take positions similar to one another. They all support internationalism in general, claiming that nations ought to cooperate, that there should be a foreign ministers' conference, that the UN will accomplish much, etc. The indices are less than the maximum possible, but they reflect the sentiment that internationalism is a good thing. European alliances are also supported in more or less the same fashion by the various party adherents. Though opposing the Schuman Plan, respondents tend to support NATO, the Marshall Plan, EDC, and European integration generally. Before 1953 they opposed British participation in a European army, and after 1953 supported it.

Table 2. INDICES OF ISSUE COHESION FOR MAJOR BRITISH AND FRENCH POLITICAL PARTIES, BY SELECTED ISSUES

QUESTIONS ON	FRENCH COM.	BRITISH PARTIES			FRENCH PARTIES				
		LAB.	LIB.	CONS.	SOC.	RAD.	M.R.P.	MOD.	GAU.
A. U.S. or U.S.S.R.	—89	—6	17	48	66	85	89	86	92
B. European alliances	—80	22	29	27	81	96	94	93	91
C. Internationalism	*	40	40	39	*	*	*	*	*
D. Armament	—2	—2	18	48	—32	—3	12	23	23
E. Far East	—100	—64	—48	—15	0	24	64	64	68
F. Colonialism	—92	—14	17	79	—10	15	30	37	*
G. Axis powers	40	†	†	†	68	76	70	63	*
Total	—71	—8	8	33	44	54	64	60	60

* Omitted: too few questions available in the French material.
† No questions on this issue in the British material.

On the other issues there is a clear difference between the Laborites and Conservatives, the latter holding opinions noticeably more favorable to American foreign policy than the former. (The Liberals consistently occupy a position between the two larger parties.) On questions specifically about the United States or U.S.S.R., for example, the Conservative Party almost always takes a more anti-Russian, less anti-American stand than does the Labor Party.[4] Similarly, the Conservatives are more in favor of armaments (a larger percentage answering "yes" on such subjects as continuing conscription, increasing the time period for such service, Britain's building an H-bomb, and increasing defense spending) than are the Laborites. On questions dealing with the Far East—largely about the Korean War, Formosa, and Red China—Conservatives and Laborites divide as sharply as they do on armaments (with index differences of 49 and 50, respectively). The Labor Party is especially more favorable to recognition of Red China and its admission into the UN than is the Conservative Party.

The sharpest difference between the parties is in the category of colonialism. The extent of the rift (ratio difference = 93) would be even greater were one to focus on the 1956 Suez adventure alone. For while the Conservatives have consistently

---

[4] Question wording, of course, forbids one to say that the Laborites are themselves *absolutely* more pro-Russian than pro-American. It does not prevent relative interparty comparisons like those made here.

supported the colonialist position—in regard to Egypt in both 1951-1953 and 1956 and in the matter of Iranian oil—the Laborites have radically shifted their stance, first supporting a tough policy against Egypt in 1951-1953 and only later turning and opposing the entry of troops into Egypt in 1956. The Suez adventure of 1956, in fact, seems to have been the *only* postwar international issue on which Laborites and Conservatives took a sharply antithetical position.

The situation is rather different when one considers the responses of the French party supporters. (In Table 2, the French Communist Party data are to the left of the British parties, the data on the other French parties to the right.) On most issues the indices for the French parties tend to be noticeably more extreme (that is, closer to +100 or −100) than for the British parties. Especially evident is the yawning chasm that separates the Communists from *all* other French parties. Even the other major left-wing party, the Socialists, is far closer to the position of the non-Communist parties than it is to its fellow Marxists.

On only one issue, treatment of the Axis powers, are the responses of the French parties relatively similar. These questions, largely from 1946 and 1947, provoke responses that are fairly consistently anti-Axis. The scores are reduced from the maximum partly because opinions about Italy are not as severe as those about Germany and partly because the left-wing parties do not entirely approve of Western policy in Germany, especially when that policy appears to be anti-Russian.

On all other issues the Communists hold strong opinions in opposition to the aims of American foreign policy. On two of these issues the non-Communist parties come down equally strong on the other side, with only the Socialists showing any tendency to waver. Regarding European alliances, for example (including a European army, the Schuman Plan, the Atlantic Pact, and the Marshall Plan), the Communists differ as much as 176 index numbers (out of a possible 200) from the Radicals and some 161 from the party nearest them, the Socialists. Similarly, in questions that specifically deal with the United States or U.S.S.R., the difference between the Communists and all other parties is sharp, and the Socialists lean far more to the right than to the left. The newly organized parties of the center and right—the M.R.P. and the Gaullists—are perhaps a bit more decisive in their answers than the more traditional parties, but these latter still incline definitely toward the United States and away from the U.S.S.R.

On the remaining issues—armaments, the Far East, and colonialism—all of them traditional stumbling blocks in French politics, only the Communists have a cohesive, well-defined position. The inability of the non-Communists to surmount these political obstacles is especially evident in the explosive area of colonialism. Questions on this topic, dealing with both Indo-China and Algeria, were easy for the Communists alone to answer. Even such a group as the moderates, who agreed fairly conclusively that Algeria must remain a department, that any means necessary should be used to achieve this end, and that colonies must be administered for the benefit of France, saw little hope that the government could solve the Algerian crisis and were unwilling to agree to new taxes if needed for that goal. In one way or another all the non-Communist parties split internally on these issues; indeed, except for a profession of faith in their own government by the Socialists, no non-Communist party gave any highly unified answer (ratio categories "1" or "10") to any question about Algeria.

With the caveats outlined in the previous section in mind, one may use Table 2 to compare the indices of issue cohesion in both Britain and France. On total issue orientation it can be seen that the French Communists have the most unified position with an anti-Western orientation of —71. On the other side, though not so extreme, are the non-Communist French parties. No British party is as firmly favorable to American foreign policy as any French non-Communist party or as unfavorable as the French Communist party.

The difference between the French and British patterns of response is quite clear. The French parties tend to be internally highly cohesive, with the extreme position of the Communist Party balanced by the relatively extreme one of the non-Communist parties. The British party responses, on the other hand, tend to be much milder, the parties within themselves not taking as cohesive a stand as the French. As for the relationship among party responses in France and Britain, in France the range (in index numbers) is 135 while in Britain it is only 41. However, if one excludes the French Communist party, then the range in France is only 20. Thus, while the over-all division among French party supporters is greater than that in Britain, this sharp split in France can clearly be traced to the existence of the Communist Party: if one ignores the Communists, he finds the French parties more similar in response than the British.

The general pattern of a sharp Communist–non-Communist division in France, with the British parties occupying an intermediary place, holds most clearly for modern, cold-war questions, that is, those that deal with the United States or U.S.S.R. or with European alliances. On the older, more traditional issues of armaments and colonialism, on the other hand, the French non-Communists take a relatively less cohesive position both internally and externally; their scores tend to fall closer to zero and to spread over a somewhat wider range than before. And on both these issues the British Conservatives give a far greater percentage of positive responses than do *any* of the French parties.[5]

Thus, the fact that the French non-Communist parties are more highly in favor of the goals of American foreign policy does not at all imply that one could place more confidence in France's standing fast in a global crisis. Not only are the Communists a large segment of the population, but the issues on which the French non-Communists are most internally divided are strategically intermeshed with those on which they are most unified. The French non-Communists may express a higher proportion of positive responses on cue-worded U.S.-U.S.S.R. questions and on the issue of European alliances than do the British party supporters, but these opinions may not be translated into budgetary actions. And the tangled colonial problems, on which the Communists alone have a pat solution while the responsible parties splinter,

---

[5]Differences in questions require caution in interpreting these results. We cannot, for example, say that the French are less colonialistic than the British Conservatives, but we can probably assert that the French are not as firmly behind the newsworthy colonial acts of their country as are the Conservatives.

especially affect all the others. For pledges to European alliances may be broken to send men to Algeria; and it is unclear how long there would be favorable French opinions toward the United States if it were to give active, or even tacit, aid to the Algerian natives.

Though the British parties severally show less internal cohesion than do the French, the British (and especially the Conservatives) have a persistent distribution of responses that probably presages a firmer foundation of support for American policy objectives than do the French replies. On all issues (except that of the Far East) the British parties have a relatively similar pattern. In France, to the contrary, the solidarity of the non-Communist parties on the "modern international issues" might very easily crumble into the continuing surge of the old "traditional issues" that never seem to ebb.

Interparty and intraparty agreement in Britain and France. If interparty agreement tends to be low when intraparty agreement is high, then we may say that the political parties tend to structure or polarize opinions rather than simply reflect a widespread national attitude. If interparty agreement tends to be moderate when intraparty agreement is moderate, then we may say that political parties tend to structure opinions but to a lesser degree. If interparty agreement tends to be high whatever the intraparty agreement, then we may say that political parties do not so much structure opinion as reflect an opinion generally held in the nation.

Agreement between the parties, as approximated by the index of interparty agreement, assumes rather different patterns in Britain and France. As Table 3 shows, on international issues British parties assume positions much more similar to one another than do those in France as a whole. If, however, one compares the distribution of British party indices with those of non-Communist French parties, then the two seem quite similar and in fact the French demonstrate rather higher agreement. Indeed, in both Britain and non-Communist France the parties show a marked consensus on international problems, nearly three-fifths of the responses in Britain and four-fifths of those in non-Communist France showing less than a quarter of the interparty deviation possible. With the French Communists averaged in, the figures change radically. Instead of piling up in the high-agreement category, some 61 per cent of the cases fall in that of low agreement. It is quite clear, then, that the low French consensus on international issues is due largely to the position of the Communist Party.

In order to study the effects of interparty agreement and intraparty agreement on one another, and thus ascertain the extent to which one may speak of a "party" position apart from a "national" position, the figures in Table 3 have been cross-tabulated against a simple index of intraparty agreement. The results are presented in Table 4 for Britain, Table 5 for non-Communist France, and Table 6 for France (including the Communists).[6]

[6]The general wedge shape of the results (i.e. results tend to fall in the upper right, center left, and lower right boxes) is caused by the partial interdependence of the two measures. For a small number of parties, for example, a moderate interparty index would be associated with a modern intraparty index. This interdependence does not, however, affect intertabular comparisons where the bias would largely be held constant.

Table 3. INDICES OF INTERPARTY AGREEMENT FOR BRITAIN, FRANCE, AND NON-COMMUNIST FRANCE

	PERCENTAGE OF QUESTIONS WHOSE RESPONSE PATTERNS FALL IN EACH LEVEL OF INTERPARTY AGREEMENT		
EXTENT OF INTERPARTY AGREEMENT	BRITAIN (N=100)	FRANCE (N=75)	NON-COMMUNIST FRANCE (N=75)
High (0.00–0.99)	59	17	79
Moderate (1.00–1.99)	29	22	18
Low (2.00–2.99)	13	61	4
Total	101	100	101

As Table 4 shows, for 33 per cent of the questions in Britain there is both close intraparty and close interparty agreement: on these questions the parties give similar and highly unified answers and may be said to reflect, rather than to generate, the national attitude. There are no examples at all of questions in which the parties rigidly oppose each other; and in only 10 per cent of the questions, many of them on the Suez crisis, do the parties come fairly close to polarizing the nation. In rough fashion, it may be stated that in about half the cases the British parties are highly agreed within and among themselves (49 per cent in the two upper right cells); in somewhat more than a fourth of the cases there is moderate agreement among the parties combined with fairly low agreement within the parties (28 per cent in the two center left cells); and in only a tenth of the cases is low interparty agreement combined with fairly high agreement within the parties (10 per cent in the two lower right cells). Only 14 per cent of the response pattern falls outside these six cells and all but 4 per cent of these are cases of high interparty agreement. In Britain, then, for the most part adherents to any one political party generally express the same sorts of opinions on international issues as adherents to another party; and a high degree of unity within a party is not so much evidence that the party tends to unify the sentiments of its supporters as that it expresses the sentiment of a highly unified nation.

Table 4. CROSS-TABULATION OF BRITISH INDICES OF INTERPARTY AND INTRAPARTY AGREEMENT
(IN PER CENT; N=100)

	EXTENT OF INTRAPARTY AGREEMENT				
EXTENT OF INTERPARTY AGREEMENT	LOW (1.00–1.99)	MODERATELY LOW (2.00–2.99)	MODERATELY HIGH (3.00–3.99)	HIGH (4.00–5.00)	TOTAL
High (0.00–0.99)	5	5	16	33	59
Moderate (1.00–1.99)	6	22	1	0	29
Low (2.00–2.99)	0	3	10	0	13
Total	11	30	27	33	101

This conclusion holds true, but in an even more exaggerated form, for that part of the French nation comprising the non-Communist parties. As Table 5 shows, here 61 per cent of the results show high inter- and intraparty agreement (as against 33 per cent for Britain); and, as in Britain, there are no examples of high intraparty agreement associated with low interparty agreement (two lower right cells). In over two thirds of the cases the French non-Communist parties are rather highly agreed within and among themselves (67 per cent in the two upper right cells); in about a seventh of the cases there is moderate disagreement between the parties and relatively low agreement within them (14 per cent in the two center left cells). Of the 20 per cent of responses outside these six cells, some 12 per cent are still examples of high interparty agreement. Like the British parties then, but in a more heightened way, the French non-Communists are vehicles for the expression of an attitude rather than the generators of that attitude. Over two-thirds of the responses bear witness to this fact, and what differences are introduced into the remaining questions by party affiliation seem weakly formed indeed.

Table 5. CROSS-TABULATION OF NON-COMMUNIST FRENCH INDICES OF INTERPARTY AND INTRAPARTY AGREEMENT
(IN PER CENT; N=75)

EXTENT OF INTERPARTY AGREEMENT	EXTENT OF INTRAPARTY AGREEMENT				
	LOW (1.00–1.99)	MODERATELY LOW (2.00–2.99)	MODERATELY HIGH (3.00–3.99)	HIGH (4.00–5.00)	TOTAL
High (0.00–0.99)	7	5	6	61	79
Moderate (1.00–1.99)	5	9	4	0	18
Low (2.00–2.99)	0	4	0	0	4
Total	12	18	10	61	101

If one passes, however, to France as a whole, including the Communists, then he finds quite a different distribution within the cells. As Table 6 shows, only 11 per cent of the responses (as against 33 per cent for Britain and 47 per cent for non-Communist France) demonstrate high inter- and intraparty agreement, whereas 43 per cent (as against zero for the other two) fall into the cell of high intraparty agreement and low interparty agreement. In France as a whole only about a sixth of the responses show high agreement within and among the political parties (16 per cent in the two upper right cells). Moderate interparty agreement combined with relatively low intraparty agreement accounts for another sixth (16 per cent in the two center left cells). But considerably more than half the responses elicited show the parties rather highly agreed within themselves but in low agreement among themselves (56 per cent in the lower right cells). Only 12 per cent of the responses fall outside these six cells, and of these only 1 per cent shows high interparty agreement.

Table 6. CROSS-TABULATION OF FRENCH INDICES OF INTERPARTY
AND INTRAPARTY AGREEMENT
(IN PER CENT; N=75)

EXTENT OF INTERPARTY AGREEMENT	EXTENT OF INTRAPARTY AGREEMENT				
	LOW (1.00–1.99)	MODERATELY LOW (2.00–2.99)	MODERATELY HIGH (3.00–3.99)	HIGH (4.00–5.00)	TOTAL
High (0.00–0.99)	0	1	5	11	17
Moderate (1.00–1.99)	3	13	1	5	22
Low (2.00–2.99)	0	5	13	43	61
Total	3	19	19	59	100

In France, then, fairly sharp disagreement between parties is the rule rather than the exception.[7] In more than half the cases the parties serve to polarize public opinion rather than convey a widely existing national attitude. Far less frequently than in Britain (16 per cent as against 49 per cent) does unity within a party merely mirror a unity felt within the nation as a whole. It is the Communist Party, of course, as a comparison of Tables 5 and 6 shows, that makes France a nation whose parties have strong opinions that rarely all run in the same direction. Whether the French non-Communists would continue to have such unified and similar opinions on other topics less ideologically connected with East-West stresses or in the absence of a pathologically different party like the Communists is, of course, quite another question.

CONCLUSION

The material presented in this article would seem to support the following conclusions:

1. Party affiliation is associated with attitudes on international questions: in both Britain and France attitudes in favor of American foreign policy objectives are most common among the parties of the right and become less common as one moves to the left.
2. British parties demonstrate a smaller range of responses on an issue than do all the French parties, but a somewhat larger range than the non-Communist French parties.
3. British response patterns are more similar from issue to issue than are the French; the latter founder especially on the more traditional issues of armament and colonialism.
4. In Britain, and more especially in non-Communist France, the parties serve by and large to express foreign policy attitudes held widely rather than to generate or polarize attitudes.

[7] Even in France, however, it should be noted, this disagreement never approaches the maximum possible: no cases exist where the index of interparty agreement lies between 3.00 and 4.50. The disagreement is sharp *in comparison to* that in Britain and in non-Communist France.

5. In France as a whole, party affiliation tends to polarize foreign policy attitudes into two fairly distinct groupings.
6. A comparison between France with and without the Communists suggests that it is Russia, explicit or implicit, that serves to dichotomize French international opinions into two ideological camps; it is, properly speaking, these two camps rather than the individual parties that polarize attitudes.

Because of limitations in method, as noted in the second section, these statements are more fittingly to be termed suggestions than conclusions. But as such they do at least form hypotheses capable of further testing. Statements 4 to 6 most particularly would seem deserving of additional investigation. Though there has been much loose and optimistic talk in political science about the effect of parties in forming opinions, the evidence presented here would seem to indicate that, in international affairs at least, parties do not perform this function too assiduously. Surely this is a topic sufficiently important to warrant further examination.

Part Three

THE FRENCH POLITICAL SYSTEM

INTRODUCTION

The chapters in this section on French politics involve comparative analyses. In the Bayeux Manifesto (Chapter 11), Charles de Gaulle outlines his philosophy of government, and contrasts his ideas with the defects of the Third Republic and of the first draft of the constitution of the Fourth Republic. This proposed new constitution had just been narrowly defeated by the French voters when he gave his speech on June 16, 1946, but a slightly modified version of the constitution was approved several months later. De Gaulle had to wait twelve years before the principles he favored were embodied in the French constitution. The Fifth Republic is clearly de Gaulle's Republic, and, whether the Fifth Republic will survive after de Gaulle or will be replaced by a Sixth Republic or other regime, the experience of Gaullist rule and the strength of Gaullist ideas will continue to have a significant influence on French politics.

Is the Fifth Republic a parliamentary or a presidential system, or a mixture of the two? In Chapter 12, Max Beloff discusses the separation-of-powers provisions in the constitution of the Fifth Republic, and compares these with American and German practices and with French historical precedents. He wrote the article shortly after the adoption of the new constitution, and his analysis is particularly interesting in the light of subsequent events. Usually when the Fifth Republic is compared with other systems and with previous regimes, it is regarded as a single entity. In Chapter 13, however, Stanley H. Hoffmann discerns two separate phases in the life of the Fifth Republic—the Fifth Republic from 1958 to 1962, and the Fifth-and-a-Half Republic" after 1962—and compares these with each other. He also speculates about future developments, and draws many comparisons between French and American politics.

The most important general question raised by the political experience of France concerns the causes of the political stability and instability of democracies. This question, of concern to Hoffmann, is systematically discussed by Eric A. Nordlinger in Chapter 14. After testing different theories against the case of France, he concludes that the thesis proposed by Michel Crozier contains the best explanation. Because the chapter deals with a general problem of democracies, of which France is a particular instance, it is implicitly comparative throughout; moreover, the author draws some explicit comparisons with Germany, Britain, *and other democratic systems. An alternative explanation of democratic stability and instability is presented in Chapter 5, in which Arend Lijphart specifically comments on Nordlinger's analysis of France.*

Chapter 15 contains probably the most famous short work on comparative political behavior: the analysis of political involvement and attitudes in France and the United States by Philip E. Converse and Georges Dupeux. The authors are experts on the politics—especially the political behavior—of the two countries, and they reach significant conclusions, by means of this comparative study, about the salience of politics at the mass level in France. It is important to note that these conclusions could be clearly established only by means of such a comparative approach.

The reader is also referred to Chapter 10, in which Morris Davis and Sidney Verba compare party affiliations and international opinions in France and Britain, and to Chapter 25, in which Harold J. Berman comments on Continental European and French legal practices.

ABOUT THE AUTHORS

CHARLES DE GAULLE *is the founder and president of the Fifth Republic.*

MAX BELOFF, *professor of government and public administration at the University of Oxford, is the author of* The American Federal Government. *Other works include* Foreign Policy and the Democratic Process, *and* Europe and the Europeans.

STANLEY H. HOFFMANN, *professor of government at Harvard University, is an outstanding expert both on French politics and on international relations. Among his many publications are* In Search of France, Le Mouvement Poujade, *and* Contemporary Theory in International Relations.

ERIC A. NORDLINGER, *of the department of politics at Brandeis University, has written on British and French politics and on the prerequisites for stable democracy. He is the author of* The Working-Class Tories.

PHILIP E. CONVERSE *is professor of political science at the University of Michigan. He has co-authored* The American Voter *and other major voting studies.*

GEORGES DUPEUX, *professor of history at the University of Bordeaux, has written* Le front populaire et les élections de 1936, *and* La société française: 1789–1960. *He has also published other works on French politics.*

CHAPTER 11

The Bayeux Manifesto

CHARLES DE GAULLE

Here in glorious and mutilated Normandy, Bayeux and its surroundings witnessed one of the greatest events in history. I can confirm that they were worthy of it. It was here that, four years after the initial disaster of France and the Allies, the final victory of the Allies and France began. The events which took place here brought conclusive justification to the effort of those who had refused to yield, and around whom the national instinct was gathered and the power of France was rebuilt.

At the same time, it was on this ancestral soil that the State reappeared. The legitimate State, because it was founded on the interest and sentiment of the Nation; the State, whose real sovereignty had emerged from war, liberty, and victory, while servitude kept only its appearance; the State, safeguarded in its rights, its dignity, and its authority, in the midst of vicissitude, destitution, and intrigue. The State, free from interference from abroad; the State, capable of restoring national and imperial unity around itself, of joining all forces of the Fatherland and the French Union, of achieving the final victory together with the Allies, of dealing as an equal with the other great nations of the world, of preserving public order, of rendering justice, and of beginning our reconstruction.

This great endeavor was accomplished outside the framework of our former governmental institutions because they had not responded to the needs of the nation and because, in the hour of crisis, they had chosen to withdraw from responsibility. Salvation had to come from elsewhere.

It came first from a group of leaders who sprang forth spontaneously from the depths of the nation and who, clearly above any narrow party or class interest, devoted themselves to the struggle for the liberation, the grandeur, and the renewal of France.

A feeling of their moral superiority, an awareness of performing a kind of religious sacrifice and example, a passion for risk and venture, a scorn for agitation, pretense, and excess, a supreme confidence in the strength and skill of its powerful conspiracy as well as in the victory and future of the Fatherland—such was the psychology of these leaders who started with nothing and who, despite heavy losses, guided and inspired the entire Empire and all of France.

The French text of General de Gaulle's speech at Bayeux on June 16, 1946, has been supplied by the French Press and Information Service, New York, and has been translated by Eva Tamm Lijphart.

But they would never have succeeded without the consent of the French masses. In fact, because of their instinctive wish for survival and triumph, the French people had never regarded the disaster of 1940 as anything but a single incident in the World War, in which France stood as the vanguard. If many were forced to yield to circumstances, the number of those who accepted them in their minds and their hearts was literally infinitesimal. Never did France believe that the enemy was anything but the enemy and that salvation was to be found anywhere but in the weapons of liberty. As the veils were torn away, the real and profound sentiment of the country became clear.

Wherever the Cross of Lorraine appeared, it caused the collapse of the platform of authority which was only imaginary, even though it appeared to have a constitutional basis. The truth is that governmental authority is valid in fact or in law only if it is in accordance with the highest interest of the country, or if it rests on the trusting support of the citizens. To build on anything else, where governmental institutions are concerned, would mean to build on sand. It would mean to risk seeing the structure collapse once more at the occasion of one of those crises to which, by the nature of things, our country so often finds itself exposed.

This is why, once the salvation of the State was assured by the victory we won and by the national unity we maintained, the most urgent and essential task was to establish new French governmental institutions. As soon as this was possible, the French people were invited to elect a constituent assembly, while attaching definite limits to its mandate and reserving the final decision for themselves.

Then, once the train was set on the tracks, I withdrew from the scene, not only in order to avoid involving in the struggle of the parties that which I can symbolize by virtue of events and which belongs to the entire nation, but also in order that no consideration toward one man, while head of the State, be able to vitiate the work of the legislators in any way.

Nevertheless, the nation and the French Union are still awaiting a constitution which is made for them and of which they can fully approve. Actually, although we may regret that the structure remains to be built, everyone certainly agrees that a success slightly postponed is more valuable than a quick, but imperfect, achievement.

In the course of a period which does not exceed twice a man's lifetime, France was invaded seven times and was governed by thirteen different regimes—to the detriment of our unfortunate people. Because of so many upheavals, poison has accumulated in our public life, and this has had an intoxicating effect on our old Gallic inclination to dissension and strife.

The unprecedented trials that we have just lived through have obviously only aggravated this state of affairs. Because of the present world situation in which the powers between which we find ourselves confront each other behind opposing ideologies, we must not permit the element of impassioned confusion to enter our internal political struggles. In short, the rivalry between the parties betrays one of our fundamental characteristics—that of always questioning everything and thus too often obscuring the highest interests of the country. This is an obvious fact which is based on the national temperament, the vicissitudes of history, and the present turmoil; but it is indispensable to the future of our country and of democracy that our governmental institutions take this fact into consideration and protect themselves, in order to preserve respect for the laws, the cohesion of the governments, the efficiency of the administration, and the prestige and authority of the State.

It is a fact that confusion in the State inevitably results in the alienation of the citizens from the governmental institutions. Thus only a single incident is sufficient to make the threat of dictatorship appear. Especially because the somewhat mechanical organization of modern society each day increases the necessity and the desirability of order in the direction and regular functioning of the machinery. How and why, then, did our First, Second, and Third Republics come to an end? How and why, then, did democracy in Italy, the German Weimar Republic, and the Spanish Republic yield to the regimes that we know? And yet, what is dictatorship but a great adventure? Its beginning undoubtedly appears advantageous. In the midst of the enthusiasm of some and the resignation of others, in the strict discipline which it imposes, with a brilliant front and with one-sided propaganda, it takes, at first, a dynamic turn in sharp contrast with the anarchy which preceded it. But it is the destiny of dictatorships to indulge in excesses.

As the citizens grow impatient and nostalgic for liberty, it becomes necessary for the dictator to offer them the compensation of increasingly greater successes at any price. The nation becomes a machine which he accelerates without restraint. In both domestic and foreign policy the goals, the risks, and the efforts gradually exceed all measure. With each step a multitude of obstacles arise, both at home and abroad. In the end the spring breaks. The magnificent structure collapses in misfortune and blood. The nation again finds itself broken and in a worse condition than it had been before the adventure began.

It is sufficient to bring this to mind in order to understand to what extent it is necessary that our new democratic institutions offset, of their own accord, the effects of our perpetual political excitement. Moreover, this is for us a question of life or death, in the world and in this century, in which the position, the independence, and even the existence of our country and our French Union are clearly at stake.

To be sure, it is the very essence of democracy that opinions are expressed, and that they endeavor, by means of the right to vote, to guide public action and legislation accordingly. But all principles and all experience also require the powers of the state—legislative, executive, and judicial—to be clearly separated and well balanced, and a national arbitration—capable of maintaining the highest degree of continuity in the midst of intrigues—to be established above political contingencies.

It is clearly understood that the final vote on the laws and the budgets belongs to an assembly elected by universal and direct suffrage. But the first actions of such an assembly do not necessarily involve perspicacity and complete serenity. Therefore it is necessary to give to a second assembly, elected and composed in a different way, the function of examining publicly that which the first has taken under consideration, of formulating amendments, and of proposing plans. The great general currents of politics may be naturally reproduced in the Chamber of Deputies, but there are also other tendencies and rights in our society.

They exist in metropolitan France. They exist, above all, in the overseas territories which are attached to the French Union by a great variety of bonds. They exist in the Saar, for which the nature of things, discovered by our victory, designates once more its place beside us, the sons of France. The future of the 110 million men and women who live under our flag lies in some form of federative organization, which will gradually become more clearly defined, but our new constitution must indicate its beginning and regulate its development.

So everything points to the institution of a second chamber, the members of which will be elected mainly by our general and municipal councils. When this chamber comes into existence, it will complement the first chamber by making it either revise its plans or examine others, and by emphasizing, in the making of the laws, that factor of administrative order which a purely political body has an inevitable tendency to neglect. On the other hand, it will be natural to include in it representatives of economic, social, and intellectual organizations, in order to make the voice of the great activities of the country heard within the State itself.

Together with representatives of the local assemblies of the overseas territories, the members of this assembly will constitute the great council of the French Union, qualified to deliberate on the laws and problems pertaining to the Union: the budget, foreign and domestic affairs, national defense, economy, and communications.

It goes without saying that the executive power cannot emanate from a parliament composed of two chambers and exercising legislative power without the danger of leading to a confusion of powers in which the government would soon be reduced to nothing but a gathering of delegations. In the present period of transition it was undoubtedly necessary for the constituent National Assembly to elect the president of the provisional government because, with a clean slate, there was no other acceptable method of selection. But this can only be a temporary arrangement. Truly, the unity, the cohesion, and the internal discipline of the French government must be sacred, or else the very leadership of the country will rapidly become powerless and disqualified.

But how could this unity, this cohesion, and this discipline be maintained in the long run, if the executive power emanated from the other power, with which it must be in balance, and if each member of the government, which is collectively responsible to the entire national representation, held his position solely as the delegate of a party?

Hence the executive power ought to emanate from the chief of state, placed above the parties, elected by a body which includes the parliament but which is much larger and is composed in such a manner as to make him the president of the French Union, as well as of the Republic. The chief of state must have the responsibility to reconcile, in the choice of men, the general interest with the direction given by the parliament; he must have the task of appointing the ministers, and first, of course, the premier, who will have to direct the policy and the work of the government; the chief of state must have the function of promulgating laws and issuing decrees, because it is toward the State as a whole that these obligate the citizens; he must have the task of presiding over meetings of the government and of exercising that influence of continuity there which is indispensable to a nation; he must serve as arbiter above political contingencies, either normally through the council or, in moments of grave confusion, by inviting the country to make known its sovereign decision through elections; he must have the duty, if the Fatherland should be in danger, to be the guarantor of the national independence and of the treaties concluded by France.

The ancient Greeks asked the sage Solon: "Which is the best constitution?" He answered: "Tell me first for which people and during which period?" Today it is for the French people and the peoples of the French Union, and during a period of great difficulty and danger. Let us take ourselves as we are. Let us take the century as it is. Despite enormous difficulties, we must achieve a profound renovation that will lead every man and every woman among us to a more comfortable, secure, and happy life, and that will increase our numbers, our strength, and our fraternal feelings.

We must preserve the freedom that was saved at the cost of so much suffering. We must ensure the destiny of France in the midst of all the obstacles arising in her path and in that to peace. We must devote all our energies among our fellow men, in order to help our poor and aged mother earth. Let us be sufficiently clear and sufficiently strong to make and observe rules of national life that will tend to unite us when division among us is a constant threat. Our whole history consists of the alternation of the immense sufferings of a divided people and of the fruitful grandeur of a free nation united under the aegis of a strong State.

CHAPTER 12

The Separation of Powers in the Constitution of the Fifth Republic

MAX BELOFF

The classic reasons for the separation of powers in constitutions has been the belief that to place all the powers of government in the same hands is to risk tyranny. Its opponents, such as those who, at the time of the making of the constitution of the Fourth Republic, sought what was in effect government by Assembly, did so in the belief that Assembly government when directed by a strong party of the Left would be the best instrument for a major transformation of

Reprinted from Max Beloff, "The Separation of Powers in the Constitution of the Fifth Republic," PARLIAMENTARY AFFAIRS, XII, *No. 1 (Winter, 1958–1959), 37–46, by permission of the author and the publisher.*

society.[1] The Constitution ultimately accepted was in fact a constitution in which the legislature retained in fact, if not on paper, all those facilities for making the lives of governments harassed and brief, that had been regarded as the bane of the Third Republic. The President was reduced on paper to the position of nominal head of State which had indeed been his rôle during most of the latter years of the Third Republic; the executive power lay with a Prime Minister and Cabinet dependent upon the National Assembly and virtually without any instruments with which to discipline its members. Ministerial instability remained the most obvious feature of the Fourth Republic as of the Third, and with far-reaching decisions having to be taken particularly in relation to the French Union and in foreign affairs, this fact had graver consequences than ever before.

In view of the fact that the inspiration of the Constitution accepted by the country in the referendum of last September came from General de Gaulle, with his well known views about the necessity of independence and energy in the executive branch of government, it is not surprising that some of its features should be governed by the intention of eliminating the instability of government characteristic of the Third and Fourth Republics. This paper is concerned only with those aspects of the Constitution that may be said to be the outcome of this particular desire on the part of its makers.

Such provisions fall under two main heads: those that enhance the rôle of the President of the Republic and in particular the wide emergency powers conferred upon him "when the institutions of the Republic, the independence of the nation, the integrity of its territory or the execution of its commitments are gravely and immediately threatened and the regular functioning of the constitutional public authorities is interrupted" (Art. 16) and the rather ambiguous terms in which he is granted the power of dissolving the National Assembly "after consulting the premier and Presidents of the Assemblies" (Art. 12); and those that define the rôle of the Prime Minister and his colleagues and their relation to the National Assembly (Titles 3 and 5)[2] to which we may add the restrictions on the duration of parliamentary sessions.

Constitutions are not made in a vacuum, and it is obvious that these arrangements not only reflect the views of General de Gaulle but are in some sense tailored to suit a President of his type and stature; an Albert Lebrun in 1940, even with the new powers of the President, might still have proved a transient and embarrassed phantom. Nevertheless constitutions may and do outlive the people and the circumstances that give birth to them and there is no reason not to try to see the new provisions against the background of constitutional experience in France and elsewhere.

In this respect the most striking feature of the Constitution is that which provides that "the office of member of the government is incompatible with the exercise of any

[1] On the making of the Constitution of the Fourth Republic see Gordon Wright, *The Reshaping of French Democracy* (London, Methuen, 1950). This also includes an English text of the Constitution itself.

[2] Quotations from the text of the new constitution are given according to the (unpublished) translation by Philip Williams and his collaborators at Nuffield College.

parliamentary mandate" (Art. 23).[3] According to the law promulgated in October in fulfilment of Article 25 of the Constitution, what is to happen is that each person in voting for a member of the Parliament shall also vote for a substitute. If the deputy or senator is nominated to the government he is replaced by his substitute for the remainder of the duration of the Assembly or of the senatorial mandate as the case may be. In other words we have here a separation of powers on the model of the Constitution of the United States where it is provided that "no Person holding any office under the United States shall be a member of either House during his continuance in office" (Art. 1. Sec. 6). The difference is that the new French constitution is seemingly more severe in that it looks as though acceptance of a post in a government, however brief its duration, would prevent any possibility of returning to the parliament until the next elections: though given the electoral conventions prevailing in the United States, the difference here is not very great. It is provided that ministers may address either house whenever they wish—thus following a modification of the American system introduced into the constitution of the Southern Confederacy though never implemented. The interdiction would seem therefore to relate primarily to their right to vote, and secondarily perhaps to their engaging in those lobbying activities which depend upon free access to those parts of the parliamentary buildings reserved for members.

It is of course the case that the American or, properly, presidential system goes further in the direction of the separation of powers in that a President cannot be got rid of (by Congress) during his term of office (except by the quasi-criminal procedure of impeachment) and that the heads of departments being his nominees can also not be got rid of, although the consent of Congress is necessary for their appointment. Under the French system, the National Assembly can still force the resignation of the government, though not of course of the President. It is true that it can only do this in specified circumstances: either by a decision of the Prime Minister to stake the existence of the government upon a vote on its programme or general policy, or else in consequence of a formal vote of censure passed under defined rules by an absolute majority of the Assembly (Art. 49). But experience in the Fourth Republic suggests that such limitations are not in themselves a guarantee of stability.[4] More important perhaps is the latitude afforded to the government under Article 38 to secure from Parliament the power to legislate "for the execution of its programme" during a "limited period" by means of ordinances even on subjects which are (according to the provisions of Article 34) normally within the domain of law. For clearly a government which asked for such powers and was denied them

---

[3] The same article also prevents a member of the government representing any economic or professional group on official consultative bodies or on the Economic and Social Council, and from being employed in any other public capacity as well as from engaging in any professional activity. But these provisions are not directly relevant to our main theme.

[4] Articles 49 and 50 of the Constitution of the Fourth Republic laid down a special procedure for putting the question of confidence and for votes of censure both of which required absolute majorities. But there were proposals for making the requirements still more stringent at the time of the constitutional reforms of 30th November, 1954 and later. See Dorothy Pickles, "Constitutional Revision in France," *Parliamentary Affairs*, VIII, 2, Spring 1955, and "The Reform of French Political Institutions," *ibid.*, X, 1, Winter 1956-7.

could hardly survive. It would also obviously have to resign if it were refused by Parliament the temporary credits that would be necessary if the annual finance bill did not become law before the beginning of the financial year to which it referred. On the other hand the power to enact the budget by ordinance if Parliament has failed to act within a period of 70 days (excluding periods of recess) would seem to be a source of strength (Art. 47).[5]

It is of course perfectly possible to devise procedures which while leaving the ministers within the legislature nevertheless make it less likely that governments will be frequently or wantonly overthrown. One way is to attempt to devise an electoral system that will diminish the number of contesting parties and so assist in producing clear majorities for governments of a particular complexion. The other is to try to restrain parliaments from overthrowing governments by insisting that the withdrawal of confidence must be accompanied by the clear expression of an alternative preference. Both these methods have been used in Western Germany in an attempt to avoid a repetition of the weaknesses of the Weimar Republic. The latter is given form in Article 67 of the basic law: "The *Bundestag* may express its lack of confidence in the Federal Chancellor only by electing a successor with the majority of its members and submitting a request to the Federal President for the dismissal of the Federal Chancellor."

The first of these options was and is hardly open to France. It is difficult to see that any electoral system—and France has tried a variety—could prevent in itself the maintenance of a multi-party system.[6] But the second which has been talked of in the past might conceivably have been tried with advantage.

On another point the new French constitution is closer to the Federal German one. The latter has a provision (Art. 68) to the effect that if the Federal Chancellor does not receive a vote of confidence when he asks for it the President may dissolve the *Bundestag* within twenty-one days unless a new Federal Chancellor has been elected by it in the meantime. The new French constitution abandons the Fourth Republic's complicated provisions about dissolution. Instead, Article 12 to which reference has already been made declares more simply that, provided one year has elapsed since the last such dissolution, "The President of the Republic may, after consulting the Premier and presidents of the Assemblies, declare the National Assembly dissolved." As already suggested the phrase about consultation is ambiguous—does the President have to take the advice of those he consults? But since (under Article 19) this is one of the occasions upon which the President acts without a ministerial counter-signature, one must presume that if he sees deadlock he is empowered to act.

With such provisions actually in the Constitution or capable of being included in some form, it is hard to see what was the precise purpose of moving the ministers out of the Parliament and preventing them from taking their places there again if

---

[5] I take it that the whole of Article 47 refers to the annual finance bill. It is very poorly drafted.

[6] This argument seems to me fully borne out by the evidence adduced in Peter Campbell, *French Electoral Systems and Elections*, 1789–1957. (London, Faber, 1958.)

subsequently defeated. In the draft of the Constitution submitted to the Constitutional Committee by the government on 29th July, 1958, the provision appears as Article 21: "No one may hold a governmental post together with a parliamentary mandate."[7] This was one of the provisions that the Committee unsuccessfully questioned in a letter written to the Prime Minister by its chairman M. Paul Reynaud on behalf of the Committee at the end of its deliberations sixteen days later.[8]

M. Reynaud wrote: "The majority of the Committee took the view that this article involved serious risks. The practical prohibition of access to government posts to political figures of weight is liable to harm parliamentary recruitment and to create between the government and the parliament an atmosphere of suspicion to which should be added the risk of a certain politicization of the upper levels of the administration." The Committee's own alternative suggestion for what it described as "the problem of the separation of the executive from the legislature" was that the minister should only be removed from parliamentary activity during his tenure of office and that he should not be replaced by anyone else.

Since this counter-proposal was rejected we must assume either that M. Reynaud's objections were thought to be invalid or that the elimination of ministers from Parliament occupied so cardinal a place in the thinking of General de Gaulle and his constitutional advisers that they were prepared to run these risks. How likely are M. Reynaud's fears to prove justified?

It is obvious from the fact that most prominent French politicians are seeking nomination in the elections now in progress (November 1958) that there has been no immediate falling-off in the appeal of a parliamentary mandate. It is of course a different question as to how this will operate in the long run. If parliamentary majorities prove and if governments in the Fifth Republic succeed each other as regularly as ever there will grow up in the country towards the end of each legislative period a body of men who are both ex-ministers and ex-deputies (or ex-senators) and who, while unable to recover their seats, have every incentive to go all out to recover their portfolios. If, on the other hand, the cabinets tend according to present American practice to be drawn largely from non-politicians M. Reynaud's forebodings will not be irrelevant. And this applies on both counts. Not only will a parliamentary career that does not make *ministrables* seem less desirable but there will be real dangers for the public service. For it is unlikely that French business can supply cabinet ministers in the way that American business can. The tendency to use civil servants for what are really political posts will be difficult to resist. No fewer than seven persons in the public service were called upon to take office in General de Gaulle's ministry formed at the beginning of June—the posts they held including such key ministries as those of foreign affairs, the interior and France *d'outre-mer*, and even if it be argued that the quasi-revolutionary situation made the circumstances exceptional, it is hard to see how nominations of this kind could be avoided in the future once the present generation of *ministrables* were used up. And it would surely make the neutrality of the civil service an impossible fiction if its members were

[7]The full text of this document was published in *Le Monde* on 30th July.
[8]Published in *Le Monde*, 17th-18th August.

habitually to serve in cabinets of a party complexion (relying that is on party combinations in the Assembly). And one could scarcely view with equanimity the prospect of their being debarred from return to official life after ministerial service—the attractions of business are strong enough anyhow. The comments that have been made on the partial obliteration of the distinction between political and administrative careers in the Federal German Republic should provide adequate warning on this score.

From the point of view of the governments themselves rather than that of the administration there is the special problem of where Prime Ministers are to be drawn from. This problem has no real parallel in the United States since there the President at least is a party leader, and chosen through familiar if to us bizarre machinery. Even with the increased powers of the French President, the Prime Minister remains the key figure of the system and cannot easily be envisaged except as a politician.

As soon therefore as one tries to look beyond the immediate crisis that gave rise to the drafting of the Constitution, one gets the feeling that in respect of the provisions with which we are concerned, there has been no real attempt to think out what they will mean in practice. They would seem to represent not so much a new solution to the problem of ministerial instability as an attempt to square the circle; to combine the dogma of the separation of powers—long assumed by General de Gaulle to be the main source of executive authority and stability where this exists—with the now entrenched republican dogma that governments must be responsible to the elected representatives of the people.

A final complication is the question of the impact upon this feature of the Constitution on the proposals for "The Community"; the group of autonomous States administering themselves and managing their own affairs "democratically and freely" which replaces the "French Union" of the Fourth Republic. It would take us too far afield to look at the whole question of the status of this "Community." But the Constitution does provide for the common handling of certain functions including "foreign policy, defence, common economic and financial policy, policy on strategic raw materials." The Executive Council of the Community provided for in Article 82 includes "the Premier of the Republic, the chiefs of government of the Member States of each Community, and the ministers responsible for the common affairs of the community." These will presumably, like the Premier, be at the same time holders of the corresponding posts within the French Republic, and it could be argued that their functions in the Community provide another reason for their being removed to some extent from direct involvement in French parliamentary struggles.

Whatever the share of American or German precedents in trying out the devices we have earlier discussed, it is still true that the easiest way to look at this problem is in the light of all the varied solutions that France itself has tried out in an effort to find the best relationship between the executive and the legislative power. No country, certainly, has a richer experience in this field. And one can only recapitulate briefly.

Under the Constitution of 1791, the Ministers who were still the King's, and suspect in consequence, were held at arm's length from the Assembly and could only address it with its permission. Under the Convention on the contrary the dogma of the sovereignty of the people's representatives was carried out to its logical conclusion. The Executive Council disappeared altogether in 1794 and political decisions were left to committees of the Convention itself. By the *Constitution de l'An III* (1795), the separation of powers was reintroduced in a radical form with the Directors elected by the legislative for fixed terms. Neither the Directory nor its ministers were responsible except in the penal sense. Under the Consulate and the Empire, the pendulum swung the other way, it now being the turn of the Executive virtually to swallow up all the legislative powers.

Parliamentary ministries begin only with the Restoration and it is then and under the July monarchy that French political institutions take on their characteristic modern forms. But ordered development was interrupted by the Revolution of 1848. The Constituent Assembly first elected an Executive Commission, a sort of Directory, with ministers as mere departmental chiefs under its authority, and with no clearly defined relationship to the Assembly. The Constitution of November, 1848, that of the Second Republic, returned to the principle of the separation of powers with a President irremovable during his short four-year term of office. The Assembly decided upon the number of the ministers and defined their functions. And (as under the Fifth Republic) they could demand to be heard in either House when they might be assisted by their officials. But there was no ministerial responsibility, since the principle of responsibility given voice in the Constitution included the Head of the State, the President himself, and was thus unenforceable.

The Constitution of 1852 was a further modification of this system in a direction favourable to the executive power and formed the basis for the institutions of the Second Empire. For the greater part of its duration, ministers could not be members of the legislature. But the pressure for liberalization brought about a modification of this position. In 1867, they were allowed to take a much less limited part in debates and were therefore subject to more parliamentary influence; and in 1869 the incompatibility between ministerial office and a parliamentary mandate was removed. The Third Republic was thus no innovator when it restored the principle of ministries drawn from the parliament and responsible to it.

It is clear enough that on the whole the French preference runs to responsible government with an executive drawn from the legislature itself. This has certainly been the characteristic system in the relatively stable eras. On the other hand, there are ample precedents both for the executive being independent of the legislature, and even dominating it, and on the other hand for almost direct assembly rule. Given this historical background it is not difficult to see how hard it must be for Frenchmen to try any device without being conscious of historical echoes. One could thus say that whereas there were overtones of the Convention about the Constituent Assembly of 1945, there are overtones of Bonapartism about the Constitution adopted in 1958. On the other hand, this precise compromise of an executive external to the legislature but responsible to it has not been tried before, and it is an open question whether its advocates or its critics will be confounded in the event.

CHAPTER 13

Succession and Stability in France

STANLEY H. HOFFMANN

There are countries in which the problem of succession involves only the mystery of one name—the name of the new leader. There are countries in which the continuity or the ending of a party's rule are involved. The French case, as usual, is far more complicated. At a time when traditional issues are fading away and political boredom seems to be settling down on the nation, French institutions continue to be one huge area of controversy and doubt. What is at stake in de Gaulle's succession is nothing less than the fate of the constitutional system which the General gave the French in 1958 and which he has already found necessary to revise in 1962.

In order to understand the nature of the issue, it is useful to first examine the system prevailing in 1958. The reasons which led de Gaulle to revamp it illuminate the problem of his succession: how can one institutionalize personal leadership in such a way that the regime will outlive the personal ruler?

THE REGIME OF 1958

The Meaning of the Constitutional Undertaking

The Constitution of 1958 consisted of two elements and tried to reach two objectives.[1]

As for the elements, one came from General de Gaulle, the other from the man who was his Minister of Justice in 1958 and who became his Prime Minister in 1959, Michel Debré. Michel Debré's scheme was what has been called "rationalized parliamentarianism." It was an attempt at establishing a genuine mixed cabinet system, i.e., one in which the Cabinet, while collectively responsible to Parliament, would nevertheless have an authority and a life of its own instead of trembling under the constant threat of parliamentary annihilation as had been the case during the Third and Fourth Republics. Since there was no chance of a stable parliamentary majority capable of guaranteeing the emergence of a solid and durable Cabinet, Debré's intention could be achieved only by institutional devices or gimmicks. In

---

[1] See the articles by Nicholas Wahl and Stanley Hoffmann in the *American Political Science Review*, Vol. 53, No. 2, June, 1959.

Reprinted from Stanley H. Hoffmann, "Succession and Stability in France," JOURNAL OF INTERNATIONAL AFFAIRS, XVIII, No. 1 (1964), 86–103, by permission of the publisher.

other words, since there was no organic salvation, there had to be mechanical solutions. There were various measures taken in order to limit the legislative and political powers of Parliament. The most spectacular was the possibility for a government to ram a law through Parliament even if there were no majority endorsing the bill. This was feasible so long as there was no majority willing to overthrow the Cabinet over the issue.

The element contributed by General de Gaulle was of a very different sort. He had always been concerned with the establishment of a strong presidency capable of acting as balance wheel for the whole governmental system. This accounts for the fairly formidable list of powers granted to the President in the text of 1958. It was this part of the constitution of 1958 which allowed some commentators to refer to it as an attempt at *néo-orléanisme*. The reference was to Louis Philippe's conception of the role of the king who, according to the doctrine of liberals such as Benjamin Constant and Guizot, was to be a supreme neutral and arbitral power dominating and regulating all others. What made the parallel even more possible was the fact that neither the king in 1830 nor the president in the 1958 version was directly selected by the people.

If one looks at the meaning of this experiment, one sees that it corresponded to a choice of medication instead of major surgery. On the one hand, the constitution still tried to reach the objectives of stability and authority through the means of a parliamentary system rather than through an abrupt switch to a presidential one. In the latter instance, the president would have been elected by universal suffrage and the Cabinet would no longer have been under the guns of the National Assembly. On the other hand, the whole constitution was based on a hope of affecting French parties, without whose reform the parliamentary ingredient of the constitution would necessarily remain somewhat artificial and fragile. Reliance was placed on the mere pressure of executive stability and the example of what executive action could accomplish, rather than on any shock techniques such as the switch to a presidential system or to repressive anti-party measures.

The Fate of the Experiment

If we look at the fate of Debré's "renovated parliamentarianism," there are two conclusions to be reached.

First, Debré's system worked only insofar and as long as the "organic" conditions for parliamentarianism obtained, in other words as long as his Cabinet enjoyed the support of a stable and coherent majority in Parliament. This majority was composed of a coalition in which the UNR (Gaullist) Party was predominant and allied to Christian Democratic and right wing partners. This coalition remained stable and coherent not because of any mutation in the mores of French political factions but because of the Algerian War. Had the alliance of these parties exploded at any time, the risk of a takeover by those very military and fascist elements from which the resort to de Gaulle had saved both the parties and the nation in May 1958 would have become overwhelming. The best evidence of the fact there had been no mutation was the disintegration of this coalition in the spring of 1962—as soon as the Algerian settlement had been achieved.

Secondly, even during the long period of Debré's majority, it had become clear that the two objectives of the Gaullist prescription had not been fully met. On the one hand, Parliament still had to be contained and compressed, and Michel Debré proved to be savagely good at doing just that.[2] Time and again the National Assembly and the Senate showed their ill will toward the mechanical limitations imposed on them by the constitution and by the rules of the two Houses, which the Constitutional Council—acting always in favor of the executive—had severely clipped. If this was the case even during France's greatest emergency, it obviously was a bad omen for the future.

On the other hand, four years of executive stability and action had had no visible healing effect on the parties, due to a number of factors. France was going through a highly abnormal and unique Algerian crisis which it was only too easy for the parties to interpret as a mere parenthesis after the closing of which they would resume their customary games. With its haughty disdain for parties, which it still treated as enemies and kept at arms' length, the regime itself was also at fault. More and more party members who had been ministers in the Cabinet were replaced by civil servants. Finally, there was the parties' own hope that once the parenthesis of the Algerian War would be closed, the very text of the constitution of 1958 could be used for a return to something close to the Fourth Republic. This explains why the old parties suddenly became the constitution's fiercest defenders when the Gaullists themselves decided to reform it in 1962.

As a result, one can say that by May or July, 1962, the medicinal approach had proved a failure. If renovated parliamentarianism had not proved conclusive, the presidential part of the Constitution had, however, become more and more predominant. In 1958 there had been a great deal of talk about the presidency as a power of arbitration to domination.

This trend was mainly the result of de Gaulle's impact. As he once said in his inimitable way, the fact that the kind of man he was and the kind of position he occupied made quite a difference. Despite the existence in the National Assembly of quasi-parliamentary conditions, i.e. Debré's coalition, it was definitely the President and not the Premier who governed. The presidency, which was supposed to serve as the spare tire for emergencies, had become the motor, and the Premier, instead of being the motor as the constitution would have had it, had become at times the fifth wheel and at times a brake. In the fall of 1962 when de Gaulle appealed to the people to ratify his constitutional revision, he defined the role of the presidency in terms infinitely more ambitious than he had in 1958.

De Gaulle's impact had been particularly heavy in three areas. First, the President had taken upon himself the interpretation of the constitution. Hence his refusal to hold any emergency session of Parliament in 1960, his extensive use of the emergency powers of Article 16, and his extraordinary resort to Article 11 instead

---

[2] See the study by J. L. Parodi, *Les Rapports entre le législatif et l'exécutif sous la Ve République*. Paris, 1962. (Fondation Nationale des Sciences Politiques, mimeo.)

of Article 89 for constitutional reform in the fall of 1962. Secondly, he had reduced the role of the Cabinet by making all important issues his "reserved domains." Algeria, of course, but also foreign and military policy remained as his own preserve, thus reducing the Cabinet to looking after what he had once disdainfully called *l'intendance*, economic, social and financial questions. Thirdly, and perhaps most importantly, he had used the referendum—which according to the constitution he could resort to only "on the proposal of the government during sessions or on joint motion of the two Assemblies"—as a personal instrument of power with which he could impose important policy decisions over the head of Parliament through carefully worded personal appeals. By connecting his own continuation in office with the adoption of the measures he wanted, he thus injected a heavy plebiscitarian element into policy-making. The importance he attached to the use of the referendum was never better shown than when he told a group of French parliamentarians that the German political system was not perfect since it did not have the referendum.

This evolution from arbitration to domination posed the problem of succession in acute terms. The trend toward presidential predominance risked being jeopardized or reversed by the very procedure for selection of a president set up in the constitution of 1958. Should the eighty thousand notables (among whom the less developed and most traditional parts of France were over-represented) select as a successor to de Gaulle one of those amiable nonentities who was likely to be acceptable precisely because he was mild, aged and noncontroversial, the powerful presidency of de Gaulle would then shrink to something like the emptiness of the Third and Fourth Republics' presidencies, despite the formal grant of powers. After all, judging from their authority on paper, the Presidents of the Third Republic should not have been mere ciphers.

This danger was suddenly brought infinitely closer by a would-be assassin's bullet in August 1962. In the famous attack at Le Petit Clamart, the bullet missed the General by only a couple of inches and something like one ten-thousandth of a second. The combination of long-term fears and imminent threats determined de Gaulle to switch from medicine to surgery. If he did not act fast, now that the Algerian War was over and the Debré majority had started to dissolve, the parties could do to de Gaulle what they had done once before in the fall and winter of 1945. At that time, they sought to erode his authority by attacking him on a terrain chosen by them, that of European and Atlantic policy, so as either to force him out or to oblige him to dissolve Parliament in circumstances most unfavorable to him and most favorable to the election of a new Parliament in which he would have had no majority whatsoever. Given this danger, he selected deliberately aggressive tactics and resorted to the kind of *Blitzkrieg* against the "parties of yesterday" that reminds one of Colonel de Gaulle's writings on offensive mechanized warfare in the 1930's. He chose to interpret the constitution in such a way as to provoke the parties into overthrowing the Pompidou Cabinet. Thus he had the double advantage of going directly to the people with the issue of constitutional reform (which was an issue chosen by him and over which the old parties had to adopt a highly unsavory position of arguing against the election of the president by universal suffrage) and of having the subsequent elections to the National Assembly centered on the issue of continuity and stability rather than on the foreign policy issues the old parties would have preferred.

FIVE AND A HALF

The New Constitutional Setup

The revision of the constitution in the fall of 1962 is important enough to allow one to say that France now lives in the Fifth-and-a-Half Republic. The design of the new system corresponds again to two objectives.

The first one is a consecration of presidential predominance both in the executive and in the system as a whole. He dominates the executive, since the premier is essentially the man chosen by the president to be a go-between in the latter's relations with Parliament. This is indeed the role Guizot had played under Louis Philippe, but de Gaulle's "néo-orléanisme" is very neo since the president is to be elected from now on by universal suffrage. The president also dominates the system as a whole, since he keeps his power to dissolve the National Assembly, a most important deterrent for protecting the premier's political life. The Houses' legislative powers remain as narrow as before, and their hold over their agenda, over amendments and over their own sessions is tenuous in the extreme. And of course the president still has the referendum and emergency powers of Article 16 at his disposal.

The second objective is to force the parties out of their grooves by making the conquest of the presidency rather than the restoration of Parliament the necessary and proper target of political life. As long as the president was merely the product of the notables, the parties could have hoped to restore Parliament to its pristine predominance. With the radical transformation in the balance of political organs, this hope now should be thoroughly in vain. In other words, the revision of 1962 was designed to blow up the road back to the Fourth Republic, which the milder obstacles of 1958 had not sufficiently barred.

The Problem of Gaullist Succession

The best way to examine the problems of Gaullist succession is to discuss first the constitutional system and later the political process itself.

Let us begin with the constitutional system as it is. The main drawback of the Fifth-and-a-Half Republic seems to be the dual executive—the co-existence of a president elected by universal suffrage and of a premier responsible to Parliament. This was the kind of system which existed in the Weimar Republic and whose catastrophic consequences are well known. Would the right of dissolution which the president enjoys suffice to deter a chaotic National Assembly in which there would be no majority capable of tolerating a Cabinet for any length of time? Or would the right of dissolution deter an Assembly, whose majority would be coherent but hostile to the president, from imposing its own premier on him? In such a case the president might find himself crippled by those constitutional provisions which continue to define quite extensively the role of the Cabinet and of the premier, particularly in the area of policy-making and in the issuance of referendums.

If the French party system remains unreformed, no presidential system may work well. After all, even if the premier and Cabinet were not responsible to Parliament, an unreformed party system still would be capable of obstructing the executive by delaying the adoption of bills in Parliament. It has been suggested that this danger could be overcome if the president relied on shifting majorities for different types of measures, as happens sometimes in the United States. This whole analysis, however, is based on a mistaken comparison of U.S. and French parties and political mores. A dual executive would undoubtedly magnify and multiply the difficulties.

Some French parliamentarians have suggested a switch to the American system— one in which there would be no dual executive but in which the president would lose his power to dissolve Parliament and in which Parliament would regain full legislative powers. It is not surprising that the pure presidential system has become the favorite of parliamentarians eager to restore Parliament to a position of greater power and prestige. After all, the U.S. Congress has a role infinitely more weighty than the French Parliament under de Gaulle. But by the same token, this is the suggestion least likely to seduce the General. The dangers of the pure American system are those of complete deadlock between the executive and the legislative and the possibility of a kind of parliamentary revenge.

The political scientist Maurice Duverger has suggested a presidential system in which deadlocks would be avoided through a resort to the voters for the simultaneous re-election of the president and election of a new National Assembly, in case of a crisis between them.[3] Duverger presents his conception as a mechanism of mutual deterrence. The trouble with it is a familiar one to students of deterrence: if deterrence fails, the mess that follows is utterly disastrous. Not only is it hard to predict what would happen if the electorate should send back to Paris the same president and the same National Assembly, still deadlocked and more at odds with each other than ever, but also the idea of resorting to elections at a time of possible national emergency is not a terribly reassuring one.

The final suggestion emanates from Michel Debré.[4] In his most recent book, the former prime minister considers that the dual executive is useful, but he, too, fears what would happen if the National Assembly disrupted the established harmony between the president and the premier. What he suggests is a further curtailment of Parliament's powers. Should the National Assembly overthrow the Cabinet, Debré would like to make it possible for the president either to ask the Assembly to reconsider and vote again, or to submit the issue at stake to a referendum.

If we turn to the political process, there are two aspects that have to be studied. Nothing has been more interesting in recent years than the way in which the average

---

[3]*La VIème République et le régime présidentiel* (Paris, 1961). Lavau's critique is in his article "Réflexions sur le régime politique de la France," *Revue Française de Science Politique*, Vol. XII, No. 4 (Dec. 1962) pp. 813–44.

[4]*Au Service de la Nation*, Paris, 1963.

citizen has lost his enthusiasm for, interest in and even patience with political parties. This has been interpreted all too often as evidence of "depolitization."[5] The truth of the matter is that the citizen has also become much more involved in large collective associations which represent neither his ideological preferences nor his general outlook on public life, but his interests as a producer, consumer and family man. The issues which these associations cope with are as public as those with which the parties deal, but in the modern industrial society of Western Europe there is a new division of attention and labor between political parties and interest groups. Political parties remain indispensable as instruments for the selection of political leaders and for the control of the administration, while the great interest groups represent the functional activities of the new society. What is original in the French case is both the belated emergence of the latter and the demotion of the former by the Fifth Republic.

As for the impact of the new political system on the parties, we have to distinguish between theoretical guesses and the conclusions that one can draw from recent practice. Theoretical arguments as to what effect the reform of 1962 will have on the parties have been running in circles. On the one hand there is Duverger's thesis of party mutation through surgical shock. Now that they will have to prepare a presidential election, French parties will see the need to reform and to merge into a much smaller number of coalitions, comparable to the very loose American parties. On the other hand there is Raymond Aron's sceptical rejoinder which one could call the hypothesis of shock absorption by party incorrigibility. He pointed to the example of Brazil, in which the existence of a presidential system has certainly not meant the disciplining of the parties, the reduction of their number or anything like the U.S. form of government. The debate can only be settled by events: the French parties are sufficiently shaken by the events of recent years to be deeply concerned with their future, as well as capable of being transformed by the new setup. And they are sufficiently incorrigible or resilient to be capable of thwarting the grand Gaullist design either *in toto* or in part.

It is precisely because nobody knows whether the shock of the new constitutional system will be enough to reform the parties that Michel Debré has suggested another shock, that of adopting the British or American type of electoral system, i.e., a single-member constituency with one ballot only (France currently has a single-member constituency system with two ballots). Debré hopes that this reform would produce a "regrouping" into a two-party system whose main beneficiary would be the Gaullist UNR and whose main loser would be the Communist Party. Maurice Duverger, on the other hand, believes that the effects of such a drastic surgical operation would not be felt immediately and that in the meantime there would be a period of very dangerous unsettled politics in which extremists on both sides might well be the winners.

Obviously, theory has to be checked against practice, and there are two experiences to which it is possible to refer. Unfortunately, neither is conclusive. The first one is the 1962 election which followed the constitutional referendum. The results were

[5] See Association Française de Science Politique, *La Dépolitisation* (Paris, 1962).

highly ambiguous. De Gaulle had obtained 64 percent of the votes cast in the referendum. A few weeks later, in the first ballot, the Gaullist party obtained slightly more than 31 percent of the popular votes—far more than the 19 percent it had gotten in the first ballot in 1958. This shows *both* that a very large section of the French electorate voted in November 1962 *as if* France had a presidential system (the voters backed the President's supporters, thus doing what the President had asked them to do), and that an equally large amount of those voters who had supported de Gaulle on the constitutional issue nevertheless voted a few weeks later for parties that had bitterly opposed him over it. In other words, the results are no more than encouraging.

The other experience concerns the behavior of the parties since November 1962. The least one can say is that the story is confusing. There are at present two crucial pivots and question marks. On the one hand there is the question of the UNR. It is trying hard to become a *mouvement gestionnaire*, i.e., the Radical party of the Fifth Republic, the party without which it is impossible to govern. This is obviously an attempt both at digging roots in the country and at exerting a major impact on all other French parties. More and more, the Gaullists talk in terms of becoming the most important French party even after General de Gaulle, the party that would be in post-Gaullist days the sun around which various satellites would turn. However, there are formidable obstacles to such ambitions. The UNR has no program except loyalty to de Gaulle; it tries hard to make a virtue out of necessity. It explains that programs were good enough in the by-gone ideological age when voters were interested in selecting a global *Weltanschauung* which had no chance of ever becoming reality. The UNR on the other hand is modern and has objectives which are pragmatic and concrete. The trouble with this argument, which certainly has a grain of truth, is that objectives are just as susceptible of not being met as the old doctrines were likely to remain up in the air. The objectives defined by the UNR so far are just as vague as the old platforms. Furthermore, the UNR until now has been not so much the inspirational force behind the government as it has been either a *masse de manoeuvre* of the executive, faithfully providing it with the necessary votes, or a kind of conglomeration of pressure group delegates haggling with the Cabinet over budgetary details which would favor special clusters of voters. Also, the UNR has not succeeded yet in giving itself a strong organization, especially at the local level, and its relations with the *forces vives* remain rather poor precisely because the latter see in it a mere instrument of the executive. One must note, however, that in so far as the UNR has become the dominant party to the right of the Socialists, it has picked up votes not exclusively on the Right but also from the Center and from the moderate Left,[6] and its emergence as the main political force on the Right implies a very drastic change in the complexion of France's Right. Whereas the latter had been largely devoted to the defense of traditional French

[6]This goes against Eugen Weber's argument of a continuous *glissement à droite* in recent French political history: *International Review of Social History* Vol. V, 1960. See on the contrary Philip Williams' analysis of the UNR in the French referendum and elections of October-November, 1962, *Parliamentary Affairs* Vol. XVI, No. 2. p. 165 ff. and René Rémond in *Esprit*, February, 1963.

society, of traditional colonialism and fearful of industrial development, the UNR, quite close in this respect to the Bonapartist candidates in the Second Empire, is interested in modernization, development, expansion and renovation.[7]

The other pivot and question mark is the Socialist Party. It finds itself at present in a highly interesting position. The non-Gaullist parties to its right, i.e., the Christian Democrats and Radicals essentially, know that they have very little chance of regaining power if they only merge among themselves while the Socialists stay away from them. The Christian Democrats in particular have multiplied, with increasing signs of impatience, their appeals to the Socialists. The Christian Democrats stress that both have the same objectives in foreign policy, i.e., Atlantic and European integration as opposed to de Gaulle's nationalism. Similarly, they point out that both parties have the same sense of priority for domestic welfare over international grandeur. The trouble is that the Communists have also been flirting with the Socialists and that at the last elections the Socialists, alone among the non-Communist anti-Gaullists, managed to pick up a very considerable number of seats due to their electoral alliance with the Communists at the second ballot. Had the Socialists chosen to ally themselves with the "democratic" Center instead, they probably would have shared the misfortunes of the MRP and the Radicals. The Socialists are entitled to hope that if they continue flirting with the Communists, especially in the new climate of international *détente*, a Socialist inevitably will be the common candidate of all the non-Gaullist parties. Such a candidate would have considerable chances of success, for the Communists obviously have no chance of electing one of *their* men and will of course prefer a Socialist to anybody further right. If the parties situated to the right of the Socialists present a joint candidate of their own against the Gaullists, or de Gaulle himself, and against the Socialists, he will "bite the dust." The common interest of all non-Gaullist parties in ejecting Gaullism from the presidency may force the parties of the Center and Right—more or less grudgingly—to accept a Socialist candidate as their spokesman. In a way, the Socialists' present dilemma is exactly comparable to the one they faced in 1943-4, when they had to choose between joining a broad French *travaillisme* and remaining a narrow socialist party working closely with the Communists. Guy Mollet's decision in the coming months may well be the same as the fateful one made by his party in the last months of the Resistance.

Thus we have no answer yet, but a new element has been injected by the man who remains the master of the game and the very skillful manipulator of the French scene, General de Gaulle. He has made it clear that he wants to be his own successor if his health continues to permit. Whether this means that he will wait until the normal expiration of his term late in 1965 or whether he will try to force the issue earlier remains to be seen and will undoubtedly be determined by circumstances. After all, young de Gaulle wrote in *The Edge of the Sword* that the best leader is the one who uses events as they come along. Should de Gaulle decide to force the issue, it is not impossible that he would tie his re-election to another

---

[7] See for instance *La Democratie à Refaire*, Paris, 1963 and the issue of *La Nef* on "Le Nouveau Contrat Social," September-November, 1963.

referendum rather than resort to the cumbersome and, in this case, comic constitutional procedure of resigning first, having the president of the Senate (who happens to be de Gaulle's arch enemy) take over during the interim, and then running for re-election. Here, we are in the realm of tactics—and de Gaulle's are as unpredictable as his objectives are clear.

Present Gaullist tactics seem to put the other parties into a dilemma. If his non-Communist opponents choose to fight him on the issues of Europe and NATO (on which they all more or less agree), then an alliance with the Communist Party becomes impossible and there will be no single anti-Gaullist candidate. In other words, de Gaulle's election will be insured. On the other hand, if they merely write a new edition of the disastrous anti-de Gaulle cartel of 1962 in order to have a single candidate—for instance, the Socialist mayor of Marseille, Gaston Defferre—it will be impossible for them to agree on any program at all except in terms so totally negative or so favorable to the Communists that de Gaulle will once again have a field day. For he could rally opinion against the incorrigible little men of yesterday, as well as characterize the opposition's candidate as a hostage of the Communist Party.

Whatever the situation will be, one thing seems clear. Despite de Gaulle's heroic attempt at establishing a political system capable of working efficiently after him, the next presidential election will hardly be a precedent for the future, if he himself is a candidate. However hard he tries to institutionalize himself, de Gaulle remains unique. The configuration of forces that will be fighting the battle of the presidential election *after* de Gaulle will hardly be predictable on the basis of the alignments for and against de Gaulle if he runs again. The very strength of his personality, his way of personalizing all issues, the manner in which the parties themselves are more and more obsessed with the single issue of preserving or ejecting him mean that the real reshaping or reshaking of the French political process will be delayed until after de Gaulle. This is not to say that his actions (whether in the form of the constitutional revision of 1962 or in the form of a campaign for re-election) do not affect the future, for they undoubtedly do; but they do not shape it once and for all.

Underneath the inconclusive agitation of the political parties we find a great number of reflections, of a much quieter and academic nature, concerning the *forces vives*. Commentators, professors, statesmen and leaders of interest groups and associations have periodically met to discuss the proper place and powers of those forces in the political system.

The limits of the problem have to be made clear. As suggested earlier, the French case is less unique than many seem to feel, because the rise of the *forces* is true of other western societies as well, especially when planning develops. What is unique is not the emergence of the new forces but the degree to which public interest in parties had dropped long before de Gaulle. This had been due to the parties' inefficiency and divisions, and the present eclipse of parties and Parliament by the spirit of Gaullism. However, the problem of the new forces is serious insofar as the lively, and to the extent that many important areas in public affairs are being dealt present parties are indeed sclerotic or anemic, whereas those forces are fresh and with by the executive directly with those new forces, thereby short-circuiting the parties. Never has this tendency been more obvious than in the fall of 1963, when the government discussed the details of its economic stabilization plan and the beginnings of a *politique des revenus* (a policy on distributing the fruits of expansion more fairly) with representatives of the various collective groups.

If one wants to draw conclusions about the role of the new forces in the French political process, there are three sets of considerations. First of all, the new forces have been very useful as economic and social issues become the dominant ones in domestic French affairs, replacing the old ideological issues. This usefulness has been all the more appreciable as the organs of consultation, in which the representatives of the new forces sit, have multiplied under the Fifth Republic. The sobering experience of participating in discussions in which plain figures matter more than figures of speech has been felt in every part of French society. Even the labor unions, despite their discontent with the economic and social policies of a regime they consider much too timid and conservative, have become acutely aware of the perils of the inflation for their own members. Similarly, the representatives of the farmers have become more and more concerned with modernization and less and less attached to their old shibboleths.

Secondly, however, the role of the new forces has been necessarily limited. In a nation in which constitutional engineering is a fine art, some grandiose schemes are flourishing. One such scheme is for a revamping of the Economic and Social Council with the intention of giving it genuine decision-making powers. Another is for the establishment of a new Senate composed of representatives of France's social and economic forces.[8] However, there are two sets of handicaps. There is the difficulty *in any country* of entrusting decision-making powers to pressure groups, even when they take a less ferociously parochial view of their problems than in the past. As Henry Ehrmann, who knows French pressure groups so well, has said, if the *forces vives* want to remain *vives*, they must remain pressure groups. And if they are primarily pressure groups, then it will be difficult to give them more than a consultative role. Then again, there is a set of especially *French* difficulties. One might call it the tension between a long tradition of group hostility to integration into the state apparatus, and the tendency of the bureaucracy to use the groups as its own agents for the implementation of policy. Each one of these tendencies, of course, breeds and strengthens the other. In recent weeks there has been quite a ritual of protest on the part of groups asserting their independence and their right of "contestation," along with equally repeated disclaimers by the government and the planning commission swearing that they have no intention of affecting the autonomy of these groups. Any attempt at developing a *politique des revenus* runs in particular against the French labor unions' resistance to any policy which seems capable of leading to an administrative or authoritarian determination of wage levels at the expense of collective agreements. The very weakness of the French labor movement, with its low membership and its three-fold division, increases the fear each fragment

---

[8] See for instance the suggestion by former Prime Ministers Pflimlin in *La Democratie à Refaire*, pp. 123–23 and Michel Debré, *op. cit.* 210 ff.

has of becoming a dupe or a stooge of the government, and consequently of losing to its rivals the support it has among the workers.

Thirdly, the role of the new forces is still minimal with respect to the indispensable blood transfusion from these forces to the parties. The latter have not been capable of attracting new men. All that is dynamic in French public life seems to go either into the new forces or into a highly efficient civil service. The remedies often suggested in this respect, such as decentralization or regional planning (in other words, changes from the bottom) seem still distant in their realization and uncertain in their effects. There are timid beginnings in the direction of regional planning, but most of the policies of regional (*aménagement du territoire*) are still determined primarily by the civil service on top. Debré's hostility in his new book to any measure that could "loosen the state" and thus weaken its authority is particularly significant, coming from a man who once was an apostle of decentralization. A change from the top, greater power of attraction exerted somehow by the parties themselves, presupposes once more a reform and reorganization of the party system. We are in a vicious circle.

Whether the blood transfusion will occur consequently depends both on the nature of the administrative reforms which General de Gaulle has hinted at in recent statements—but given his *mystique de l'Etat*, how far can one expect them to go—and on the sense of responsibility that the parties, including the UNR, will show after the closing of the Gaullist era.

HUNCHES AND HEDGES ABOUT THE FUTURE

The political scientist who sees the future as obscure has two ways of dealing with uncertainty. Either he can build models, or developmental constructs, or scenarios and thus give rein to his scientifically playful instincts. Or else he can say that he does not know. In this instance, it is better to conclude with a few tentative remarks. They refer to the political situation of France after General de Gaulle (whenever that may be), since the real test will come not at the time of the next presidential election if de Gaulle is a candidate, but only after his disappearance from the political scene.

The French political problem is *sui generis*. The main difficulty continues to be the translation of the deep social and economic changes that have revolutionized French society into the present political system.[9] Specifically, the party system remains the chief bottleneck. So far, the shocks administered to it from above have been much smaller than those suffered by political forces in totalitarian or authoritarian regimes (Nazism, Fascism, Bonapartism), for in those instances the period of oppression was long enough and the effects of suppression deep enough to either eliminate the old forces altogether or prepare the clandestine emergence of new ones, or both. Also, the transformations below have not yet reached the political forces.

In the present situation, with the disappearance of the threat from the extreme Right, the lowering of political tensions and a new constitutional experiment, there is one hope that deserves discussion: the "American illusion" of many French social

[9] See my remarks in *In Search of France*, Cambridge, 1963.

scientists. They tend to make two mistakes. They exaggerate the merits of the U. S. system. Professor Georges Vedel has even gone so far as to state that the American presidential system allows a full discussion before the electorate of the issues between the candidates—a point which hardly corresponds to reality. Indeed, one might almost say that the American system of government shares with the Third Republic in its heyday the characteristic of being so full of checks, balances, brakes and inhibitions that only a prosperous and relatively secure nation could afford the waste of energy involved. Also, the writers underemphasize the purely French obstacles to the functioning of a system such as that of the United States.

There has been no sufficient discussion yet of one important problem: the relation of the presidential candidates to the parties (or coalitions), especially in the first presidential election after General de Gaulle's removal from the political scene. It is a crucial issue both because of its precedent-making value and because the "reform" of the parties may still be incomplete. The French tradition has been one of either parliamentary domination or extra-parliamentary "personal power." The new system tries to break away from the tradition, by increasing the distance between the president and the representatives, as through direct popular election. It tries, as well, by making saviors unnecessary now that the executive is granted sufficient authority or, if one prefers, now that the savior is being institutionalized.

However, the very differences between French and American parties which many French authors minimize may thwart this design. U.S. parties have neither the organization, doctrinal history, nor the parliamentary traditions of French ones. Besides, there are only two of them, despite the strange and strained analysis by French zealots of U. S. presidentialism who pretend that there are actually a hundred American parties, two in each state, that get together into two coalitions only at election times. French parties have a far more centralized structure, which makes it very difficult to compare any of them with the situation of that American party which has lost the presidential election and which finds itself without any nominal or real leader. Also, despite the decline of dogmas and programs, the French parties still try to preserve from their past a certain distinctive flavor which will save them from the loss of identity which mergers would imply and may ultimately force them to accept. Finally, there is in their relations among themselves and in their relations with the executive such a long record of extremism and *politique du pire,* developed throughout the long periods of parliamentary predominance, that the transition to pragmatism, goodwill and good faith may be anything but brief.

Consequently, relations between the presidential aspirants or nominees to their parties in the U. S. are very different from what they may be in France. In a U. S. convention what is at stake is the temporary control of a party by a candidate; the party is moved, not mover. In France, what may be at stake, if parties remain unreformed at the start, is the control of the candidate by the party or party coalition. De Gaulle's strength is due to his extra-party position and to his personality. Will the popular election of his successor, by what may be a narrow margin, and the broad grant of powers to the presidency (on paper) be enough to overcome party attempts to tie down the candidates by promises and commitments? If not, the tradition may continue even within the new framework, and the presidents may be either hapless prisoners of puny parties or Bonapartist types. *Le pire n'est pas toujours sûr,* but nor is smooth success.

If one turns to the political climate as a whole, one also has to hedge. On problems of substance, the gap seems to be narrowing. Ideologies continue to disintegrate even in the field of foreign policy where (leaving aside the problem of the Communist Party) one notices that even the non-Gaullist parties tend to dispute de Gaulle's methods much more than de Gaulle's objectives. This is a point overlooked by far too many American commentators who delude themselves by believing that de Gaulle's successors will be more tractable in Atlantic affairs. It is significant that the opposition at present attacks much less the content of de Gaulle's domestic and foreign policies than their style, which is inextricably tied to the General's personality, and his order of priority among the several objectives which he and they accept. In this last respect there is therefore much less of a difference between France and her European neighbors, or between France and the U. S. today than there has been in the past.

On the other hand, built-in obstacles to a French version of the "American image" still remain. First, the constitutional problem is not settled. As indicated above, de Gaulle himself may have further reforms in mind. Should an anti-Gaullist either win the next presidential election against the General himself or succeed the General after the latter's second term, it is quite possible that the various non-Gaullist parties could unite at least on some constitutional reforms designed to curb the presidency and restore the powers of Parliament. Secondly, the existence of a Communist Party, which remains the second party in popular support and certainly the one with the biggest membership and best organization, makes the Anglo-Saxon analogies (so dear to certain French authors) questionable and party regroupments difficult. Thirdly, the domestic conservatism of the present Cabinet, which does not even have the merit of great administrative efficiency that distinguished the Debré era, tends to revive traditional left-wing, popular front type reflexes. Thus we have the paradox of the Left representing tradition today far more than a Right shaken and transformed by the changes in France's traditional society, in France's role in the world and in the nature of the parties that obtain conservative votes.

Probably because of the many tribulations of recent years, French writers now tend to indulge in another illusion—what one might call the Quest for the Perfect Regime or the Illusion of Political Idyll. Both with respect to constitutional machinery and with respect to the integration of the "new forces" into the political system, too many people expect too much and invent ingenious schemes for perfect, frictionless harmony. These put far too much emphasis on procedure and too little on substance, such as on those issues that might still disrupt the most clever techniques of government.[10] Here again, Lavau is the voice of wisdom:[11] No constitutional

---

[10] This is particularly true of recent discussions on making French planning more democratic. Very much has been said about including more economic, social, regional and political representatives into the procedure. Very little has been said about either the possibly cacophonic results of such an orchestration or about the substantive objectives that ought to be pursued.

[11] Lavau, *op. cit.,* p. 841.

machinery can be so good as to work all the time. And when it works too smoothly, sometimes nothing comes out of it. The U.S. is an example of the first instance, England today perhaps a good example of the latter. A return to modesty of expectations might be a prelude to reasonable success and an antidote to new disillusionment. The crucial problem is not whether the system will be perfect, but on what kinds of issues it will be stalled, and whether they will affect its survival. Perfection requires an "organic" solution, and it is not yet in sight. Seymour Lipset has recently written a book about the U.S. as "the first new nation." Insofar as many of the new nations, whose societies are much less advanced than France's, suffer from political uncertainties comparable to those of France, one might well ask in the same vein of forced modernism whether France will not remain the last new nation.

CHAPTER 14

Democratic Stability and Instability: The French Case

ERIC A. NORDLINGER

I

The outstanding characteristic of the French political system is its historical instability. Constitutional monarchy was overthrown by a revolution, replaced by a republic, which in turn quickly evolved into a dictatorship, and when it too was dismissed by an armed uprising, the interminable squabbles among the monarchist factions allowed another republic to come into existence by default. But for an "accident" of history this republic too would have given way to a dictatorship through the bloodless medium of the coup d'état, but while the republic tottered on in the interwar period the life-span of its governments was calculated in terms of months rather than years, and with its "collapse" under the coup de grâce of military defeat a new dictatorship immediately sprang up to take its place, to be succeeded by another republic lasting for thirteen years amid constantly recurring cabinet crises, then falling in the wake of an eminently successful revolution, out of which emerged the present regime. Here we have what sociologists might label the "institution-

Reprinted from Eric A. Nordlinger, "Democratic Stability and Instability: The French Case," WORLD POLITICS, *XVIII, No. 1 (October, 1965), 127–57, by permission of the publisher. This article is a review of Michel Crozier,* THE BUREAUCRATIC PHENOMENON: AN EXAMINATION OF BUREAUCRACY IN MODERN ORGANIZATIONS AND ITS CULTURAL SETTING IN FRANCE, *translated by the author (Chicago: University of Chicago Press, 1964).*

alization of instability," interpreted by a number of leading writers on French politics as the product of a deep-seated conflict between the "two Frances," whether these two political subcultures are viewed as the parties of *mouvement* and of *l'ordre établi*,[1] or as the "administrative and representative traditions."[2]

Nor is this cleavage confined to the political realm. It is well grounded in two opposing and mutually exclusive cultural edifices. To quote Siegfried, "Party divisions . . . are based on opposing conceptions of life, and the instinctive personal reaction of like or dislike toward a certain social order."[3] Traditional France, favoring order at the expense of liberty, the Catholic Church at the expense of the republics, following and pushing forward the "caesars" of France in the face of democratically constituted authority, can most succinctly (for our purposes) be subsumed under the phrase "the France of hierarchical order." The France of the Left, with its clarion call for liberty and equality, for decades working toward a levelling state socialism, believing in the camaraderie of the working class, can most aptly be characterized as "the France of idealistic egalitarianism."

Adherence to the doctrine of hierarchy is manifested, for example, in attitudes toward the equality of the sexes. When asked in a national survey whether or not they believe in the equality of salaries for men and women doing the same jobs, the proportion of respondents saying "yes" steadily decreases as one moves from Left to Right on the political spectrum.[4] Two conflicting orientations toward authority are also exhibited in the work-place, the meeting place of the bourgeoisie and the workers. Desiring equality in their social relations, the workers manifest similar attitudes toward employer-employee relations. As one French worker put it: "The worker wants equality on the job because he practices equality with his wife and with his pals and children."[5] Though wanting equality for themselves, almost all the workers interviewed by Cantril and Rodnick had the feeling that they were being looked down on by the *patron*, who viewed them as objects with which to make money. Nor were these feelings entirely illusory, for the same authors found that *patrons* of all types believed in a social hierarchy with themselves as the self-appointed elite. And it is just this attitude that is largely implemented by the bourgeoisie in their economic establishments.[6] In France the common pattern of

---

[1] François Goguel, *La politique des partis sous la troisième République* (Paris 1946).
[2] Nicholas Wahl, "The French Political System," in Samuel H. Beer and Adam B. Ulam, eds., *Patterns of Government* (New York 1958).
[3] André Siegfried, *France: A Study in Nationality* (New Haven 1952), 26.
[4] *Sondages*, No. 17 (1948), 222.
[5] Hadley Cantril and David Rodnick, *On Understanding the French Left* (Princeton 1956), 14.
[6] *Ibid.*, 18, 67. Roy Lewis and Rosemary Stewart describe the French businessman as "paternalist and autocratic, treating his employees as *'mes enfants'* . . ." (*The Managers: A New Examination of the English, German, and American Executives* [New York 1961], 186). It is this hierarchical orientation among the employers that is responsible for the deep division between employers and employees. Given their elitist assumptions the employers have not been able to bring themselves to accept the legitimacy of trade unions in anything resembling a graceful fashion. Not even the crisis period of the "phony war" in 1940 dissuaded the employers and the Daladier government from mounting an unrelenting attack upon the unions' organizing and bargaining rights.

employer-employee relations is almost simplistically military in nature.[7] The *patron* exercises practically all authority, without even the presence of an American-type foreman through whom authority is mediated. Nor does the *patron* seek advice or approval from his workers, except in those instances in which effective unions force him into this position. It is then no wonder that the worker feels himself to be acted upon, consequently developing an intense desire for liberty.[8]

No doubt the disparate political orientations dividing the France of hierarchical order from the France of idealistic egalitarianism are of fundamental importance in accounting for the system's constitutional and governmental instability. The differing conceptions of the state and its relation to society held by the two Frances continually call into question the legitimate form of the state as prescribed by the constitution; and their near balance of strength, combined with the ostracism of the extreme Left and extreme Right from both the government coalitions and the "loyal" opposition, places a heavy burden upon cabinet government. Yet what makes the mutual antagonism so intense and pervasive is the solid congruence of these two conceptions of government with their accompanying attitudes toward religion, the family, social equality, and the relationship of man to society, resulting in two remarkably resilient cultural edifices supporting the two conflicting views of the ideal political system. It is just this structural rigidity—this congruence of political and non-political attitudes —that severely militates against the two political subcultures' becoming inclusive rather than exclusive toward each other.

However, notwithstanding the extensive validity and explanatory power attaching to this interpretation of French political instability, there are three reasons for thinking it inadequate: first, the dichotomous argument is too simplistic, particularly as modernization and industrialization have erased the acute distinctiveness of the two cultures, leaving only vastly depleted political troops as representatives of the traditional France of hierarchical order;[9] secondly, the argument is difficult to apply to the actual workings of the governmental system—a system founded on excessively pragmatic bargaining techniques rather than on the clash of ideologies;[10] and thirdly, with the publication of Michel Crozier's exceptional study, a more powerful, elegant, and contemporaneously valid interpretation of the French political system is available.

II

Crozier is concerned with two problems: a novel and persuasive theoretical analysis of bureaucracy, and an analysis of the interconnections between a bu-

---

[7] Georges Lasserre, "Le Monde ouvrier dans la société," in André Siegfried, ed., *Aspects de la société Française* (Paris 1954), 122.

[8] In a 1955 survey of manual workers, the questions were asked: "Are there ideas to which you are strongly attached in the political and social spheres? What are these ideas?" About twice as many workers selected "liberty" rather than "betterment of workers' conditions," "equality," or "social justice" (*Les Temps Modernes*, July 1955). These data sharply underline the workers' desire for liberty since more than twice as many workers prefer this amorphous entity to specific and concrete economic betterment and social justice.

[9] For an excellent series of essays on political, social, and economic change in contemporary France, see Stanley Hoffmann and others, *In Search of France* (Cambridge, Mass., 1963).

[10] The place of ideology in the political system is discussed in Section IV of this review article.

reaucratic system of action and the social and cultural setting in which it functions. The author's analytical skill is not evidenced only in the execution of these two objectives. His work must be accorded additional praise for the "sociological imagination" shown in developing the rather standardized original data concerning the functioning of two bureaucratic organizations into a study that should be most useful to political scientists, sociologists, and students of French society. In the present review article, the discussion will be limited to Crozier's interpretation of French authority relations and the application of his model to the political system.

Just as a country's political culture is embedded in the larger social culture, authority relations are set in the matrix of social relations. Thus before characterizing French authority relations *per se*, it would be well first to look at Crozier's comments regarding a broader range of interpersonal relations. In his survey data and research into the operation of two bureaucratic organizations, Crozier found three typical patterns of action that are also manifested throughout French society: "the isolation of the individual, the predominance of formal over informal activities, [and] the isolation of the strata" (p. 214). For example, notwithstanding the clerical workers' need for social support (most of the workers having come to Paris from the provinces in order to find work), only rarely did any of them report having any friends at the work-place. And even those who did have friendships found that these did not develop into strong and lasting relationships. Moreover, there were few associations of any sort that could serve as instruments of social integration. The trade unions were fairly active, but membership in these organizations remained on a purely formal level, entailing a minimum of participation. And even when friendships and informal groupings (e.g., cultural, educational, or social activities) did exist, their membership did not cut across the various formal strata within the two bureaucratic organizations.

According to Crozier, these patterns of action are rational responses for the attainment of the clerical workers' values. Contrary to the assumption of American social scientists that people want to participate in their organizations, Crozier correctly points out that "It is a partial view indeed which expects people to be always eager for participation. People are very ambivalent towards participation. . . . On the one hand, people would like very much to participate in order to control their own environment. On the other hand, they fear that if and when they participate, their own behavior will be controlled by their coparticipants. It is far easier to preserve one's independence and integrity if one does not participate in decision-making. By refusing to be involved in policy determination, one remains much more free from outside pressure" (p. 204). It is for this reason that the French tend to insist upon strict equality between members of the same strata; protection is afforded the members of the strata by the imposition of this equality, which prevents interference from higher authority by a defensive banding together whenever a superior attempts to exercise his formal authority. In addition, it is recognized that separate informal activities could very well weaken the cohesion of the formal group. It is this protective rationale that supports the sterility of French interpersonal and intergroup relations. (This analysis could be taken one step further, to show that it is a deep-seated insecurity among many sections of French society that leads to the assumption of a protective posture in their social relations.)

Crozier goes on to cite the work of Jesse R. Pitts, who characterizes both informal and formal French associations as "delinquent communities." These are defined by "jealous equalitarianism among the members, difficulty in admitting newcomers, and a conspiracy of silence against superior authority. They do not deny authority, however. Indeed they are incapable of taking initiatives except in interpreting the directives of superior authority and accommodating themselves to those interpretations. In an effort to create for each member a zone of autonomy, of caprice, of creativity, these peer groups thrive on the unrealism of the authority's directives."[11] French associational activities thus maintain a negative solidarity—a solidarity against authority. Their rabid egalitarianism prevents the emergence of peer-group leadership, or even acceptance of the authority of formal superiors (to whom it might be easier to accord this prerogative, since peer-group jealousy would not be an issue).

The "delinquent community" is thus a protective device vis-à-vis external authority and is at the same time an efficient system for making it impossible for a member of the peer group to attain a leadership position. As a number of sociologists have pointed out (among them Wylie, Bernot, and Blancard), this strong tendency to prevent any group member from raising himself above the others, and to see initiative as bossism, is a common French behavioral characteristic. "Non-participation is then a rational response if people want, above all, to evade conflict situations and to escape dependence relationships. Strata isolation, focusing on rank and status, and the impossibility of informal grouping across strata all stem from the same difficulties. All these traits ultimately refer to the basic cultural conditions predetermining the possible scope of authority relationships" (p. 220).

Authority relationships are marked by the avoidance of face-to-face dependence relations (*l'horreur du face à face*) and the avoidance of open conflicts between groups that directly confront each other. "Authority is converted, as much as possible, into impersonal rules. The whole structure is so devised that whatever authority cannot be eliminated is allocated so that it is at a safe distance from the people who are affected" (p. 222). This pattern of authority stems from two contradictory goals: the extremely high value placed upon personal independence—or in reverse fashion, the abhorrence of personal dependence relationships—and the recognized need for a powerful directing authority if any cooperative activity is to succeed. The French try to reconcile these two goals by allowing for a centralized authority, while at the same time insisting upon impersonal, protective relationships, i.e., individual and strata isolation to prevent interference from above. This reconciliation may be a satisfactory one for the individual, but, as will be seen later, it is dysfunctional for the efficient operation of formal and informal systems of action. For what the French are doing is balancing (rather than integrating) absolutist conceptions of authority. Central authority is invested with supreme power, yet the people formally subject to that authority refuse to give up even a modicum of their independence so that the authority's directives can be carried out. Both superiors and subordinates claim for themselves an absolute right that cannot be shared or compromised.

[11]"Continuity and Change in Bourgeois France," in Hoffmann and others, 259.

Such a system of action cannot handle moderate and steady change. It produces a cyclical alternation between periods of unrelenting and stubborn routine and periods of crisis when change finally becomes inevitable. Crozier believes this pattern of action to be "a distinctly French feature, inasmuch as it relies on the complex model of individual isolation, lack of communication between strata, and avoidance of face-to-face dependence relationships" (pp. 224-25). Since authority cannot be shared or diffused while dependence relationships are not easily accepted, impersonal rules and centralization offer the only escape. Consequently, power tends to be pushed further and further upward until, even if the subordinates were to permit interference by their leaders, the leaders could not act effectively because of the distance separating the order from its execution. Thus continuous leadership is impossible. The system cannot adapt to change; when change does come it must be in the form of a crisis. Although the people at the top "are all-powerful because they are at the apex of the whole centralized system, they are made so weak by the pattern of resistance of the different isolated strata that they can use their power only in truly exceptional circumstances" (p. 225).

III

Up to this point discussion of Crozier's analysis has centered upon French patterns of interpersonal and intergroup relations in the abstract, although the analysis was originally formulated primarily on the basis of two case studies. In order to provide his interpretation with a further empirical grounding and to show that the patterns of action just described are ideal-typical for all French institutions, Crozier applied his model to four important institutional systems: the educational system, the colonial system, the industrial relations system, and the "politico-administrative system." This reader, for one, is satisfied that with the possible exception of the colonial system Crozier's model finds its validation in each of these four areas. Here, however, the discussion will be confined to the political system.

Crozier's analysis of the political decision-making process in terms of three interrelated subsystems—the administrative, the policy-making, and the revolutionary grievance-settling subsystems—brings coherence into what many other students of French politics have analyzed either in an unsystematic or in a legalistic-institutional fashion. Moreover, Crozier's scheme is not confined to a static analysis of the political system; it also leaves room for a dynamic interpretation.

The administrative subsystem handles the routine, already established programs of government. It is afflicted with two characteristic weaknesses of the ideal-typical French decision-making process. The men who actually make the decisions reside at an excessive distance from the people to whom the decisions are applicable and the reality to which they refer. Secondly, the difficulties of coordination, which are omnipresent in all organizations, are heightened in this instance by the civil servants' rigidity (their absolutist conception of authority) in relations with groups inside and outside their organizations, while the fear of face-to-face dependence relationships leads them into a panicky fear of overlapping spheres of authority. The outcome is a self-perpetuating circle in which further centralization appears to be the only answer to the bureaucrats' unwillingness to share authority and their abhorrence of overlapping responsibilities. In turn, this structural factor (overcentralization) reinforces the upward spiral of responsibility, increases the frustration and lack of initiative found at the lower levels, and leads to further isolation of the various strata. Given these disabilities, the administrative subsystem is able to act only as an instrument of order, rather than as one of change and innovation.[12] Consequently, when issues unquestionably demanding change arise, the administrative subsystem must resort to the next highest level, the policy-making subsystem.

The parliament and the governments it produces would appear to be omnipotent, considering the absence of any institutions that could challenge their preeminent authority. Yet again, this seeming omnipotence goes hand in hand with actual powerlessness. The civil service has taken over responsibility for the country's actual problems at the local and regional levels, while the deputies join in unproductive clashes over the divisions separating Frenchmen from one another. Since these issues have not been handled effectively by the deputies, they have had to be settled by changes of government rather than by the writing of legislation. This cyclical pattern has produced "the predominance of the government, since matters could be settled only at the government level, and its helplessness, since parliament has been the constant arbiter of intragovernmental struggles and has yielded to the government only for as long as necessary to handle the crisis it has been unable to solve alone" (p. 256). Or to summarize the situation by adapting one of Stanley Hoffmann's felicitous comments: French governments do not govern; they occupy power.

Crozier is quite correct in disregarding the presumed structural factors as the causes of the government's ineffectiveness and instability, and in turning instead to the cultural characteristics of the political class. But here one is best off following Melnik and Leites in their description and detailed analysis of the men of the policy-making subsystem.[13] In the first place, both the *piétaille* (the backbenchers) and the *ministrables* (the group of potential ministers) subscribe to an isolationist view of the parliament. "There is in the Parliament a tendency to ignore the outside world and concentrate almost fanatically on its internal affairs." And elsewhere in their study, Melnik and Leites can write that parliament members carry about with them an "impression of the Parliament as existing outside of space and time, cut off from the outside world."[14] There also exists in the parliament an extreme egalitarianism stemming from an intense desire to be independent, to be free to act in whatever manner personal advantage may dictate.[15] Further, the deputies manifest a strong urge to put off action until the last possible moment; they will not act until

---

[12] This is true at the prefecture level as well as at the highest level of the *Grand Corps*, whose major function is to control and oversee the work of the lower echelons.

[13] Constantin Melnik and Nathan Leites, *The House Without Windows* (Evanston 1958). It is recognized that this study is based upon analysis of a single event—the process by which René Coty was elected President of the Fourth Republic—and that the conditions under which the election took place in Versailles differed from those prevailing in the Assembly and Senate: the absence of tradition-honored procedures for electing a President and the secret ballot are just two factors that differ from the conditions under which prime ministers were invested. Nevertheless, the generalizations cited here are thought to be almost equally applicable to the politicians in their usual habitat.

[14] *Ibid.*, 16, 66.   [15] *Ibid.*, 97.

impelled to do so by an external force. This predisposition is expressed in the familiar parliamentary dictum that "It is only with your back to the wall that you can make a proper choice."[16] Clearly these types of behavior are largely responsible for the ineffectiveness and instability of the parliamentary system.

But the most characteristic and significant feature of parliamentary behavior is a rebelliousness against authority, an inability to allow authority to others or to engage in collective action entailing superordinate-subordinate relations. The premier is "tolerated only when he is at everyone's disposal: resistance (or an error) on his part provokes demands for his resignation. The [ensuing] hubbub provides a good deal of pleasure, for it enables the individual to rebel against authority in an atmosphere of communicative good fellowship."[17] The political malaise produced by this cultural foundation is poignantly expressed in one of Raymond Aron's quotable comments: "The Republic was so afraid of great men that it was forced, from time to time, to have recourse to saviours."

Since the policy-making subsystem is unable to furnish neither leadership nor a minimum amount of authority, it is not able to find or impose solutions upon the problems gripping the society. The result is "escalation": a third subsystem has necessarily evolved to resolve the conflicts that should have found their solution in the legislature or the executive. Referring to the Fifth Republic, Hoffmann could write that "never before has resort to violence been so widespread and treated so casually. The ungodly spectacle of party squabbles and cabinet crises had been eliminated, only to be replaced by an even ungodlier one. If the institutions of yesterday were too close to a shaky ground, those of today are too far removed, and dissent tends to express itself through direct action—strikes, plots, bombs and coups."[18] It is this institutionalized resort to violence that Crozier calls "the revolutionary grievance-settling subsystem." Nor is this mode of action confined to a particular time or to a particular social class. When other avenues have failed to provide complete satisfaction, resourt to direct action has been a common French reaction. Since the end of the nineteenth century, workers, farmers, civil servants, shopkeepers, students, and *colons* have all resorted to this style of decision-making; and its acceptance by the French is perhaps best illustrated by the leniency with which the courts treated the perpetrators of terrorism and rebellion under the Fifth Republic.

IV

It was originally suggested that Crozier's interpretation is preferable to the standard argument regarding French political instability delineated at the outset. But what of the other competing hypotheses that are said to explain political instability? Additional weight is given Crozier's argument when it is seen that these competing hypotheses are untenable or that they can be subsumed under Croziers' wide-ranging interpretation.

[16]*Ibid.*, 34.
[17]*Ibid.*, 108.
[18]"Paradoxes of the French Political Community," in Hoffmann and others, 94.

One of the political scientists' more popular whipping boys for the instability of French governments is the multiparty system. It is undoubtedly true that the presence of more than two parties can make the task of forming and supporting stable and effective cabinets a more difficult one; and in a multiparty system it is also more difficult than in a two-party system to aggregate interests. Yet this is hardly an adequate explanation for governmental instability in France, especially since there are a half-dozen or more European countries that have or have had multiparty systems supporting stable and effective coalition governments.

It may be, however, that it is not the multiparty system alone but a confluence of the multiparty system and the ideological inundation of French politics that is responsible for the political system's weaknesses, with ideologism aggravating the structural disabilities of a fragmented party system. This argument finds support in a comparison of successful and unsuccessful multiparty systems: in Austria, Norway, Denmark, Switzerland, Belgium, and Holland, where dogmatic ideological stances are infrequently assumed, the existence of more than two parties has not served to undermine the system by preventing the emergence of effective coalition governments, while in Weimar Germany and contemporary Italy, the conflict of mutually exclusive ideologies has eliminated this possibility.

Notwithstanding the force of this argument and the many writers who have attributed France's political ills to the diffusion of conflicting ideologies, the evidence for this proposition is far from conclusive. Almost all those writers who point to the political parties' ideological rigidity as the impediment to the development of a pragmatic politics based on compromise have conveniently overlooked the way in which the game of politics is actually played in France. Although ideologism pervades the parties' electoral and propaganda efforts, this public ideological posturing of French politicians does not prevent them from playing out their game of compromise in the Assembly and its *couloirs*. In fact, the political class thinks of compromise as a positive principle of action, with parliamentary activity largely revolving around nonideological squabbles in which pork-barrel legislation, private gain, political advantage for the individual deputy and his party, and personal vendetta take pride of place. Nor does the ideological argument hold much water when it is recalled that in the Third and Fourth Republics the least ideologically inclined party—and that is just about the most positive statement that can be made about the Radical Socialists—provided the largest number of prime ministers. Furthermore, the old ideological debates over education, clericalism, foreign policy, and nationalization lost most of their dogmatic intensity after 1954, yet the stability and effectiveness of the system did not vary concomitantly with the softening of ideologies.[19]

A third argument regarding the instability of French government demands a good deal more attention to uncover its inadequacies. Kornhauser's theory of mass society is not directly a theory of stable democracy.[20] Yet in the French case it could be

---

[19]This is not to say that ideological predispositions—with their affinity for deductive systems of thought in which the real world can often be ignored—have disappeared.

[20]William Kornhauser, *The Politics of Mass Society* (Glencoe 1959).

viewed as such since the rise of mass movements was frequently the immediate cause of the downfall of republics and governments. Although there is some debate whether the Third Republic "collapsed" or whether its demise was simply due to military defeat, in either case the proliferation of mass movements in the 1930's made a signal contribution. Two rapidly growing parties on the Left—the Communists and Socialists—were able to mobilize a mass following, while the nationalist and semi-Fascist Right was spewing out a host of mass movements, at the head of which stood the Croix de Feu. And in the 1950's the Fourth Republic had to contend with an ugly Poujadism, to be succeeded in time by a successful Gaullist revolution. There is thus certainly more than one connection between the existence of mass movements and the functioning of the political system. But can Kornhauser's theory, which attributes the rise of mass movements to the absence of voluntary associations mediating between the masses and the elite, account for the rise of mass movements in France?

So far as this writer is aware, there is only one nationwide sample of Frenchmen who were asked about their membership in voluntary associations. It was found that 41 percent of the French belong to at least one such organization.[21] To be meaningful, this figure must be related to comparable data for other countries. In their comparative study, Almond and Verba found that 57 percent of the Americans, 47 percent of the British, 44 percent of the Germans, and 29 percent of the Italians were association members.[22] When the French figure of 41 percent is inserted into this grouping, it is seen that the data hardly support Kornhauser's theory. The French figure is significantly closer to the British and German ones than it is to the Italian; yet it is the British system that has been historically stable, as is the present German regime, something that cannot be said of either the French or Italian systems. In short, if a low rate of group membership accounts for the rise of mass movements, as claimed by Kornhauser with specific reference to the French case, then one would expect the French figure to be closer to the Italian than to the British and German ones. Yet just the opposite is true. Consequently, if Kornhauser's thesis were to be accepted, it would be necessary to argue that it is simply the difference between the three and six percentage points separating the proportion of group members in France from the proportion of group members in Germany and Britain respectively that accounts for the rise of mass movements in the former but not in the latter two countries.[23]

Such an argument is patently inadmissible; clearly extensive group membership is not sufficient protection against the rise of mass movements. An addition must therefore be made to Kornhauser's theory. Presumably it is not the number of organization members alone that is significant; the nature of that membership and possibly the nature of the organizations' goals are also relevant.

[21] Arnold M. Rose, *Theory and Method in the Social Sciences* (Minneapolis 1954), 70.
[22] Gabriel A. Almond and Sidney Verba, *The Civic Culture* (Princeton 1963), 302.
[23] And even these differences may not be reliable since the French survey was carried out in the early 1950's and the Almond and Verba surveys were done some six years later, when it is reasonable to expect that the continued industrialization of France had led to an increase in the proportion of group members.

As we have seen, despite the relative extensiveness of French group life, the country has witnessed the rise of a host of mass movements. The same is also true of Weimar Germany. Although no quantitative data are available for Weimar, it is clear that there was a proliferation of voluntary associations; dramatic societies, choral groups, gymnastic clubs, walking clubs, and card-playing and drinking fraternities dotted the landscape.[24] These two exceptions to Kornhauser's theory—and they are critical exceptions since his data refer almost exclusively to European countries—have two characteristics in common. Every observer of French social life has been impressed by the passive and casual nature of group membership. And the same was true of Weimar's group life. The members joined, but they did not generally participate in running their organizations. While French group members refuse to recognize the authority of association leaders, the Germans hardly thought to question their authority. Yet in both instances the result was a passive membership. Though based on only two cases, the conclusion seems to be warranted that extensive organizational membership alone, without a concomitant involvement, does not offer sufficient protection against the growth of mass movements. Moreover, it will be seen (though limitations of space and subject matter do not permit such a demonstration here) that each of the arguments put forward by Kornhauser in support of the theory linking group membership and mass-movement politics is unsatisfactory (logically and empirically) if the size of group membership alone is considered.

A second possible addition to Kornhauser's theory might also be suggested. Arnold Rose has noted the preponderance in France of what he calls "expressive associations," which meet only for social and cultural purposes, over "social influence organizations," which are directed toward public purposes.[25] Again, in Weimar Germany the situation was basically similar. There too the expressive associations greatly outnumbered those organizations whose goals entailed a form of community action. The exceptional ease with which these associations were *gleischgeschaltet* (the process by which Nazi leaders were substituted for the original ones) was at least partly related to the nonpublic and nonactivist goals of these groups. This is to say that expressive associations are generally less able to prevent the flowering of mass movements than are social influence organizations. Certainly if the goals of voluntary groups were added to Kornhauser's theory its validity would be enhanced. In this instance, it would lead to a better understanding of France's mass-movement politics. As it stands now, the theory is inadequate as an explanation for the rise of mass movements in France, and by extension, for its political instability.

[24] Writing of the *fin de siècle* period, before further urbanization and industrialization led to an even greater increase in the number of voluntary organizations, Robert Michels noted that "the German worker, as his wages have increased, has acquired the disease which is in the blood of the German petty bourgeoisie, the club-mania. In every large town, and not a few small ones, there is a swarm of working-class societies" (*Political Parties* [New York 1959], 289).

[25] Rose, 73. Charles Bettelheim and Suzanne Frère come to the same conclusion in *Une Ville Française moyenne: Auxerre en 1950* (Paris 1950), 252.

V

Crozier's analysis of French political instability was said to be more adequate than the generally accepted interpretation of a culturally divided France, while a number of alternative explanations have also proved to be unsatisfactory. There remain a handful of hypotheses, not directly touched upon by Crozier, that appear to be eminently valid in accounting for the disabilities of the political system. The purpose of this section is to demonstrate that these hypotheses can easily be encompassed within the Crozier interpretation—that they stand at a lower level of generality than does Crozier's. The article will then, it is hoped, have shown Crozier to be a leading guide for students of French politics, for not only is his interpretation the most general and coherent one available, but it is also able to subsume those valid arguments put forward at a less general level, while competing hypotheses are shown to be untenable.

The French are well known for their *incivisme*, a quality that has a political and a nonpolitical meaning. In one sense it is a mixture of a diffuse social negativism, a refusal to cooperate with others or to accept responsibility, and a fierce clinging to individual independence—a string of attitudes that might best be characterized as "privatization." It is thought that this cultural trait prevents the formation of the active group life necessary for a two-way flow of communication and influence between the governed and the governors. On a political plane, *incivisme* refers to a singular distrust of politicians and a pervasive suspicion of government. In the popular Radical doctrine erected by Alain, democracy becomes nothing more than the control of the governors by the governed—a system of surveillance rather than representation. Surely Crozier's analysis suggests a connection between both types of *incivisme* and French attitudes toward authority. On the one hand, the fear of face-to-face dependence relations prevents the French from engaging in a personally meaningful group life since this would entail the exercise of informal authority and its acceptance by the group members. On the other hand, the absolutist conception of authority as something that is to be hoarded and that cannot be shared accounts for a large part of the hostility toward politicians and government.

Another widespread interpretation of the political system's malaise focuses upon the political alienation of Frenchmen—their refusal or inability to invest their connections with the regime with any positive meaning, and the related tendency to take direct and extra-legal action against it. French politics is largely played out at two levels—that of the individual citizen and that of the center—and the interconnections between the two are few in number and tenuous at best. The prefecture system is primarily organized around the execution of orders and the forwarding of communications from the center, while the political parties suffer either from the lack of any organization to speak of or from the failure to provide pyramidal structures connecting the local level to the national level through intermediate bodies. It is this overcentralization that is largely responsible for the Frenchman's political alienation; he sees little or no connection between the casting of his vote or the statement of his interests and the decisions produced by the governmental system. There is consequently little reason for him to participate in an orderly democratic politics.

And as Crozier has pointed out, French attitudes toward authority impel the system toward overcentralization, in turn creating a diffuse political alienation with its dysfunctional consequences for the system's stability. At the same time, the citizens themselves are responsible for their lack of meaningful contact with the political system. Having elected representatives with orders to oppose governmental authority, to defend their *situations acquises,* they prevent the emergence of a government effective enough to make contact with the people. Furthermore, the political class is opposed to any direct contact between a strong leader and the people —something that would mitigate the effects of overcentralization. Unlike legislators in the United States, where Presidential fireside chats are an accepted commonplace, the deputies bitterly resented the attempts of Doumergue in the Third, Mendès-France in the Fourth, and de Gaulle in the Fifth Republic to establish a direct connection, over the heads of the elected representatives, between themselves and the electorate.[26] Needless to say, this reaction is securely anchored in the deputies' determined refusal to accord the executive more than the minimum of authority necessary to survive—temporarily.

From a functionalist perspective, the difficulties of the French political system can also be attributed to the loud and coarse style of interest articulation and the weak processes of interest aggregation characterizing the system. Adhering to an absolutist conception of authority—an authority that they alone possess—the citizens and their interest groups state their desires not as requests but as demands upon the government—and exaggerated demands at that, which, if they are not completely fulfilled, are considered to be unfulfilled. Moreover, the absence of a respect for authority prevents the formation of a well articulated system of interest aggregation, in which all would accept the decisions of the persons in authority rather than continually appealing for redress to competing centers of power or resorting to direct action. Interest groups operate almost solely on the national level, the place where power resides, without extending organizational roots down to the regional and local levels. These organizational lacunae also contribute to the irregularities of the interest-aggregation process, given the tenuous system of communications between the pressure-group leaders and the people for whom they are presumably speaking.

Nor does the unorganized state of the political parties offer a remedy for the unedifying spectacle of a disorderly pressure-group politics. It is a primary function of political parties to aggregate diverse interests so that they can be introduced into the policy-making process in an orderly and properly processed manner. Yet the absence of channels extending from the grass roots to the center removes the possibility of the parties' effectively fulfilling this function. Then, too, the lack of discipline within the parliamentary hemicycle and the dependence of most deputies upon nonparty sources for electioneering expenses make the deputies fair game for the pressures exerted and blandishments offered by the various interest groups,

[26]Philip Williams, *Crisis and Compromise* (London 1964), 440.

resulting not only in an ineffective system of interest aggregation, but also in what Gabriel Almond has called poor "boundary maintenance" between the society and the polity. The interest groups have managed to insert themselves directly into the Assembly through their control and direct representation in the *groupes parlementaires d'études*, causing a marked confusion between the interest-articulation function, the interest-aggregation function, and the rule-making function. Because the interest groups have gained a direct influence on legislative activity, their interests cannot be articulated and aggregated in an orderly fashion; instead, uncompromised or "raw" claims are introduced directly into the legislative process where it becomes more difficult to handle them in an efficient and orderly manner. Surely it is unnecessary to indicate further how each of the factors contributing to the weakness of the interest-articulation and interest-aggregation functions is closely related to French attitudes toward authority.

VI

If a single variable—in this case, a society's attitudes toward authority—is powerful enough to account for the broad operational contours of its political system, is it not possible that the authority variable can also account for the stability and instability of other political systems? Does not the French case suggest that in attempting to account for democratic stability and instability on a universal scale a society's authority relations are the key explanatory factor? For one thing, a stable democratic government must be able to strike a workable balance between its two functions, representing and governing the electorate. An imbalance between the two will lead either to unstable and ineffective government or to inauthentic democracy in which the political elite is able to disregard the desires of the electorate. It would seem that the crucial determinant in patterning the extent to which a government is led to balance these two, frequently incompatible, functions is the attitude toward authority held by both the political class and the electorate. Secondly, the authority dimension in a country's political culture appears to be an exceptionally powerful explanatory variable, able to subsume other types of attitudinal, behavioral, and structural factors under its umbrella. It therefore seems reasonable to believe that if various attitudes toward authority could be analytically structured, a theory of stable democracy utilizing them as the independent variable would appear eminently plausible on at least the two following grounds: Attitudes toward authority are intimately related to the stability, efficiency, and authenticity of democratic systems; and because this single variable is able to encompass other types of empirically well-founded hypotheses, whatever validity attaches to these narrower hypotheses can be logically transferred to the authority variable.[27]

[27] Such a theory is developed in the present writer's *The Working-Class Tories: Authority, Deference, and Stable Democracy* (Berkeley, 1967). Pp. 234–50 enlarge upon some of the points made in this essay, and integrate them within a theory of stable democracy.

CHAPTER 15

Politicization of the Electorate in France and the United States

PHILIP E. CONVERSE AND GEORGES DUPEUX

The turbulence of French politics has long fascinated observers, particularly when comparisons have been drawn with the stability or, according to one's point of view, the dull complacency of American political life. Profound ideological cleavages in France, the occasional threat of civil war, rather strong voter turnout, the instability of governments and republics, and the rise and fall of "flash" parties like the R.P.F. in 1951, the Poujadists in 1956, and the U.N.R. in 1958 have all contributed to the impression of a peculiar intensity in the tenor of French political life.

It is a sign of progress in the study of political behavior that such symptoms no longer seem to form a self-evident whole. We feel increasingly obliged, for example, to take note of the level in the society at which the symptoms are manifest. Most of our impressions of the French scene reflect only the behavior of French political leadership. Growing familiarity with survey data from broad publics has schooled us not to assume perfect continuity between the decision-making characteristics of a leadership and the predispositions of its rank and file. The extremism of the military elite in Algeria or ideological intransigence in the French National Assembly are in themselves poor proof that the shipyard worker in Nantes has political reflexes which differ from those of the shipyard worker in Norfolk.

We feel increasingly obliged, moreover, to discriminate between some of these well-known symptoms of turbulence, for they no longer point in a common direction as clearly as was once assumed. Two signs which unquestionably reflect mass electoral behavior in France provide a case in point. Turnout levels in France are indeed high relative to those in the United States,[1] suggesting that, in the politically indifferent strata of the electorate where nonvoting is considered, political motivations are more intense. On the other hand, we now doubt that the rise and fall of "flash" parties are parallel symptoms of intense involvement. Rather, it seems likely that such episodes represent spasms of political excitement in unusually hard

---

[1] They are not, of course, outstanding against the backdrop provided by other Western European nations.

Reprinted from Philip E. Converse and Georges Dupeux, "Politicization of the Electorate in France and the United States," PUBLIC OPINION QUARTERLY, *XXVI, No. 1 (Spring, 1962), 1–23, by permission of the authors and the publisher.*

times on the part of citizens whose year-in, year-out involvement in political affairs is abnormally weak.[2] Obviously, for France and the United States, the basic traditions of a two-party or a multiparty system affect the likelihood that the flash party phenomenon will occur. But other things being equal, it seems that such phenomena are hardly signals of long-term public involvement in politics but betray instead a normal weak involvement. The durably involved voter tends toward strong partisan commitments, and his behavior over time stabilizes party fortunes within a nation.

Other less direct indicators add doubt as to the high involvement of the broad French public. Demographically, French society differs from the American in its lesser urbanization and lower mean formal education. Intranational studies have persistently shown higher political involvement among urban residents and, more strongly still, among people of more advanced education. While cross-national extrapolation of such data may be precarious, it does leave further room to question our intuitive impressions.

We intend in this paper to examine comparative data on the French and American publics in an effort to determine more precisely the locus of Franco-American differences in these matters.[3] We shall consider the locus in qualitative terms, covering an extended series of political characteristics which run from expressions of involvement, acts of participation, and information seeking to orientations whereby the voter links party alternatives to the basic ideological issues in the society. We shall throughout maintain an interest as well in a vertical locus of differences. That is, we shall think of the two electorates as stratified from persistent nonvoters at the bottom, through the large middle mass of average voters, to citizens who engage in some further partisan activity, and thence by extrapolation to the higher leadership whose highly visible behavior is so frequently the source of our cross-national impressions. Such extrapolation is necessary, of course, because it is unlikely that the handful of "activists" whom we can distinguish at the top layer of both national samples include more than one or two persons who have ever had any direct hand in a leadership decision of even a parochial party organization or political interest group.

INVOLVEMENT, PARTICIPATION, AND INFORMATION SEEKING

While a relatively large number of comparisons may be drawn with regard to simple manifestations of political involvement in the two countries, these comparisons vary widely in quality. Broad differences in institutions and political

[2] For a fuller discussion, see Angus Campbell, Philip E. Converse, Warren E. Miller, and Donald E. Stokes, *The American Voter*, New York, Wiley, 1960, Chap. 15.
[3] The French data were gathered in three waves of a national cross-section sample in the fall of 1958, during the constitutional referendum launching the Fifth Republic and the ensuing legislative elections. The study was jointly supported by the Conseil Supérieur de la Recherche Scientifique, the Rockefeller Foundation, and the Foundation Nationale des Sciences Politiques. The American studies over six elections have been conducted by the Survey Research Center of the University of Michigan, under grants from the Rockefeller Foundation and the Carnegie Corporation. Informal cross-national collaboration prior to the 1958 French study led to a French interview schedule permitting more rigorous comparative analysis than unrelated studies usually offer.

practices in the two societies can serve to channel public interest in different directions. The French political poster, often a full-blown campaign document, is addressed to other goals than the American political billboard, and hence the reading of such posters in the two societies is in no sense comparable activity. Similarly, the national control of the domestic airwaves in France means that two media of communication are given a totally different cast than in the United States. This fact, coupled with reduced access to radio or television sets in France, renders the attention paid by the two publics to such political broadcasts fundamentally incomparable. Or, in a different vein, certain manifestations of involvement are known to vary widely in their frequency within a nation from one type of election to another, or for the same type of election between periods of crisis and troughs of routine politics. While an extended American time series has provided some useful norms, these were more difficult to find for the French data. In general, then, we shall elaborate upon only a few of the most solid comparisons, referring summarily to the flavor conveyed by other, looser comparisons.

Given the broad institutional differences between the two societies, it might seem useful to draw contrasts between self-estimates of psychological involvement between the two nations, however differently institutions might channel the ultimate behavioral expressions of such interest. While the data permit a number of matches between questions on political interest, posed at comparable times with comparable wording and with superficially comparable alternatives, one hesitates at comparisons which depend on crude "amount words" such as "very," "fairly," and the like. Cautiously, however, it may be observed that Americans gauge their interest in their elections at a rather higher level than do the French. Two to five times as many French respondents indicated that they were "not at all" interested in the 1958 elections as is the tendency for Americans with regard to their presidential elections; three to four times as many Americans say that they are "very" interested. Distributions from France in the more normal political year of 1953 show slightly higher levels of expressed interest, but even this distribution fails to approach the most unenthusiastic American distributions collected at the time of off-year congressional elections. For what it is worth, then, it is relatively hard to get French citizens to confess much interest in their elections.

More solid are comparisons of reported acts of political participation selected as involving comparable motivation in the two systems: membership in political organizations, attendance at political rallies, and attempts to influence the political choice of others through informal communication.[4] As Figure 1 suggests, the cross-national similarities on these items are impressive. Furthermore, we can examine such additional points as the number of meetings attended by those French

---

[4] Of these three pairings, the first is technically the weakest. The American item asks about membership in "any political club or organization," while the French item focuses directly on political party membership, although the term "party" may be rather broadly construed in France. Furthermore, there were a substantial number of refusals to answer this membership item in France. These refusals have simply been removed from the calculations, since such treatment leaves the gross rate on the upper side of the range that informed estimates have suggested for total party membership in France, after realistic appraisal of the memberships claimed by the parties.

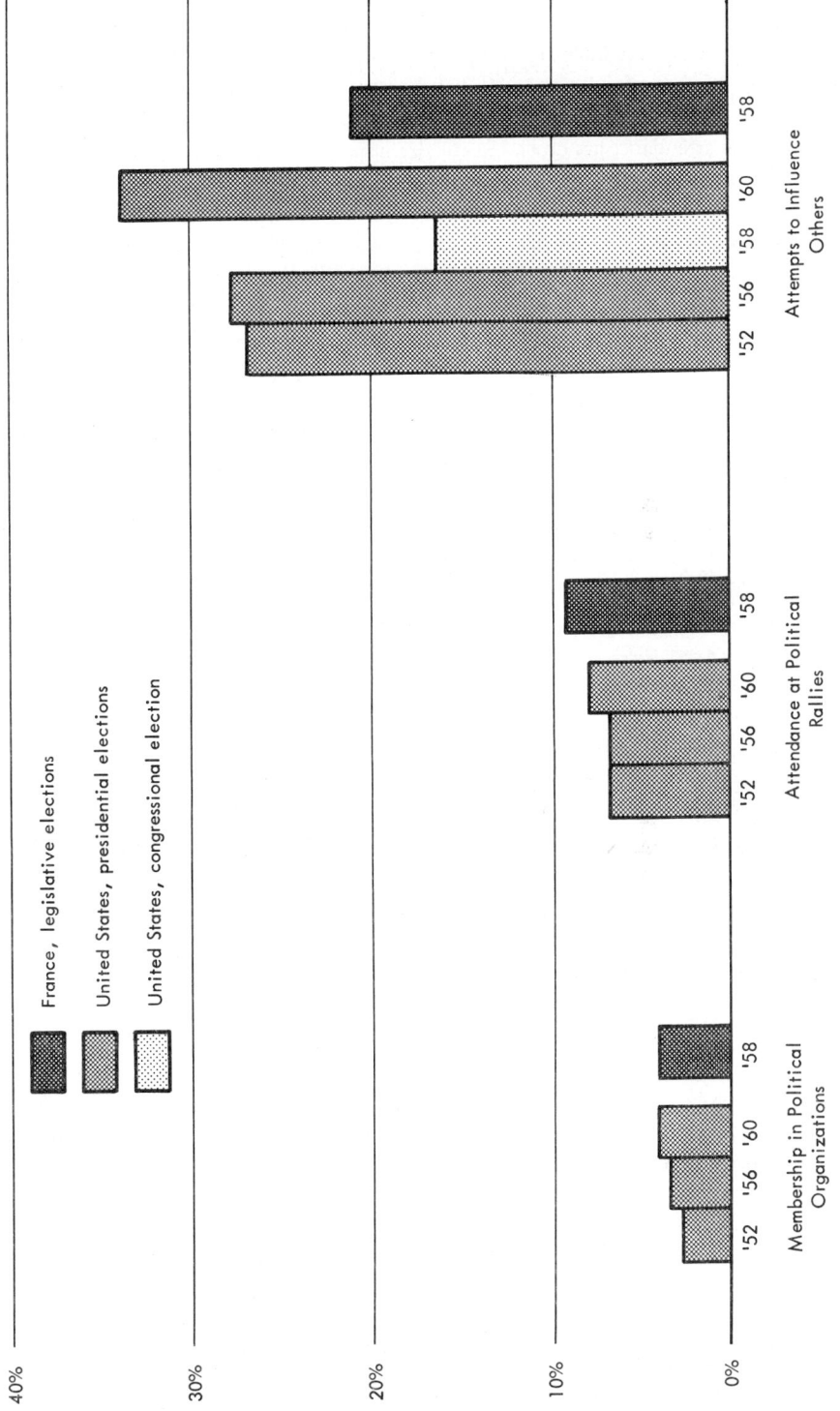

Figure 1. Rates of several forms of political participation, France and the United States.

and Americans who do attend political gatherings. Graphs of the frequency of attendance by attender are almost indistinguishable between the two countries. The mean number of meetings attended among those who attend at all is in both cases a slight fraction over two. In sum, it is hard to imagine that any slight divergences in rates of attendance are crucial in any dramatic differences between the two systems.

Data were collected in both countries as well with regard to dependence on the mass media for political information. Here one of the most excellent bases for comparison which the data afford is provided by reports of the regularity with which political news about the elections was followed in daily newspapers. Although the structured alternatives again involve "amount words," there is a more tangible standard for responses implicit in the rhythm of newspaper production. "Regularly" or "never," when applied to readership of daily papers, has a common meaning in any language. Furthermore, we find empirically that responses to the newspaper questions show much higher stability for individuals over time than the direct interest questions. It is clear that we are dealing here with stable habits which are reliably reported.

When we compare distributions from the two countries (Table 1), there seems little doubt of higher American devotion to newspapers as a source of political information. Furthermore, the French citizen appears also to monitor other media for political information less. Thus, for example, he is less likely to have a television or radio set than his American counterpart; but even among those French who possess sets, attention to political broadcasts is markedly less than for comparable Americans. As we have observed, these latter differences are not in themselves proof of lesser French motivation, since the choices offered through the airwaves are not comparable cross-nationally. But such facts, along with further comparisons as to magazine reading, indicate that lower French attention to political material in the newspapers is not compensated for from other sources. Since elite political competi-

Table 1. FREQUENCY OF NEWSPAPER READING FOR POLITICAL INFORMATION
(IN PER CENT)

	FRANCE, 1958		UNITED STATES, 1960	
	POST-REFERENDUM	POST-ELECTION	POST-ELECTION	
Regulièrement	19	18	44	Regularly
Souvent	12	10	12	Often
De temps en temps	25	29	16	From time to time
Très rarement	19	21	7	Just once in a great while
Jamais	25	22	21	Never
	100	100	100	

tion in France, even when reduced to simplest terms, is considerably more complex than two-party competition in the United States, it is ironic that the French voter exposes himself less faithfully to the flow of current political information.

Education being a strong determinant of all these information-seeking activities, it is of interest to control the substantial Franco-American differences in level of formal education. For Figure 2 we have applied a single integer scoring to the five response categories of Table 1 and extracted means within education categories, the latter having been carefully tailored on the American side to match French intervals with regard to simple number of years of formal education. While the two curves do not match precisely, the main departures lie at the extremes, where case numbers are

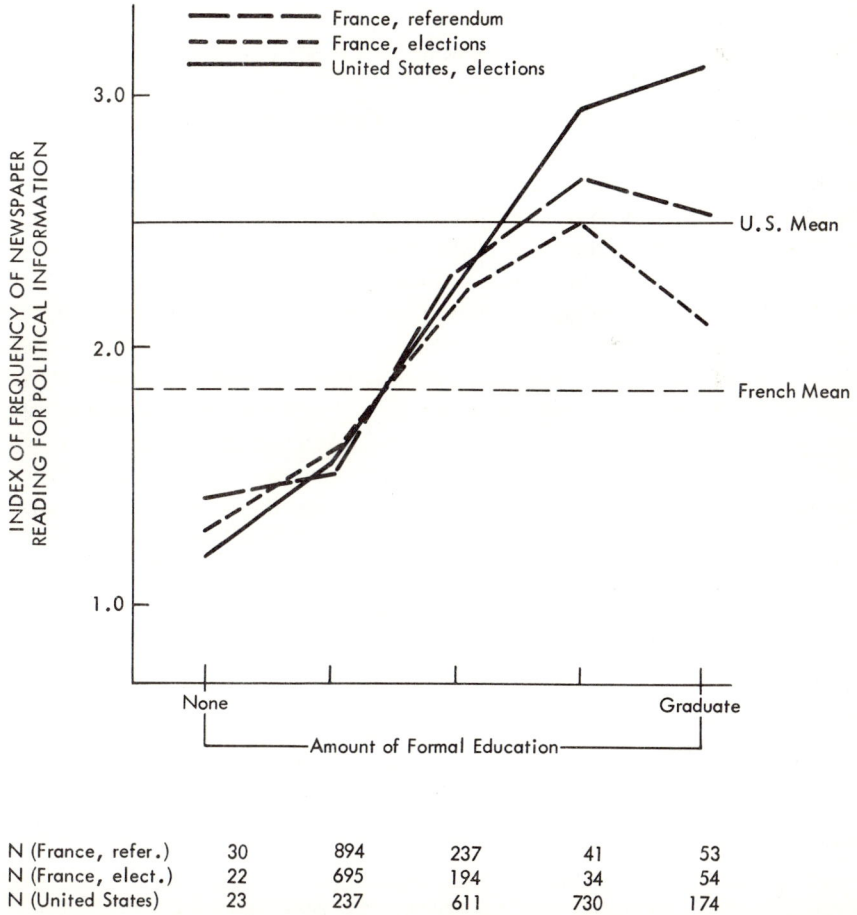

Figure 2. Frequency of newspaper reading for political information in France and the United States, by education.

N (France, refer.)	30	894	237	41	53
N (France, elect.)	22	695	194	34	54
N (United States)	23	237	611	730	174

lowest, and hence where sampling error is bound to disperse results.[5] Where more than 200 cases are available from both sides, the estimates show a most remarkable convergence.

As we see, there are strong cross-national differences in total distribution of regularity of newspaper reading (as represented by the distance between the horizontal lines in Figure 2). But these differences very nearly disappear (the general proximity of the two slopes, and their essential identity when case numbers are sufficient), with education controlled. The gap in total news reading for the two electorates comes about, then, simply because the American cases are loaded heavily to the right side of the graph, while the center of gravity of the French education distribution is to the left, or low, side. The capacity to move cross-national differences out to the marginals in this fashion not only strengthens presumption of common causal factors, but also is a reassuring anchor in the unknown waters of cross-national research, where the basic comparability of data must be held to special question. We shall see a more dramatic demonstration of this circuit of empirical reasoning below.

By way of summary, then, comparisons up to this point create a general sense of Franco-American similarity, with occasional mild divergences suggesting stronger American political involvement. The locus of these differences in the vertical sense is interesting. Let us bear in mind that most of these involvement actions in both societies stand in a scalar relationship to one another, in the Guttman sense. That is, the party member who passes the "hardest" of our items is very likely to have attended meetings, monitored the media, and so on. In this sense there is an underlying involvement dimension represented here. Furthermore, we have established cutting points on this dimension in quite different ranges of the continuum, at the high end for party membership and meeting attendance, but much deeper in the middle mass where information seeking and expressions of involvement are concerned. In a rough way, we may observe that the American data seem to show somewhat higher motivation in the middle ranges, with cross-national differences narrowing near the very top, and perhaps even showing a slight French advantage. Interestingly enough, this pattern would describe as well the cumulative frequency distributions expressing differences in years of formal education in the two countries. This identity is, of course, no proof that education accounts for the involvement divergences. But it does remind us that these patterns, if they differ cross-nationally at all, may partake of the sharper discontinuities in France between a tiny elite and the remainder of the population that one suspects for a variety of characteristics, and can readily demonstrate for education.[6]

---

[5]The decline in reading among the most educated French citizens approaches statistical significance and is currently unaccounted for. We have been able to show that these people are not substituting political reviews and weekly magazines for daily news reading. The educated elite which follows the reviews also reads the newspapers faithfully; the remainder which fails to attend to the newspapers ignores political magazines as well.

[6]We shall not treat Franco-American differences in vote turnout, save to observe that they are probably more institutional than motivational. American registration requirements in many states are such that an American is persistently confronted by an institutional barrier

PARTISAN ORIENTATIONS

The gross similarities between the two publics in apparent political interest do not, to be sure, remove the possibility that the Frenchman in his interested moments may respond to politics in much different terms than his American counterpart. Actually, when we consider the character of partisan ties felt by citizens in the two countries, we strike upon some contrasts of great magnitude.

If Americans are asked to locate themselves relative to the American party system, 75 per cent classify themselves without further probing as psychological members of one of the two major parties, or of some minor party. In France, somewhat before the elections, less than 45 per cent of those who did not refuse to answer the question were able to classify themselves in one of the parties or splinter groups, while another 10 to 15 per cent associated themselves with a more or less recognizable broad *tendance* ("left," "right," a labor union, etc.). The cross-national differences of 20 to 30 per cent are sufficiently large here to contribute to fundamental differences in the flavor of partisan processes in the two electorates. For a long time, we wrote off these differences as products of incomparable circumstances or of reticence on the part of the French concerning partisanship, most of which was being expressed not as refusal to answer the question, but as some other evasion. As we grew more familiar with the data, however, these differences took on vital new interest.

The hypothesis of concealed partisanship was very largely dispelled by a close reading of the actual interviews. It is undeniable that nearly 10 per cent of the French sample explicitly refused to answer the question, as compared with a tiny fraction in the United States. However, we have already subtracted this group from the accounting. Beyond the explicit refusals, the remarks and explanations which often accompanied statements classified as "no party," or as "don't know which party," had a very genuine air about them which made them hard to read as hasty evasions. No few of these respondents were obviously embarrassed at their lack of a party; some confessed that they just hadn't been able to keep track of which party was which. The phrase "je n'y ai jamais pensé" was extremely common. Others indicated that they found it too hard to choose between so many parties; some indicated preferences for a specific political leader but admitted that they did not know which party he belonged to or, more often, had no interest in the identity of his party, whatever it might be. Others, forming a tiny minority of the nonparty people, rejected the notion of parties with some hostility.

It became clear, too, that people reporting no party attachments were distinct on other grounds from those who willingly classified themselves as close to a party. On our vertical involvement dimension, for example, they tended to fall in the bottom

---

which is rarely erected in France. It can be argued most strongly that the act of getting somewhere to register demands higher political motivation than getting to the polls on Elec- registration barriers as change in residence, failure to renew on time, etc. If such reports are credited, the registration toll in the United States would easily make up the apparent Franco-American differences.

stratum of the least involved, just as the paper-thin stratum unable to choose a party in the United States consists heavily of the least involved. Demographically, these nonparty people were disproportionately housewives, poorly educated, young, and the other familiar statuses which tend to be uninformed and uninvolved.

Among actual party identifiers in France there was further interesting variation in the character of the party objects to which reference was made. A very few linked themselves with small new ideological splinter groups which had developed during the political crises of 1958. For these people, it was not enough to indicate that they felt closest to the Radical-Socialists, for example: they had to specify that they were Mendesists or anti-Mendesists, Valoisiens, and the like. Most identifiers suffered no difficulty in seeing themselves as "Radical-Socialists," completely shattered though the party was. Others, perceiving the system even more grossly, linked themselves only with a broad *tendance*. On involvement measures these groupings showed the expected differences: the grosser the discrimination, the lower the involvement.

In other ways as well it was clear that the extreme ideological fractionation of parties in France has few roots in the mass population, members of which simply pay too little attention to politics to follow the nicer discriminations involved. When asked whether the number of parties in France was too great, about right, or too few, 97 per cent of those responding said there were too many parties, and less than 1 per cent said there were too few. In response to an ensuing question as to the desirable number of parties, the mean of the number seen as optimal was 3.5 for the handful of adherents of the new ideological splinters, 3.0 for the partisans of the traditional mass parties, and less than 2.8 among those who had formed no party attachments. Perhaps the most apt expression of the problem of partisan fractionation and discrimination came from the naïve respondent who opined that France should have two or three parties, "enough to express the differences in opinion."

The fact that large proportions of the French public have failed to form any very strong attachments to one of the political parties should not be taken to mean that these people are totally disoriented in the French party system. In particular, a sensitivity to the gulf separating the Communist Party from the remainder of French parties does pervade the mass public. There seems to be less confusion as to the identity of the Communist Party than for any of the other parties; and for the bulk of non-Communists, the Communist Party is a pariah. There are some nonidentifiers who appear to shift from Communist to non-Communist votes with abandon, and were all of these votes to fall to the Communists in the same election, the Party would undoubtedly exceed its previous high-water mark in its proportion of the French popular vote. At the same time, however, one cannot help but be impressed by the number of respondents who, while indicating they were not really sure what they were in partisan terms, indicated as well at one point or another in the interview that they were not only non-Communist but anti-Communist. In other words, were the descriptions of party adherents to proceed simply in terms of a Communist, non-Communist division, the proportion of ready self-classifications would advance considerably toward the American figure, and would probably exceed that which could be attained by any other two-class division in France.

Nevertheless, the limited party attachments outside the Communist camp in France retain strong theoretical interest, as they seem so obviously linked to a symptom of turbulence which is clearly not an elite phenomenon alone—the flash party. With a very large proportion of the electorate feeling no anchoring loyalty, it is not surprising that a new party can attract a large vote "overnight," or that this base can be so rapidly dissolved. Furthermore, there is a problem here that is peculiarly French, in that the low proportion of expressed attachments cannot simply be seen as a necessary consequence of a multiparty system per se. Fairly comparable data from Norway, where six parties are prominent, show party attachments as widespread as those in the two-party United States.[7]

The French sample was asked further to recall the party or *tendance* which the respondent's father had supported at the polls. Here the departure from comparable American data became even more extreme (Table 2). Of those Americans in 1958

Table 2. RESPONDENT'S CHARACTERIZATION OF FATHER'S POLITICAL BEHAVIOR, BY COUNTRY, 1958
(IN PER CENT)

	FRANCE	UNITED STATES
Located father in party or broad *tendance*	25	76
Recalled father as "independent," "shifting around," or as apolitical, nonvoting	3	6
Total able to characterize father's political behavior	28	82
Unable to characterize father's political behavior	68	8
Father did not reside in country or was never a citizen		3
Did not know father; question not asked about father surrogate	4	6
Refused; other		1
	100	100
(N)	(1,166)	(1,795)

having had a known father who had resided in the United States as an American citizen, thereby participating in American political life, 86 per cent could characterize his partisanship, and another 5 per cent knew enough of his political behavior to describe him as apolitical or independent. Among comparable French respondents, only 26 per cent could link fathers with any party or with the vaguest of *tendances* (including such responses as "il a toujours voté pour la patrie"), and another 3 per cent could describe the father's disposition as variable or apolitical. In other words, among those eligible to respond to the question, 91 per cent of Americans could characterize their father's political behavior, as opposed to 29 per cent of the French.

It goes without saying that differences of this magnitude rarely emerge from individual data in social research. And they occur at a point of prime theoretical interest. We have long been impressed in the United States by the degree to which

[7] Angus Campbell and Henry Valen, "Party Identification in Norway and the United States," *Public Opinion Quarterly*, Vol. 25, 1961, pp. 505–525.

partisan orientations appear to be passed hereditarily, from generation to generation, through families. It has seemed likely that such transmission is crucial in the stability of American partisan voting patterns. Therefore, we find it startling to encounter a situation in which huge discontinuities seem to appear in this transmission.

What do the French responses concerning paternal partisanship really mean? As best we can determine, they mean what they appear to mean: the French father is uncommunicative about his political behavior before his children, just as he is more reserved in the interviewing situation than Americans or Norwegians. It seems highly unlikely, for example, that Franco-American differences in recall represent French concealment: large numbers of the French willing to speak of their own party preference are unable to give the father's preference of a generation before, and explicit refusals to answer, while attaining 10 per cent or more where own partisanship is at stake, are almost nonexistent for paternal partisanship.

Furthermore, we have come to reject the possibility that the bulk of the Franco-American difference is some simple consequence of the more fluid and complex French party system. Responses to a similar question in the Norwegian multiparty system look like our American results, and not like the French. Nor is there reason to believe that the Frenchman has trouble finding comparable modern terms for the party groupings of a generation ago. As we have observed, the respondent was invited to give a rough equivalent of his father's position in terms of *tendance*. Moreover, where there are any elaborations of "I don't know" captured in the interview, the consistent theme seemed to be that the respondent did not feel he had ever known his father's position ("je n'ai jamais su"; "je ne lui ai jamais demandé"; "il ne disait rien à ses enfants"; "il n'en parlait jamais"). Finally, if special problems were occasioned on the French side by the changing party landscape over time, we should certainly expect that older French respondents would have greater difficulty locating their fathers politically than would younger respondents. They do not: the tabulation by age in France shows only the slightest of variations attributable to age, and these lend no support to the hypothesis (e.g., slightly less knowledge of father's position for children under thirty) and are variations which may be found in the comparable American table as well.

If we accept the proposition, then, that there are basic discontinuities in the familial transmission of party orientations in France, all of our theory would argue that weaker party attachments should result in the current generation. The data do indeed show a remarkable association between the two phenomena, once again involving differences of 30 per cent or more. Both French and Americans who recall their father's partisanship are much more likely themselves to have developed party loyalties than are people who were not aware of their father's position. Of still greater importance are the more absolute Franco-American similarities. Setting aside those people whose fathers were noncitizens, dead, apolitical, or floaters, or who refused to answer the question, we can focus on the core of the comparison (in per cent) [see p. 189].

Where the socialization processes have been the same in the two societies, the results in current behavior appear to be the same, in rates of formation of

	KNOW FATHER'S PARTY		DO NOT KNOW FATHER'S PARTY	
	FRANCE	U.S.	FRANCE	U.S.
Proportion having some partisan self-location (party or vague *tendance*)	79.4	81.6	47.7	50.7
Proportion that these are of total electorate	24	75	63	8

identification. The strong cross-national differences lie in the socialization processes. In other words, we have come full circle again: we have encountered large national differences but have once again succeeded in moving them to the marginals of the table. This is our best assurance that our measurements are tapping comparable phenomena.

Partisan attachments appear therefore to be very weakly developed within the less politically involved half of the French electorate. While undoubtedly a large variety of factors, including the notoriety which the French parties had acquired in the later stages of the Fourth Republic, have helped to inhibit their development, more basic discontinuities of political socialization in the French family appear to be making some persisting contribution as well.[8] Of course, similar lack of party attachment does occur among people indifferent to politics in the American and Norwegian systems as well; but the strata of unidentified people are thinner in these systems and do not extend greatly above that layer of persistent nonvoters which is present in any system.

The link between an electorate heavily populated with voters feeling no continuing party attachments and a susceptibility to "flash" parties is an obvious one. It must be recognized at the outset, of course, that such phenomena arise only under the pressure of social, political, or economic dislocations occurring in some segment of the population, thereby generating an elite which wishes to organize a movement and a public which is restive. This means that even a system highly susceptible to such phenomena is not likely to experience them when it is functioning smoothly:

---

[8] Among other factors, an alleged paucity of voluntary associations acting vigorously to mediate between the mass of citizens and centralized authority in France has often been cited as a crucial differentium in the quality of the political process between France and the United States. See William Kornhauser, *The Politics of Mass Society*, Glencoe, Ill., Free Press, 1959. If such differences do exist, they may well have some bearing on the prevalence of partisan attachments, for it is clear intranationally, at least, that high rates of participation in nonpolitical voluntary associations and strong partisan attachments tend to co-occur at the individual level (although it is much less clear whether this represents a causal progression or two aspects of the same stance toward community life). In other contexts, however, it has been argued that ostensibly nonpolitical associations of mass membership in France tend to play more vigorous roles as parapolitical agents than do comparable associations in the United States, which so often tend to regard political entanglement with horror. Both views have some appeal on the basis of loose impressions of the two societies, and are not in the strictest sense contradictory. However, their thrusts diverge sufficiently that a confrontation would seem worthwhile if either can be borne out by any systematic evidence. Where grass-roots participation in expressly political associations is concerned, we have seen no notable differences between the nations in either membership rates or rates of attendance at political gatherings.

their prevalence in postwar France cannot be divorced from the severe dislocations the society has been undergoing. Once misfortunes breed discontent, however, the proportion of partisans in an electorate is a datum of fundamental significance. One cannot fail to be impressed by the agility with which the strong partisan can blame the opposing party for almost any misfortune or deny the political relevance of the misfortune if some opposing party cannot conceivably be blamed. Hence, where partisans are concerned, misfortunes do relatively little to shift voting patterns. Independents, however, have no stake in such reinforcements and defenses and move more massively in response to grievances. In France, the institutions which conduce to a multiparty system make the organization of new party movements more feasible from an elite point of view than it is likely to be under two-party traditions. At the same time, the presence of a large number of French voters who have developed no continuing attachments to a particular party provides an "available" mass base for such movements. This available base is no necessary concomitant of a multiparty system, but is rather a peculiarity of the current French scene.

PARTIES AND POLICY CONTROVERSY

Whatever differences exist in partisan orientations, no assessment of politicization would be complete without consideration of the manner in which ideological conflict is worked out through the party system. If parties are recognized at all in the classical view of democratic process, they are relegated to a distinctly secondary position: they are means to policy ends, and should be judged by the citizen accordingly. In this light, the number of Americans with strong party loyalty and a poor sense of what either party stands for in policy terms represents a distinct perversion of the democratic process. In this light, too, weaker partisan orientations in the French populace might simply mean a relegation of party to second rank, with a primary focus on policy goals.

At an elite level, of course, there are distinct Franco-American differences in the phrasing of the means-end relation between party and policy, and these contrasts weigh heavily in our impressions of differences in quality of political process between the two systems. That is, while French political elites are not insensitive to party formations as instruments toward policy goals, the fact remains that parties are split and reshaped with relative freedom in order that the party may be the purest possible expression, not only of the politician's position on a single basic issue dimension, but of the total configuration of positions adopted on cross-cutting issue dimensions. On the American side, remarkable policy accommodations are made to preserve the semblance of party unity, and party competition for votes "in the middle" leads to a considerable blurring of interparty differences on policy. The crucial role of basic political institutions in stimulating either multipartite or bipartite trends has often been discussed, and whether French elite activities would survive long under American ground rules is a moot point. We may consider, however, whether the ideological clarity or intransigence associated with French political elites and the policy compromise or confusion which characterizes the American party system reflect properties of their mass publics.

Data have been collected in both countries concerning reactions to a variety of issues confronting the two systems. While both sets of items must be regarded as only the crudest samplings of hypothetical issue universes, selection on both sides was performed in an attempt to tap some of the most basic controversies of the period. In France, three items were devoted to the classic socio-economic left and right, with one concerning the role of labor and the other two the relative roles of government and private enterprise in housing; two more involved the clerical question; a sixth item had to do with military expenditures and national prestige; a seventh concerned the freedom of the press to criticize the government. Of eight American questions, two dealt with social-welfare legislation and a third with the relative role of government and private enterprise in housing and utilities, covering the classic right and left; two more dealt with the government's role in racial matters (FEPC and school desegregation); and three others were concerned with the internationalist or isolationist posture of the government in foreign affairs. All questions were in Likert scale form.

We shall focus upon three properties of these issues which we can more or less crudely measure in the two countries: (1) the degree to which public opinion is sharply crystallized on each issue; (2) the degree to which opinion within the two publics is polarized on each; and (3) for each issue, the degree to which individual opinion is associated with partisan preference.[9] Assuming the items do give fair coverage to most primary issue dimensions in the two nations, we are interested to see if opinion in France at a mass level appears more sharply crystallized or polarized, and to assess the manner in which policy concerns are linked with party preference. As before, we shall distinguish layers of both populations in terms of partisan involvement. At the top, we isolate as political "actives" those people who were either party members or reported attending two or more political rallies in the respective election campaigns, a group which amounts to 5 to 7 per cent within each population and hence is sufficiently large for analysis. We also continue to distinguish between party identifiers (three-quarters of the American population, but half of the French) and nonidentifiers.

In both nations, the issue items were asked again of the same respondents after an

---

[9] Of these three properties, polarization is most dependent on question wording. It is measured by the standard deviation of the response distribution after the five steps of the scale have been assigned simple integer scores. The statistic takes high values (e.g. over 1.50) only when the distribution of opinion is relatively U-shaped. Party-relatedness is measured by a rank-order correlation between the respondent's partisan position and his issue position. In the United States, the Democratic Party was presumed to be the more liberal on domestic issues and the more internationalist in foreign affairs, and respondents were arrayed from a Democratic to a Republican pole on the basis of party loyalty for identifiers, or patterns of 1956–1958 vote for nonidentifiers. In France, a panel of expert observers arranged the many parties or fractions thereof on a socio-economic left-right continuum and again on a continuum from clerical to anticlerical. The second was used to array respondents for the two religious issues; the first was used for the other five issues. Once again, nonidentifiers were located on the basis of reports of 1956 and 1958 votes. All rank-order correlations, including those used for the crystallization measure discussed in the text, are tau-betas, based on tables of equal rows and columns. See Hubert M. Blalock, *Social Statistics*, New York, McGraw-Hill, 1960, pp. 321ff.

interval of time. We take as a measure of crystallization of opinion the rank-order correlation between the two expressions of opinion. There is a good deal of internal evidence to suggest that "change" in opinion between the two readings is almost never a matter of true conversion, but rather represents haphazard reactions to items on which the respondent has never formed much opinion. With minor exceptions, there is no significant change in the marginal distributions of the tables, despite the high turnover of opinion. There is a persistent relation between the proportions of people who confess they have no opinion on any given issue and the amount of turnover shown by those who do attempt an opinion. As one might expect, too, there is a tendency for high crystallization, high polarization, and high party-relatedness to co-occur, despite intriguing exceptions. Clearly both publics are more likely to have arrived at stable prior opinions on some items than on others, and this degree of crystallization has an obvious bearing on the vitality of the role the issue dimension may play in partisan choice.

Unfortunately, the magnitude of these turnover coefficients may not be compared cross-nationally, since the interval between tests averaged little more than a month on the French side but ran twenty-six months on the American side.[10] Nevertheless, as Table 3 indicates, the level of these coefficients is by any standard remarkably low in both populations. Taken as test-retest reliability coefficients, they would send the psychologist in search of a better measuring instrument. After all, on an item where the stability seems relatively high (freedom of the press), less than eight Frenchmen in ten take the same side of the issue twice in a five-week period, when five out of ten would succeed in doing so by making entirely random choices. On the other hand, while more routine measurement error certainly imposes a rather constant ceiling on these coefficients which may not greatly exceed .8, the further incapacity of the two publics to respond reliably to these items must be considered a substantive datum of the first water. For if these items, reduced to an unusually simple vocabulary, fail to touch off well-formed opinions, the remoteness of both publics from most political and journalistic debate on such dimensions is obvious. It is not as though the items presented new controversies on which opinion had not yet had time to develop. With few exceptions, they have been the basic stuff of political disagreement for decades or generations. Opinions still unformed are unlikely to develop further.

In this light, then, it is interesting to compare the stability over time of reactions to parties with the stability of responses to these "basic" controversies shown in the

[10] In American panel studies we are beginning to fill in a picture of the manner in which these coefficients erode over time. For example, coefficients after four years show almost no decline from their two-year levels, and it seems likely that, in the infrequent instances where opinions on these issues are truly crystallized, they are subject to little change. As the test-retest interval changes, we may suppose that the coefficient declines very rapidly in the brief period in which respondents forget their previous answers and hence are obliged to "guess again," and then stabilizes at a hard core of well-informed opinions. The French interval was so brief, however, that it is hard to imagine that the coefficients had yet dropped to their stablized level. We would hazard the loose judgment that the French coefficients lie about where one would expect were they destined to decline to American levels in a comparable period of time.

Table 3. SELECTED ISSUE CHARACTERISTICS IN FRANCE AND THE UNITED STATES

	CRYSTALLIZATION		POLARIZATION		PARTY-RELATEDNESS		
	TOTAL SAMPLE	AC-TIVES	TOTAL SAMPLE	AC-TIVES	ACTIVES	IDENTI-FIERS	UNIDEN-TIFIED
France:							
State support of religious schools	.65	.74	1.54	1.62	.58	.39	.32
Strikes by government employees	.52	.69	1.60	1.70	.59	.31	.22
Current threat posed by clergy	.47	.80	1.32	1.64	.56	.34	.19
Freedom of press	.47	.68	1.50	1.60	.39	.22	.13
State responsibility for housing	.42	.34	1.39	1.45	.38	.13	.08
Level of military expenditures	.34	.46	1.36	1.56	.33	.18	.17
Private responsibility for housing	.28	.35	1.26	1.60	.37	.22	.04

	CRYSTALLIZATION		POLARIZATION		TOTAL SAMPLE ACTIVES	NON-SOUTH		
	TOTAL SAMPLE	AC-TIVES	TOTAL SAMPLE	AC-TIVES		AC-TIVES	IDENTI-FIERS	UNIDEN-TIFIED
United States:								
Federal school integration action	.42	.47	1.69	1.72	.00	.12	.07	—.06
Federal guarantees of employment	.35	.49	1.45	1.55	.16	.30	.19	.03
Federal FEPC	.34	.34	1.41	1.55	.00	.14	.06	.01
Federal aid to education	.34	.54	1.09	1.60	.16	.29	.21	.16
General isolationism-internationalism	.33	.25	1.48	1.48	.16	.06	.03	.04
Deployment of U.S. forces abroad	.28	.10	1.23	1.25	.07	.04	.05	—.02
Government vs. private enterprise in power and housing	.25	.41	1.37	1.45	.21	.27	.18	.21
Foreign aid	.24	.31	1.36	1.38	.11	.10	.02	—.05

first column of Table 3. This assessment is rather difficult on the French side in view of the frequent indeterminacy of party locations; however, it seems that, in a comparable period of time, affective reactions to the parties are more stable than issue reactions even in France. In the United States, we know that partisan reactions show dramatically greater stability than the issue responses. Most important, perhaps, is the failure of data in Table 3 to support an image of the mass French public as remaining aloof from party sentiments while hewing dogmatically to ideological goals. Beyond the political actives, stability of issue opinion seems unimpressive, and, for the majority of French voters without party attachments, the articulation of party choice with any of the issue dimensions covered here is slight indeed (Table 3, final column).

While the instability of opinion in both nations is of primary interest in Table 3, several further comparisons may be summarized. The major cross-national contrast comes in the party-relatedness column, where French actives and partisans show much higher coefficients than their American counterparts. The most obvious American phenomenon which blunts interparty policy differences is the disparity between Southern and non-Southern wings of the Democratic Party. While setting aside the Southern Democratic rank and file does not remove the perceptual problem posed for Northern Democrats who may find the top leaders of their party at odds on many issues, we complete this exercise in Table 3 to show that, even for actives, this regional limitation does not begin to bring the American coefficients up to the French level. While the higher French coefficients are no statistical necessity, it is likely that, in practice, closer party-relatedness is inevitable in the multiparty system. The interparty differences in opinion among French *partisans* appear to lie in about the same range as those found in Norway.[11] However, as we have seen, party attachments are more prevalent in Norway than in France; when the unidentified enter the French electorate in an actual vote, it is likely that individual issue opinions receive less clear expression across the electorate as a whole than is the case in Norway.

Beyond this primary contrast, Table 3 is impressive for its cross-national similarities. Actives in both countries show more highly crystallized opinions, and usually more polarized opinions as well, although American actives differ less sharply and consistently from their mass public than do French actives. In neither country do identifiers differ reliably from nonidentifiers with regard to crystallization or polarization of opinion. In both countries, however, there are quite reliable differences in party-relatedness, not only between actives and the remaining 95 per cent of the population, but between identifiers and nonidentified. In other words, while the partisan manner of relating to the political process makes little difference in basic opinion formation save for the extremely active, the translation of these attitudes to some kind of party choice seems increasingly haphazard as party attachments become weaker.

Throughout these comparisons, however, we may remain struck by the fact that the "slope" is steeper on the French side: the differences between actives and mass are large relative to those in the United States. From the upper end of this steep

[11] Campbell and Valen, *op. cit.*

slope, one might wish to extrapolate to the sharp and rigid cleavages on policy matters for which French elites are noted; for our purposes, it is sufficient to observe that these cleavages blur rapidly and lose their tone in the mass of the French electorate.

Finally, it should be observed that the issues seem to sort themselves into two rough categories in both nations: (1) emotional-symbol issues involving some of the more gross group conflicts within the two societies (racial in the United States, religious in France, along with items which touch in a direct way upon labor as an interest group), which show relatively high crystallization and polarization; and (2) more complex questions of relations between the state and private enterprise which, along with all foreign policy issues, tend to be less crystallized.

These differences in crystallization are scarcely surprising, as the objects and means involved in the second group of issues are clearly more remote from the common experience of the man-in-the-street. Yet the pattern is ironic, for the issues which show a stronger resonance in both mass publics tend to be those which both elites make some attempt to soft-pedal, in favor of direct debate over such more "ideological" matters as arrangements between state and private enterprise. The more resonant issues are not dead, of course, and are used for tactical advantage by elites in both countries. Calculations of vote gain are made in the United States on the basis of the religion of the nominee, and the clerical question in France has been resuscitated repeatedly as a handy crowbar to split apart government coalitions. At the same time, however, there is genuine elite effort to keep such cleavage issues in the background: the American public is told that religion is not a proper criterion for candidate choice, and the battleground for elite debate on the racial issue is usually displaced quite notably from race itself in the modern period. Similarly, much sophisticated French opinion has for some time argued that even the secondary role which the clerical question has been playing in elite debate exaggerates its importance.

Given this common background, the different manner in which the two types of controversy weave into partisan choices in the two countries is fascinating. In France, there is fair coincidence between the ordering of issues in terms of party-relatedness and the ordering on the other two properties. The clerical questions, for example, are highly crystallized and polarized, and show high levels of party-relatedness as well. The structure of party competition is such that, elite values notwithstanding, these emotional cleavages achieve prominent partisan expression. Such is not the case in the United States: there is little coincidence between the party-relatedness of issues and the other two properties. Indeed, the racial issue finds little clear party expression, while the "elite" issue concerning government and private enterprise, one of the least crystallized issues, is at the same time one of the most party-related across the full electorate.

Where mass or elite control of issue controversy is concerned, then, the two systems have rather paradoxical outcomes. By conception, the French party system is geared to elites, encouraging them to a multi-faceted ideological expression which is too complex for most of the public to encompass. At the same time, the multidimensional clarity of party positions serves to return a measure of control to part of the public, for the more involved citizens can single out certain dimensions to reduce the system to manageable simplicity. These reductions are naturally made in terms of issues which are more resonant in the public, even if these are not the dimensions which the elites might wish to stress. The American system is less elite in conception; it is sufficiently simple in its gross characteristics that it is easier for the common citizen to follow it with only limited attention. But this simplification requires great blurring of party differences across most of the universe of possible issues, and the differences which are maintained are those which the competing elites select as battlegrounds. Hence, control of controversy which can be given partisan expression is, paradoxically, more nearly in elite hands.

CONCLUSIONS

We have attempted to sort through a number of those characteristics of French politics which add up to vague impressions of intense French politicization, in order to identify more precise loci for Franco-American differences. It appears likely that the more notable of these differences stem from the actions of elites and require study and explanation primarily at this level, rather than at the level of the mass electorate. While certain peculiarities reminiscent of French political elites are visible in the most politically active twentieth of the French population, these peculiarities fade out rapidly as one approaches the more "representative" portions of the broad French public.

It is unlikely that the common French citizen devotes any greater portion of his attention to politics than does his American counterpart, and he may well give less. His behavior is constrained within a much different set of political institutions, and these differences have important consequences for the character of his political behavior, including the opportunity of closer articulation between any crystallized opinions he may hold and an appropriate party instrument. However, the data give no striking reason to believe that the French citizen, either through the vagaries of national character, institutions, or history, is predisposed to form political opinions which are more sharply crystallized or which embrace a more comprehensive range of political issues than do comparable Americans. On both sides, opinion formation declines as objects and arrangements become more remote from the observer; and much of politics, for both French and Americans, is remote. Hence the proliferation of choices offered by the multi-party system is itself a mixed blessing: it is capitalized upon only by the more politically interested segments of the electorate, and appears to represent "too much" choice to be managed comfortably by citizens whose political involvement is average or less.

Over the range of characteristics surveyed, only one striking difference at the level of the mass public was encountered which seemed more uniquely French than the multiparty system itself. There is evidence of a widespread absence of party loyalties, a phenomenon which can be empirically associated with peculiarities in the French socialization process. This characteristic has obvious links with the major symptom of French political turbulence, which is based on the behavior of the mass population rather than that of elites—the current availability of a mass base for flash party movements under circumstances of distress.

Part Four

THE POLITICAL SYSTEM OF THE GERMAN FEDERAL REPUBLIC

INTRODUCTION

The founding fathers of the Bonn Republic were not only the members of the German Parliamentary Council who, under Konrad Adenauer's leadership, drafted the Basic Law—but also the Western Allies, and especially the United States, whose deliberate aim was to build a stable and democratic Germany after the Second World War. It is therefore appropriate to devote the first chapter on Germany to an analysis of German politics by James B. Conant, the last United States High Commissioner for Germany (1953–1955) and the first United States Ambassador (1955–1957) to the sovereign Federal Republic. In his speech to the Bar Association of New York City on April 28, 1954, Conant compares the Bonn Republic with the Weimar Republic, which he knew from personal experience, and with the Communist regime in East Germany.

A comparison of the highly unstable and ineffective Weimar Republic and the stable and relatively effective Bonn Republic also constitutes the topic of Chapter 17. Charles E. Frye subjects the political parties and pressure groups of these two vastly different German regimes to a thorough analysis, providing another good example of how the comparative method can lead to general propositions. Because the Weimar and the Bonn Republic are both "Germanies" and culturally similar in most respects, the analysis can focus on the relationships between parties, interest groups, and the general political system, while other factors are kept constant.

The Bonn Republic cannot be fully understood without considering the legacy of Weimar. Nor can it be understood without considering the deep and continuing effect of the division of Germany. In Chapter 18, Ralf Dahrendorf compares the "new Germanies," West and East. He emphasizes the wide divergence of the developments in the two parts of Germany, speaking not only of two different political systems but also of two radically different societies.

Chapters 19 and 20 contain broad comparisons of Germany with several other countries. Taylor Cole compares the constitutional courts of Germany, Italy, and Austria. He also calls attention to the medieval natural law tradition and more recent European precedents, and to the parallel of judicial review in the United States. It is to be noted that West Germany has gone further in making a supreme court the guarantor of constitutional provisions than has any other European country. In Chapter 20, Giuseppe Di Palma compares political disaffection and *participation in Germany, Britain, Italy, and the United States. His striking conclusion is that all four countries are quite similar in this respect: Social and demographic factors are related to party preference, but the general orientation toward the political system does not have much influence on party choice. Also, the more positively the citizen is inclined toward the system, the more he will tend to participate—regardless of party. To the extent that differences among the countries do appear, Germany is more like Britain and the United States than like Italy.*

ABOUT THE AUTHORS

JAMES B. CONANT *served as professor of chemistry at Harvard University from 1919 to 1933, and as president of Harvard from 1933 to 1953. He was the American diplomatic representative to Bonn, first as High Commissioner and then as Ambassador, from 1953 to 1957. He has written extensively on the subject of education, and is also the author of* Germany and Freedom.

CHARLES E. FRYE *is associate professor of political science at Bryn Mawr College. His special interest is German politics, and he has made a thorough study of the Weimar Republic.*

RALF DAHRENDORF *is professor of sociology at the University of Konstanz. Among his published writings are* Class and Class Conflict in Industrial Society *and* Society and Democracy in Germany. *He has recently entered German party politics, and has become a leader of the Free Democrats.*

TAYLOR COLE, *professor of political science at Duke University, and president of the American Political Science Association in 1958–1959, has a special interest in European comparative government and in Canadian politics. He is the author of* The Canadian Bureaucracy *and the editor of* European Political Systems.

GIUSEPPE DI PALMA, *of the department of political science at the University of California in Berkeley, is an expert on comparative political behavior. His paper included in this book will also appear as a part of a larger volume on sources and implications of political participation in Western democracies.*

CHAPTER 16

The Foundations of a Democratic Future for Germany

JAMES B. CONANT

I propose tonight to speak to you as a reporter, though perhaps some of you would prefer that I assume the role of prophet. For I have found in private conversations in the last few days that many people are more interested in what is going to happen in Europe in the coming months than in an analysis of the situation that now exists. This is particularly true in regard to the plans for the European Defense Community whose fate depends on the vote of the French Assembly when the ratification of the EDC treaty comes up for debate next month. I have repeatedly said that to my mind there is no practical alternative to the EDC and therefore I believe the EDC treaty would be ratified by all six nations. I repeat that statement again; but beyond that, I intend to resist the temptation to indulge in a prophecy tonight. Years ago when I was a chemistry professor I learned a lesson in this regard. Like many teachers of chemistry in college, I used to employ the device of lecture-table experiments often involving explosions to keep my class awake. However, the experiments were not always successful; the predicted explosions sometimes failed to occur. After one such fiasco, an elderly gentleman in the audience who was there as a listener came up after the lecture and offered the following advice: Young man, it is always better to speak after the event as an historian rather than before the event as a prophet. Following this sound precept, I propose to place my remarks about Germany in an historical framework this evening.

I can do this the more readily because it so happens that as a young chemist I was in Germany for 8 months 7 years after the end of World War I. As United States High Commissioner for Germany, I once again entered that country 7 years after the end of another world war. Therefore, in the past 12 months or more there have constantly come to my mind comparisons between what I saw and heard in Germany in 1925 and what I have seen and heard in the past year. As a young professor of chemistry, I traveled widely in 1925 visiting the various universities, but my conversations with my contemporaries were by no means confined to technical subjects. For those were days of considerable political excitement in the Weimar

Reprinted from James B. Conant, "The Foundations of a Democratic Future for Germany," DEPARTMENT OF STATE BULLETIN, XXX, No. 777 (May 17, 1954), 750–55.

Republic. The first President, Friedrich Ebert, had just died and the electoral campaign to choose his successor was in full swing. I discussed with my acquaintances quite frankly the past, present, and future of Germany and Europe. In the course of such informal discussions in the pleasant quarters of a *Weinstube* or *Bierhalle* I received a fairly accurate impression of what the people of Germany were thinking in 1925. I have tried to repeat this experience by making rather extensive trips throughout Germany as United States High Commissioner, including many off-the-record discussions with small groups. In addition there are other ways of assessing public opinion, by indirection so to speak, and we have an excellent staff not only in Bonn but in the consulates scattered throughout Germany. Therefore I think that the comparisons I shall make between Germany in 1925 and Germany in 1954 are based on fairly reliable information.

First of all I am tempted to compare the attitude of the citizens of the German Federal Republic today toward their Government with the reactions of the citizens of the Weimar Republic to the democratic institutions of that time. And the difference is very great. In 1925 a considerable proportion of those people with whom I talked were either indifferent to or hostile to the principles on which the Weimar Republic was founded. It was not a question of being members of the opposition party, not a matter of party politics, but a question of fundamental loyalty to the then newly established republican institutions. It seemed to me at that time that the new governmental structure of Germany had not won the loyal support of many influential sections of the German people. This was in part because of the failure of the Western democracies to give encouragement and support to those elements in Germany which were trying to build a democratic government. These democratic elements were opposed by German conservative and reactionary forces who had never accepted the military defeat of World War I as final and who therefore refused to break with the imperialistic past. Practically from the beginning of the Weimar Republic, the official German government found itself competing for popular support with an opposition which was a shadow system consisting of antidemocratic elements whose purpose was to achieve a nationalistic restoration and who were unscrupulous in the choice of their means. Those who had created the new constitution were rarely in full political control of the Weimar Republic and partly for this reason failed to educate the people and, above all, the youth to accept and support the democratic system of government. I think I am not simply writing history backward when I say that I came away from Germany at that time with a feeling of a lack of confidence in the ability of the Weimar Republic to weather any storms that might be ahead. And after a quick trip to Germany in 1930 when the shadow of Hitler was already on the wall, I returned in a mood of pessimism about the future of the German nation.

Today the situation is quite otherwise. I am referring to something deeper and more significant than the fact that the electorate returned Chancellor Adenauer's own party to the lower house of the Federal legislature with a majority and his coalition with a two-thirds majority. This result of last September's election is of major

THE FOUNDATIONS OF A DEMOCRATIC FUTURE FOR GERMANY

importance in assessing Germany today and was a tribute to the effective leadership of Chancellor Adenauer and the work of his Cabinet during the first 4 years of the existence of the Federal Republic. But what is even more significant is the fundamental attachment of the German people irrespective of party to a federalized republican form of government based on democratic principles; one manifestation was the failure of either right radical parties or the Communists to place a single member in the Bundestag. The German people appear to have broken with their undemocratic past. Conservative and liberal elements have jointly created a democratic constitution and all parties are loyally supporting the new political system. Such opposition as exists today is not directed against the principles and structure of the new Republic but against certain policies of the Government.

GERMANY LOOKS TOWARD THE FUTURE

A second major difference between 1925 and 1954 is the attitude of the Germans toward the immediate past and their hopes for the future. Seven years after the end of World War I one could hardly discuss any political problem in Germany without becoming involved in an endless debate about the origins of the world war, who had in fact won it or lost it, and the role of the founders of the Weimar Republic in the disturbances which followed on the heels of the armistice in November 1918. The stab in the back legend about the Liberals and the Socialists confronted one at every turn. Indeed, this deep concern with the immediate past led many Germans to distrust the Weimar Republic and to hate the democratic and socialist parties. Today, one very rarely hears any discussions of the events of 1933 to 1945. I won't say that there may not be groups of former Nazis here and there who look back with nostalgia and possibly with satisfaction to the days when they were in power, but the results of the last election show that the overwhelming majority of the German people are now repudiating the extremists of both the right and of the left. In fact the leaders of the major parties in the coalition as well as of the Social Democrats (the opposition) are men who do not hesitate in their public speeches to condemn the Nazi regime and the internal as well as the foreign policy of Hitler.

But in general, the eyes of the Germans today are focused not on the past but on the future; and this future they envisage as something different from anything in their past. If one defines a progressive as a man who looks toward a new and better future and a reactionary as one who looks longingly to the past, then I think it would be fair to say that the prevailing attitude in the German Federal Republic today is a progressive attitude. Certainly there are few reactionaries who are longing to turn back the clock of history.

When I first arrived in Germany early in 1953, I was amazed to find how widely the plans for European integration were being discussed and with what degree of confidence the German leaders looked forward to the development of a new Europe. To be sure, the spokesmen for the Social Democratic Party, the opposition party, oppose the formation of a European Defense Community, but being democratic and oriented toward the West, even they have their plans for close military and economic cooperation between a large group of Western European nations.

The city of Passau last summer arranged a festival of which the main theme was the development of a European community of nations. Anyone who had suggested an assembly to talk about European integration in 1925 would surely have been declared a visionary fanatic. It may be that the German enthusiasm for the ideal of a united Europe is somewhat less today than it was 18 months ago; the slowness with which the ratification of the EDC treaty has proceeded has had a somewhat chilling effect on the enthusiasm of some of the most European-minded leaders of German opinion. It is also true, and in view of the unprecedented character of the project not surprising, that difficulties are now beginning to appear in connection with the Coal and Steel Community which are being given considerable publicity, but in spite of both these negative factors, it seems that the ideal of a new sort of future for Europe still has great vitality for a surprising number of people in the Federal Republic. Reports which have come to me from several sources indicate that the young people of Germany are, for the present at least, really enthusiastic about going forward with plans for a close military, economic, and political integration of the six nations who signed the EDC treaty and are now part of the Coal and Steel Community. Moreover, the agreement just signed by this Government to extend a loan of $100 million to the Coal and Steel Community, I hope will galvanize German and West European interests and prove a timely shot-in-the-arm for what is still the most important single venture in the field of European economic cooperation. The year 1954 appears to be one of those years in European history when there is a tide running in a direction which we Americans can only regard as being the right direction. Whether this tide will be taken at its flood is still admittedly uncertain.

Each one of you has probably his own version of the history of the last 50 years and is ready to defend his own particular thesis as to the origins of World War I, the failure of the Versailles Treaty, the rise of Hitler, and the subsequent disaster of World War II. My own interpretation is that the Weimar Republic was founded on shifting sands. The violence of the years 1919 and following, in which German assassinations, street fighting, and putsches played an important role, furthered the growth of political reaction. Many turned their eyes to the nationalistic and militaristic ideals of the period from 1870 to 1914. Therefore when the forces created by the great social catastrophe of inflation and unemployment and the failure of the victors in World War I to carry out a wise and prudent policy staged a series of revolutionary political events, nationalistic and militaristic ideals—reaction in short—came to the fore embodied in the person of Adolph Hitler.

I recall this bit of history to your minds for the purpose of contrast. If I am right, the number of Germans today who envisage the future of their country in terms resembling the period of imperialistic glory is very small. Tonight, I refuse to be prophet, therefore if some of you are inclined to say, ah!, but the German mood may change, I can only repeat that the difference in attitude between 1954 and 1925 is a difference not of degree but of kind. And after all, we mortals can only predict the future in terms of the facts of the present and the past.

THE DIVIDED WORLD

Of course the fundamental difference in Germany and throughout the free world today between the present agonizing period in which we live and the relatively tranquil times of the 1920's reflects the basic fact that we live in a divided world in an atomic age. For the people of Germany since 1945, the existence of a divided world has been ever present before their minds. Since the Berlin blockade, no German could question the fact that the Iron Curtain was being moved westward to the line of the Elbe River. This fact, coupled with the utter destruction of most German cities and the complete collapse of all German governmental structures on VE-Day in 1945 has meant that the Germans since the end of World War II have been literally struggling for their existence. The day-to-day task of merely staying alive and attempting to reconstruct some kind of order from the ruins of their nation occupied all their energies until very recent years. Therefore, the great difference in attitude between Germany today and 30 years ago, one may well say, is a consequence of the total defeat of Germany in World War II followed by the decision of the Western allies to prevent the sovietization of all of Germany by the Russians during the occupation period. Certainly the military events of 1944 and 1945 and the East-West diplomatic struggle between the occupying powers of the period 1945 to 1949 set the scene for Germany today and for some years to come.

When we talk of Germany, it is well to bear in mind that there are three Germanys: The Federal Republic comprising that portion of the former German Reich lying in the occupation zones of the British, the French, and the Americans, including some 50 million inhabitants; the Russian Zone with its 18 million Germans lying to the East; and the city of Berlin, the Western sectors of which are an island of freedom deep in the heart of the Soviet-occupied territory. When I have been speaking about the attitude of the Germans, I have referred to the attitude of the citizens of the Federal Republic and West Berlin. What the people living in the Soviet Zone feel about the past and future can be deduced from the evidence supplied by thousands of refugees. But the tragic fate of these 18 million Germans is one of the brutal facts of history which stand before the eyes of the fortunate Germans who live in the Western Zones. The dramatic events of June 17 last year underline the plight of the East Germans. They also demonstrated their courage and their desire for freedom.

In the almost 9 years that have elapsed since the end of World War II, the 50 million Germans in the Federal Republic have been able to reconstruct a free democratic form of government which I hope and believe will soon be essentially sovereign. The cultural life of Western Germany, thanks to the wisdom of the British, the French, and the Americans, is again beginning to flourish in an atmosphere of democracy and freedom. As to the physical rebuilding of West Germany, that is a fact so striking and so well known as to require no underlining to this audience. Thanks very largely to American aid, first by special appropriations and then through the Marshall plan, but thanks also to the energetic and skillful use of those funds by the Germans, German industry has revived and the cities are in process of being rapidly rebuilt. A traveler through Western Germany today will find all the signs of a prosperous, stable, industrialized society and, unless our economic experts are completely wrong, the prospects for the continued satisfactory development of industry and commerce in Western Germany are excellent indeed. The currency is stable and the relation between the banking system and the government is such as to insure a stable financial policy. The attitude of the economic advisers of Chancellor Adenauer is very much on the side of the American concepts of initiative and free enterprise. There is further evidence of the growing stability and health of the German economic and financial situation which is of particular interest to Americans. The Federal Government, in recent months, has found it possible to eliminate restrictions on the import from the United States of nearly 3,000 commodities, many of which are of considerable importance to our agriculture. It has also substantially reduced restrictions on the transfer of earnings on investments in Germany of U.S. residents and on transfer of so-called "blocked mark accounts."

CONTRAST BETWEEN EAST AND WEST ZONES

Contrast all this with what has been going on in such cities as Leipzig, Dresden, and the Soviet sector of Berlin. Here a puppet government was installed by Soviet fiat in 1949 and later given the appearance of constitutionality through sham elections in 1950 which favored the Communists. While the degrees of the severity of the regime have varied from time to time, the characteristics of a totalitarian state have been present from the day of the surrender of Germany in 1945. Indeed, and this is important, you must remember that great numbers of the inhabitants of this eastern part of Germany can never recall a time when they have not lived either under the totalitarian rule of the Nazis or the tyranny of the Soviet occupying forces. The economic situation reflects the attempts of the Soviets to push their system westward to the Elbe River; the farms have been collectivized and the stores and industries largely nationalized. Those who have visited the cities in the Russian Zone tell me the physical contrast between the East and West is so evident as to be shocking.

Certainly as I myself have seen so often in Berlin, there is a great distinction between West Berlin, our side of the fence, with its well-stocked shops, well-dressed inhabitants, motorcars, new or rebuilt libraries, churches and theaters, and its general air of freedom, and East Berlin, the Soviet side of the fence. When one enters the Soviet sector, one sees drabness and depression. In spite of the much vaunted Stalinallee, a workers' housing development built along Moscow lines, not a great deal of reconstruction has taken place. Above all else one is oppressed by the atmosphere of police control and austerity.

The control in East Berlin and the East Zone is actually the control of Moscow. The appearances could lead a naive observer to think that the German Communist regime were masters in their own house. This facade has been redecorated recently by the proclamation of the sovereignty of the puppet government. The Soviet maneuver has fooled no one and no standing will be accorded to the regime by the free nations of the world.

THE REFUGEE PROBLEM

That there have been two million refugees from the Russian Zone to the Federal Republic in the last 3 years will surprise no one. The stream is continuing at the rate of some 20,000 a month. For the last year the Russians have permitted relatively free travel between their zone and the rest of Germany—I emphasize the word *relative* for the number of ports of entry are few indeed and the traffic is strictly controlled. The border between the Russian Zone and the Federal Republic is marked by barbed-wire fences, a plowed strip, and armed guards at every turn. But the relatively free travel means that as many refugees now come across the border as through Berlin. This continued influx of several hundred thousand a year added to the 10 million refugees already in West Germany presents the Federal Republic with a serious problem.

In recent months, as you are well aware, the Russians have been doing all in their power to attempt to raise the prestige of their satellite government, the so-called German Democratic Republic referred to colloquially in Germany as the Pankow regime. Mr. Molotov at the Berlin Conference asked for the representatives of this government to come to the conference. When challenged by Mr. Dulles as to the legitimacy of this government and twitted about the forced election methods used to choose the Legislative Assembly in his zone, he unashamedly defended the Soviet concept of free elections. Both in his remarks and subsequent articles in the East Berlin press, the system of elections with the help of Soviet-controlled unity lists was defined as the "only free and democratic " method of choosing representatives. The results of such elections (farcical from our point of view) were contrasted with what happens when in Soviet terminology militarists and capitalists are allowed to compete as they did in the elections of September which resulted in the return to power of Chancellor Adenauer in the Federal Republic.

SOVIET INTENTIONS

More than one observer of the Berlin Conference has drawn the conclusion from Mr. Molotov's amazingly frank attitude that he was quite unwilling to consider proposals for free elections in all of Germany first of all because he did not wish to relinquish his control of the Russian Zone, and furthermore because he had his eyes fixed on the ultimate control of all of Germany itself. I shall long remember his cynical contempt for democratic procedures when he warned us: "We must not be carried away by parliamentary formalities and the organizational and technical aspects of this matter"; "this matter" happened to be the idea proposed by the United States, Great Britain, and France and desired by all Germans of holding free and democratic elections throughout Germany. But Mr. Molotov thought that we were "carried way by formal constitutionalism." It seemed to some of us that he was anxious to support his puppet regime in the Russian Zone for several reasons. First, for the sake of the prestige of those Germans who had cast their lot in with the Russians; second, because he needed to support the prestige of the satellite governments in Poland, Czechoslovakia, and nearby lands; and third, because he was preparing his case for the extension some day of his electoral methods to all of Germany. He seemed to have his eye on a future which would come when the wedge he was trying to drive between the Western allies finally found a weak spot and opened enormous cracks. That Mr. Molotov and his colleagues in the Kremlin may dream such dreams at the present moment may seem fantastic. However, whether it may sound impossible to us here and whether it is disbelieved by every German matters little to the men of the Kremlin. Mr. Molotov may assume that some day the present American foreign policy will weaken and our economic structure collapse, that the American military forces will no longer consider Europe the outpost of their own defense, that the French and the Germans will renew their old hostility and be ready to stab each other in the back, that the whole free world will go through a major depression with a consequent vast unemployment in Western Germany. Under such a set of circumstances, the masters of the Soviet Union would be indeed in a position to talk about the German problem in far different tones from those we heard in Berlin last January.

Let me remind you that tonight I have promised not to be a prophet. My last few sentences have dealt only with what may well be a prophetic vision in the eyes of the dwellers of the Kremlin. If I am at all right in this supposition, then the task for us in the free world is to do all in our power to prevent the future resembling in any way that which Mr. Molotov and his associates may hopefully have in mind, and in this regard the United States and West Germany at present see eye to eye. For, as I have already reported, the German leaders seem anxious to work for some type of European integration; among the youth of Germany today the ideal of a new type of European community has a powerful hold. One need not be a prophet to say that the future of free Europe depends on the future relations of Germany and France; in spite of many discouragements of the past few years, I believe the signs are still predominantly favorable for continued progress toward European cooperation.

INTEGRATION WITH WESTERN EUROPE

The policy of Chancellor Adenauer is a policy of integration of West Germany with Western Europe. He regards such integration as a necessary step toward the reunification of Germany in peace and freedom. Contrary to what some of his political opponents maintain he believes there is no antithesis between unification and European integration. Rather, he and his associates believe that the Russians made it plain at Berlin that until the West proves itself to be strong and united, the Russians will not forego their ambitions to move the Iron Curtain further westward and will not consider relinquishing their hold on the Russian Zone of Germany. Following his line of thought, one could look forward to the day when a reunited Germany can become one of the stalwart nations in a new type of free Europe, a free Europe which can face boldly the totalitarian challenge from the East.

This vision of the future which appears to be in the minds of the leaders of the

Federal Republic today may be regarded by some of you as an illusion. I know the fear of a revitalized Germany exists.

Many people in the United States are apprehensive about what the new Germany will do. They have seen or heard or read of the teeming energy of West Germany in 1954. They only ask themselves, "Will this powerful new nation prove a stabilizing influence or will it as it has twice in the memory of most of us draw us into a holocaust?" I have said I was not going to be a prophet but I do venture to summarize my previous diagnosis: Germany today is unlike Germany either in the 1920's, the 1930's, or before World War I. There are a number of powerful political personalities in different parties working toward a close cooperation with the West and strong believers in a peaceful and democratic Germany. These men need the help and understanding of the freedom loving people of this country and the European nations. The error of the victors in the 1920's must not be repeated. The recent declaration of assurance of Great Britain and the United States, contingent, of course, on the realization of the European Defense Community, renders it clear that the new Germany will be a firmly integrated member of the free world. I know no better answer to those who raise questions about the future of Germany after its sovereignty is restored.

CHAPTER 17

Parties and Pressure Groups in Weimar and Bonn

CHARLES E. FRYE

I

In the last forty years, Germany has had three radically different political systems. In each case, the party system, better than any other single index, reflects the style of politics of that period. The highly splintered, multiparty system of Weimar mirrors perfectly the extreme ideological dissension and radicalism of postwar German politics. The one-party system of the Third Reich epitomizes the attempt to destroy the individual's traditional social ties and then to absorb him totally in a coordinated movement. Finally, the two-party system of Bonn reflects the growing social and political consensus concerning the more pragmatic and concrete political goals of post-Hitlerian Germany. Although these three political systems are intimately related, the main question for us is why the democratic party systems of Weimar and

Reprinted from Charles E. Frye, "Parties and Pressure Groups in Weimar and Bonn," WORLD POLITICS, *XVII, No. 4 (July, 1965), 632–55, by permission of the publisher.*

Bonn are so different. The party systems are unintelligible, however, without an understanding of the patterns of pressure-group politics as well.

One measure of the differences between the two democracies is simply the number of active political parties. At any given election to the Reichstag, thirty-odd political parties were likely to be competing in Weimar; less than a third that number competed at the last elections in Bonn.[1] But by themselves these figures might be misleading, for although more than thirty parties in Weimar had candidate lists, rarely more than a dozen of them won seats to the Reichstag. Never more than six parties (sometimes as few as three) united to form coalition governments. Social Democrats (SPD), German Democrats (DDP), and Centrists (Z), for example, united to form the Weimar Coalition of the Constitutional Assembly and the early republican years.

Today in Bonn one also finds a coalition government—formed of the Christian Democratic Union, its Bavarian affiliate, the Christian Socialist Union (CDU/CSU),[2] and the Free Democrats (FDP). Despite their relative similarity in this respect, however, in actual functioning the governments of Weimar and Bonn are worlds apart.

The CDU singlehandedly won a majority of seats in the lower house of parliament in 1953, something no other German party in free elections had ever done, and followed that by victory in 1957[3] with an unprecedented absolute majority of the popular vote. Moreover, whether the CDU has commanded a majority of the seats in the Bundestag or not, the Social Democrats have dominated the opposition; and the opposition, it should be remembered, is nearly as important as the government in determining the style of politics in a parliamentary system. In Weimar, there were never fewer than five parties in opposition—two of them always at the extremes, which is not the case in Bonn. How can one account for these differences? Why did the number of parties in Weimar tend to multiply whereas in Bonn it has constantly decreased?

Or, taken from another point of view: German farmers pressed their demands upon the government through three pressure groups in Weimar.[4] But there was also the German Farmers' Party (DBP); and the German National People's Party (DNVP), heir to the conservative tradition in Weimar, was in large measure a party of farmers—aristocratic Junkers. Today most German farmers have united in the *Deutsche Bauernverband* and, with good reason, this group is held to be among the most powerful in Bonn.[5] Yet, despite persistent demands for the creation of a

---

[1] For a general descriptive account of German parties since their beginnings, see Ludwig Bergstraesser, *Geschichte der politischen Parteien in Deutschland* (10th ed., Munich 1960).

[2] I consider the CDU and CSU as one party and for brevity's sake usually refer to it hereafter as the CDU.

[3] U. W. Kitzinger's *German Electoral Politics* (Oxford 1960) is a study of that election.

[4] *Reichslandbund, Bauernverein, and Bauernschaft.* The best descriptive account of these and other pressure groups early in Weimar is Edgar Tatarin-Tarnheyden, *Die Berufsstaende* (Berlin 1922).

[5] "Around the political power potential which the *Deutsche Bauernverband* represents, there has gathered an economic concentration of power which in breadth, rationality, and certainty of achieving its goals has no equal in West Germany" ((Kurt Pritzkoleit, *Maenner Maechte Monopole* [2nd ed., Duesseldorf 1960], 61).

farmers' party, they until now have remained organizationally active within the bounds of the *Bauernverband*. Of other economic associations that were or became active parties in Weimar but have become mere pressure groups in Bonn, there are at least two instances—the *Wirtschaftspartei* and the *Handwerkerpartei*. In Bonn itself the remnants of the Refugee Party, Bavarian Party, Economic Reconstruction League, and South Schleswig Voters' League, among others, are forceful reminders of the continuous threat to such narrowly oriented parties. Eleven parties won seats to the Bundestag in the election of 1949; only six remained in 1953, and by 1961 the number was three—in each case, the victims being splinter parties.[6] But why have associations that were active as parties in Weimar contented themselves, or had to content themselves, with the status of pressure groups in Bonn?

It is the main hypothesis of this article that these two kinds of association are interrelated. A political party system affects the activities of pressure groups (their goals and the means of attaining them) at the same time that pressure-group patterns condition the functioning of political parties. The closeness of the ties between parties and pressure groups, as well as the kinds of ties that exist, is directly related to the number of parties in a system.[7] Party and pressure-group activities also interact in other respects. At what point pressure groups will tend to bring pressure, how dogmatic and persistent they are likely to be, and whether or not parties will be able to compromise their programs are the kinds of questions that cannot be settled without an understanding of party and pressure-group patterns.[8]

The consequences of the kinds of relationships that exist between parties and pressure groups in modern democracies run deep. Political party and pressure-group activity is, in fact, a primary determinant of the stability and effectiveness of a political system. Political effectiveness presupposes, among other things, a degree of stability that highly fragmented multiparty systems such as Weimar or the Third and Fourth Republics in France have never achieved. And for the study of these relationships, Weimar and Bonn make a particularly good case because there are more constants and relatively fewer variables than in many cross-national studies. Yet the differences could hardly be sharper. Ultimately, the question is why Weimar was so unstable and ineffective as a political system and why Bonn has been so stable and relatively effective.

II

Perhaps the most striking fact about parties in Weimar was the narrowness of their appeals; the whole tendency was for them to become representatives of particular interests.[9] The closeness of their affiliations can be seen most clearly among the small *Splinterparteien* in Weimar—Party of House and Land Owners; People's

---

[6]Kitzinger, 170–97, discusses the efforts of the minor parties in the 1957 election.

[7]Compare Gabriel A. Almond, "A Comparative Study of Interest Groups and the Political Process," *American Political Science Review*, LII (March 1958), 275–76.

[8]For a statement of determinants of pressure-group politics, see Harry Eckstein, *Pressure Group Politics* (Stanford 1960), 15–39.

[9]See, for example, Karl D. Bracher, *Die Aufloesung der Weimarer Republik* (3rd ed., Villingen 1960), esp. 65–69.

Coalition of the Victims of Inflation; Non-political List of War Victims, Work Invalids, and Welfare Recipients.[10] But it is hardly less striking among the larger parties—Social Democrats and labor, German Populists and the upper middle class, National Populists and Junkerdom. In fact, the only party to escape the bonds of really narrow identification was the Center party, and it was almost exclusively Catholic.

Despite the large number of parties in Weimar and the specificity of their appeals, party programs did overlap—particularly at the center of the political continuum.[11] The economic position of the German Democratic Party, for example, was very similar to that of the Stresemann-led German People's Party (DVP); and the Center party was famous (or infamous, depending upon the contemporary's point of view) as the party that could find a community of belief with several other parties. In Prussia, which comprised approximately two-thirds of both the people and territory of Germany, the Center party maintained a very stable coalition with the Social Democrats throughout the Republic.[12] It also participated in every coalition in the Reichstag, although after 1920 those coalitions always included at least one rightist party and excluded the SPD an even dozen times. As the Republic became older, this area of overlap moved constantly to the right.

That parties in multiparty systems tend to make narrower appeals tells us something very important about their relations with interest groups—namely, that parties and interest groups in a multiparty system are more closely integrated than parties and interest groups in a two-party system. They become more interdependent because the relative importance of each particular interest to the life of the party increases as the party appeal becomes more narrow. By tying a given interest group more closely to itself, a party is seeking to compensate, as it were, for its inability to integrate a large variety of interests. It becomes increasingly dependent upon particular interest groups for funds and electoral support, while the interest groups are increasingly dependent upon it for furthering their interests. When this interdependence is complete, one has interest groups fully politicized—as a *Wirtschaftspartei*, a *Bauernpartei*, a *Handwerkerpartei*.

Several forces were operating to counteract the narrowness of these identifications. First of all, no party could hope to join a coalition, let alone win an election, if it did not succeed in bringing large numbers of people into its fold.[13] This meant an appreciable advantage for the Social Democrats. On the strength of worker class-consciousness, they could direct their appeal to one particular class and still win

[10]The *Deutsche Allgemeine Zeitung*, May 9, 1928, gives a list of the parties competing in the 1928 election.

[11]James K. Pollock, Jr., "The German Party System," *American Political Science Review*, XXIII (November 1929), 878–79.

[12]Several people have commented upon this startling incongruity. Allan Bullock, for example, has written that "In Prussia . . . the State Government enjoyed a stability which made it the bulwark of democracy in Germany . . ." (*Hitler: A Study in Tyranny* [New York 1961], 121). See also Arnold Brecht, *Federalism and Regionalism in Germany* (New York 1945), 20–23. But no one has adequately accounted for the disparity between Prussia's stability in Weimar and, despite its compromising such a large part of Germany, its instability in the Reich.

[13]The SPD won the largest plurality of any democratic party in the Republic with 37.9% of the vote in 1919. It dropped to 21.6% in the next election and never again reached 30%.

a sizable representation. To attract a comparable following, any bourgeois party would have had to extend its appeal across a broad front of social and economic interests. For as the large number of parties competing among the bourgeoisie suggests, middle-class identification was much more ambiguous. Moreover, when the identification between the party and its social groupings became too tenuous, every party faced the threat that another party might be formed or that the voters might switch to a party more specifically directed to their social and economic interests. A case in point is the National Populists, who split at least three times in the course of their brief history—each time sending out another party shoot.

Second, and more important, that many pressure groups could not afford to put all their eggs in one party basket helped to loosen the ties between parties and particular groups. Pressure groups could exert a significant influence in a coalition government only by operating across several party lines. The national organizations of industry, trade, and agriculture, for instance, cultivated relations with all the bourgeois parties and contributed heavily to their campaign funds.[14] Furthermore, the strongest professional groups were represented in the Reichstag without regard to party lines.[15] Important as these affiliations were to the individual parties and pressure groups, they also had an enormous impact upon the political system as a whole.

Cross-fertilization was the latent function of this pressure-group activity. Without a web of contact and affiliation that allowed interchange of information and ideas, it is inconceivable that there would have been a political system at all—multiplicity of self-contained entities within a state, to be sure,[16] but little or nothing to bind them together and relate them to one another. Pressure-group activity across party lines helped to keep those lines from hardening more rapidly than they did, and it prevented the parties from completely isolating themselves behind their ideological walls.[17] As a consequence of the immobilization of the Reichstag under Bruening after 1930, this extensive pattern of interchange and communication collapsed; the Republic itself would soon follow.

III

With the declaration of the Republic, it became absolutely essential for all German parties to develop organizations capable of enlisting the support of the voting masses.[18] Only the SPD among pre-republican German parties had set up its

[14] Bracher, 199–200.
[15] Ernst Saemisch, "Wer vertritt das Volk?" *Die Tat*, XXI (March 1930), 932.
[16] For an apt description by a contemporary of the tendency toward party isolation, see Carl Schmitt, *Positionen und Begriffe im Kampf mit Weimar-Genf-Versailles 1923–1939* (Hamburg 1940), 189.
[17] Werner Liebe states, for instance, that "Economic cross-ties, as with the *Reichslandbund*, the *Reichsverband der deutschen Industrie* and the *Deutschnationalen Handlungsgehilfenverband* . . . bound the National Populists more and more strongly to the German People's Party and to the Center party. It furthered the process of inner accommodation of the moderate circles on the right with the bourgeois Republic" (*Die Deutschnationale Volkspartei, 1918–1924* [Deusseldorf 1956], 105).
[18] Max Weber, "Parlamentarisierung und Demokratisierung," *Gesammelte politische Schriften* (Tuebingen 1958).

own bureaucracy; and the strength of its organization (like the weakness of the bourgeois organizations) reflected the role it had always wanted the people to play. But it was not alone in having a large following. After 1890, the *Volksverein fuer das katholische Deutschland* had served to deliver masses of votes to the Center party, although neither it nor many of its members were directly affiliated with the party.[19] The Center party itself had only a small membership, consisting chiefly of party notables.

In contrast to the SPD, the prewar bourgeois parties, liberal and conservative, were decentralized parties of notables (*Honoratiorenparteien*).[20] It was to them that the advent of the Republic presented the greatest organizational challenge. The Center party rationalized its organization and gradually took over the functions of the *Volksverein* itself. But, characteristically, the most truly liberal party of the Republic, the German Democratic Party, failed to meet the challenge and soon began its rapid decline. While its neighbor to the right, the German People's Party, was somewhat more successful, the greatest change came from the quarter where one might least have expected it—from the Prussian conservatives and the DNVP. Overnight the National Populists had created an organization that would increase their stature considerably by 1920 and in representation make them second only to the Social Democrats in 1924.[21]

Part of the impetus for this change came from the Christian Socialists, one of the five parties that had united in the DNVP in 1918. They had also enjoyed the fruits of an extraparty mass organization (*Bund der Landwirte*) before the war.[22] Formed in 1893, this association, which concentrated its strength in western and southern Germany, was estimated to have a peak membership of between 200,000 and 300,000; and although it too was formally independent of party, in fact it worked intimately with the Christian Socialists and was indispensable to them. These mass organizations had helped prepare the parties for the important changes they faced in Weimar. In several short years, the German democratic parties were to develop "a highly integrated, smoothly working party machinery which has an equal in England but no superior anywhere in the world."[23]

The very excellence of the bureaucratic organizations, however, tended to make the parties in Weimar undemocratic, unwieldly, and unresponsive to popular opinion. They were often criticized as parties of aged functionaries divorced from the people.[24] And despite other dramatic changes, the parties in Bonn seem to have changed little structurally. Perhaps today the change in electoral law from the "pure" list system has decreased the power of the party *Apparate*,[25] but the German parties

[19] Thomas Nipperdey, *Die Organisation der deutschen Parteien vor 1918* (Duesseldorf 1961), 281.
[20] *Ibid.*, 42.
[21] Liebe, 30ff.
[22] Nipperdey, 249.
[23] Pollock, 865.
[24] Compare Herbert Sultan, "Zur Soziologie des modernen Parteiensystems," *Archiv fuer Sozialwissenschaft und Sozialpolitik*, LV (1926), 109; Oswald Spengler, *The Decline of the West* (2 vols., New York 1928), II, 457; Ottmar Buehler, *Die Reichsverfassung vom II. August, 1919* (2nd ed., Berlin 1921), 158.
[25] See Kitzinger, who gives considerable weight to the local constituency in the all-important selection of candidates (p. 64).

still continue to be bureaucratically strong—particularly the SPD. The Bonn parties are also similar to those in Weimar in that their membership is made up of a relatively small percentage of the voters and it consists predominantly of older people. Duverger gives the following percentages of party voters who are members of some European socialist parties: Germany 9.1 per cent (1949); Sweden 35.5 per cent (1948); Norway 25.6 per cent (1949); France 8.4 per cent (1946); Austria 37.9 per cent (1949).[26] As far as one can tell from the few studies we have, there would not seem to be any important organizational differences either between the highly bureaucraticized interest groups in Weimar and those in Bonn.[27]

IV

The intimacy of the ties between many pressure groups and parties in Weimar meant that each party could transmit the demands of its component groups directly into the legislative process. There was no need to try to harmonize or compromise them first with other conflicting demands within or outside the party, because the possibility of such conflict had already been minimized. Existing differences among various interests in society had been hidden, as it were, by segregating them behind party walls. Labor and management, Catholic and Protestant, farmer and entrepreneur did not have to iron out their differences within a party, for each party tended to represent non-conflicting interests—which is not at all to say that there was no internal party conflict; it is to say that it had been minimized.

As potential conflict was reduced *within* parties, it became all the more real *among* parties in the Reichstag, at least as long as that body continued to legislate. (With its complete paralysis after 1930, the only recourse for conflicting interests was to the party battle leagues, the *Kampfbuende.*) In the Reichstag, the hitherto unchecked competing demands entered a common hopper. Between the articulation of interests and their introduction into the legislative process, no intermediary stage had existed where differences of opinion had had to be curbed and interests compromised. Perhaps more important, life in the social and political institutions of Weimar (in the authoritarian family, schools, churches, political parties, military organizations) had not prepared the politicians for the reconciliation and compromise they faced in the Reichstag. But if settlement was to be reached within the polity, it had to occur in the legislature or, beyond it, in the administrative hierarchy. The paucity of legislation indicates how much compromise of this kind ever really took place.[28]

It is said that pressure groups concentrate their activities upon positions of power in a political system.[29] In fact, centers of pressure-group activity are considered to be

[26] Maurice Duverger, *Political Parties* (New York 1954), 95. Perhaps in partial explanation of this disparity, Erich Reigrotzki has suggested that "The parties in Germany according to this [the frequency of membership in groups] have the character not of mass organizations, but much more so of local groups [or] clubs of notables . . ." (*Soziale Verflechtungen in der Bundesrepublik* [Tuebingen 1956], 191).

[27] Rupert Breitling has pointed out in this connection that in general he believes the German groups, though less numerous, "have a higher degree of organization, a more monopolistic position, and are more privileged" than their counterparts in the United States (*Die Verbaende in der Bundesrepublik* [Meisenheim am Glan 1955], 2).

[28] Fritz Poetzsch-Heffter, "Vom Staatsleben unter der Weimarer Verfassung," *Jahrbuch des oeffentlichen Rechts der Gegenwart*, XIII (1925), 216.

[29] Almond, 278.

reliable indices of the loci of power in a political system.[30] Yet in Weimar, where the Reichstag was practically immobilized, pressure groups directed much of their activity to that body. The Reichstag, of course, had served as a center for pressure-group activities early in the Republic primarily because the identification between parties and pressure groups was so close. Also the Reichstag did pass some legislation prior to 1930. Extensive so-called "Aid to the East"—that is, to the Junkers—came as late as 1930.[31] Moreover, because the bureaucracy was ostensibly responsible to the Reichstag, pressure upon the administration would more likely be effective coming through a party than directly from a pressure group. Identification with a party also gave the pressure-group officials access to governmental agencies and circles, and the Reichstag gave them a national platform from which to broadcast their goals. Today, in Bonn, it is not that pressure groups do not wish to have the official status of parties[32] or at least to have the advantages which go with it. Many of them do, but conditions do not permit it. In Weimar they did: "In 1928 the Wulfmeyer family formed a party which was called 'Law and Renter's Protection Party.' As the *Deutsche Allgemeine Zeitung* remarked: 'Here is real anarchy—every family its own party.' "[33]

The very narrowness of interests being represented in the Reichstag conditioned that body's deliberations in several important respects. It increased the number of issues before the Reichstag by staggering proportions and overburdened that body to the point of collapse. And because the members of parliament could never transcend their differences, it also prohibited meaningful debate. Discussion in the Reichstag tended to be about matters of little consequence or *Weltanschauungen*.[34] And if discussion of middle-range issues was rare, even rarer was a piece of middle-range legislation, for the parties in Weimar could not reconcile their differences. Obviously, if a party has but one interest to represent, it is going to be much more dogmatic and uncompromising in that interest's behalf than if it represents a variety of interests. That one interest is its *raison d'être*. A party with a number of interests can afford to win a little here, lose a little there, and still maintain its integrity.

The consequences of differences in the breadth of party appeals are manifest in the differences between the extreme splinter parties and the parties which had some hand in governing Weimar. Because Weimar was a highly pluralistic and stratified society —and a republic—to have an effective parliamentary party, each of the five or six major parties had to make relatively broader appeals and attempt to aggregate at least several interests. In the Empire, it had sufficed for the German Conservative

---

[30]Eckstein, 16.

[31]Bruno Buchta, *Die Junker und die Weimarer Republik: Charakter und Bedeutung der Osthilfe in den Jahren 1928–1933* (Berlin 1959).

[32]The Basic Law of Bonn broke with the German constitutional tradition in officially recognizing parties (Art. 21) as having a role in the formation of the will of the people. For a criticism of this change, see Werner Weber, *Spannungen und Kraefte im westdeutschen Verfassungssystem* (Stuttgart 1951), 20–23. Also Gerhard Leibholz, *Der Strukturwandel der modernen Demokratie* (Karlsruhe 1952).

[33]Pollock, 882.

[34]Gerhard A. Ritter, *Deutscher und britischer Parlamentarismus* (Tuebingen 1962), 45.

Party to restrict itself to the Junkers of East Elbia;[35] one of the first actions of the conservative party (DNVP) in the Republic was to extend its appeal geographically. That in turn necessitated a broadening of its social appeal. Any party that had expectations of participating in a government had to attract as many voters from as wide an area as possible, but the minor parties did not even make any pretensions of inclusiveness. That is the point—they were not interested in governing. They were merely carrying their particular interests into the governing process. They were pressure groups turned parties.

The most broadly based party in Bonn, the Christian Democratic Union, is a coalition of a wide variety of social, economic, and geographic elements: farmers, workers, professional people, entrepreneurs, refugees, Bavarians, Rhinelanders.[36] But what is most striking about its make-up, what would have been inconceivable in Weimar, is its union of Catholics and Protestants—in almost equal proportions.[37] Possibly as anomalous would have been the recent Social Democratic overtures to German farmers. No party in Bonn, however, can hope to succeed to power without embracing all, or nearly all, major social groupings. The mellowed tone of debate in the Bundestag reflects that change.

V

Widespread in Weimar was the charge that the Reichstag was not legislating in the interests of the community. It was not without substance. The Reichstag could not ordinarily reach any agreement on important policy issues when mere discussion of them was sufficient to cause a coalition split. But for that very reason, it was all the more urgent that every interest group not only be heard but satisfied—and satisfied in its boldest demands. Had there been more general legislation on the many pressing issues in Weimar, the need for each group to pursue its own policies on each issue would not have been as great, nor would the differences among groups have been as clear and distinct.

Between the area covered by party ideologies, on the one hand, and the everyday business matters of the legislature, on the other, there was an immense array of middle-range problems which the Republic had faced ever since its inception: schools, taxation, unemployment and other economic matters, social welfare, and farming—the list was interminable. But the parties had shown themselves to be impotent in dealing with most concrete problems. As a result, the Reichstag increasingly lost power and stature to the executive.[38] It was not that the Reichstag refused the programs submitted to it; most frequently it never had to make a choice. Only three of the seventeen different governments to be formed prior to Bruening in 1930 had to dissolve according to constitutional provision—two because of votes of

[35]Hans Booms, *Die Deutschkonservative Partei* (Duesseldorf 1954), 7–8.
[36]Although the data are now over twelve years old, useful in this connection is Friedrich A. von der Heydte and Karl Sacherl, *Soziologie der deutschen Parteien* (Munich 1955), esp. 288–93.
[37]Thomas Ellwein, *Klerikalismus in der deutschen Politik* (Munich 1955), 99.
[38]Poetzsch-Heffter, 131.

censure and one because of a vote of lack of confidence.³⁹ The remainder collapsed from inability to reach agreement in the Cabinet—before the matter ever came up for a final vote in the Reichstag.

Near immobilization of the Reichstag early shifted the balance of power to the President and the bureaucracy in Weimar. Of 1,150 pieces of legislation enacted between 1920 and 1924, two-fifths came in the form of emergency statutes from the President.⁴⁰ Among them were some whose enactment was crucial: approximately forty economic emergency laws, for example. This helped not only to decrease the size of the legislature's workload, but also to relieve it of some pressure-group activity. It simultaneously overburdened a bureaucracy which had been schooled rigorously in objectivity.⁴¹ And unfortunately the National Economic Council, which could have served as a channel for much of this activity, was born lame.

The Prussian civil service had always prided itself on being apolitical, and yet the bureaucracy in Weimar was being drawn inexorably into the very heart of the political process.⁴² Aside from hampering effective administration, this pressure increased the tensions and strains of the civil service as a class and alienated it further from the Republic. Nevertheless, after 1930, when the Reichstag had been shunted out of the governmental process, there was nowhere else pressure groups could turn.

VI

From the point of view of a two-party system, Almond suggests, parties and pressure groups in highly fragmented multiparty systems are not markedly different in function.⁴³ Considered in their own right, however, the parties and pressure groups of Weimar were quite different: parties tended to be seats of ideology; pressure groups, associations for the representation of concrete interests. If there had been no such role differentiation, it would have been unnecessary to have both a party and several pressure groups operating in the same area at once, as the German Farmers' Party and the *Bauernverein, Landbund,* and *Bauernschaft* did or as the *Handwerkerpartei* and the various artisan pressure groups did.

Parties in Weimar served as the official keepers of ideological seals,⁴⁴ and pressure groups existed for the articulation of specific determinate demands. The bond between them came, of course, in that party ideology was a rationalization of the particular interests. Labor unions (SPD) and Marxism, for example; Junkers

---

³⁹ Poetzsch-Heffter, "Uebersicht ueber die in den Jahren 1924–1928 gestellten Vertrauens- und Misstrauensantraege," *Jahrbuch des oeffentlichen Rechts der Gegenwart,* XVII (1929), 106ff.

⁴⁰ Poetzsch-Heffter, "Vom Staatsleben unter der Weimarer Verfassung," 216.

⁴¹ Theodor Eschenburg discusses and denounces the increasing pressures upon the bureaucracy in Weimar and particularly in Bonn (*Herrschaft der Verbaende* [Stuttgart 1955], esp. 13–16).

⁴² Ritter talks of "Die der Buerokratisierung der Politik parallel laufende Politisierung der Buerokratie..." (p. 51).

⁴³ Almond, 275–76.

⁴⁴ Werner Conze, "Die deutschen Parteien in der Staatsverfassung vor 1933," in Erich Matthias and Rudolf Morsey, eds., *Das Ende der Parteien 1933* (Duesseldorf 1960), 9.

(DNVP) and Prussian conservatism; *Grossbuergertum* (DVP) and capitalism; Catholics (Z) and Catholicism. German parties had traditionally been charged with the exposition of a philosophy.[45] It perhaps strikes us as strange that their contemporaries exhorted them to become more rather than less philosophical.[46] Nearly every party commentator decried the increasing ties of parties with particular interests and their decreasing emphasis on philosophy.

As one might expect, there are significant differences between interest-group activities in Weimar and Bonn. Not only were more groups probably active in Weimar, where German society was more highly stratified and heterogeneous, but also relatively more interest groups became politicized as pressure groups. It was much more important, as we have seen, that each group pursue its own interests politically because the general policy machinery had been incapacitated. This urgency also affected the tone of pressure-group activity: it made the groups more dogmatic and adamant in their demands. Although pressure groups had focused their activities on the Reichstag at the outset simply because they were so closely identified with the parties, they steadily shifted the center of their activities to the executive and the administration. In Bonn activity tends to be directed at the party hierarchies[47] and the cabinet or shadow cabinet, rather than at Bundestag deputies and the administrative bureaus.[48]

The primary distinction, however, in party and pressure-group activity lies between the *parties* in Weimar and Bonn. Weimar parties could not seek out a common denominator among a broad range of interests as their successors in Bonn do. But they did formulate policies for the narrow range of interests they represented. The DNVP, for example, had a tariff policy and an agricultural policy—policies geared, in large part, to Junker interests and derived from a conservative philosophy. The Social Democrats also had a policy regarding tariffs and agriculture, but it was based on the interests of labor and deduced from Marxism. The Center party also had a policy on tariffs and agriculture—a Catholic one.[49] In other words, it was not the absence of policy or program ideas, but their ideological tone, their irreconcilability, their mutual exclusiveness that differentiated them from the programs of the parties in Bonn. Thus, in differentiating parties and pressure groups in two-party and multiparty systems, it seems more appropriate to distinguish between systems by the way in which policy is formulated, rather than by the presence or absence of programs.

Parties in modern states also serve as the channel for the recruitment of political

[45] Sigmund Neumann, *Die deutschen Parteien* (Berlin 1932), 10–11; Gerhard Ritter, "Das politische Parteiwesen in Deutschland," in his *Lebendige Vergangenheit* (Munich 1958), 63.

[46] Emil Lederer, "Das oekonomische Element und die politische Idee im modernen Parteiwesen," *Zeitschrift fuer Politik*, v (1912); Fritz van Calker, *Wesen und Sinn der politischen Parteien* (Tuebingen 1928), 18.

[47] Breitling, 91; Karl W. Deutsch and Lewis J. Edinger, *Germany Rejoins the Powers* (Stanford 1959), 90; Otto Stammer, "Interessenverbaende und Parteien," *Koelner Zeitschrift fuer Soziologie und Sozialpsychologie*, IX (1957), 595.

[48] See especially Eschenburg.

[49] Wilhelm Mommsen and Guenther Franz, *Die deutschen Parteiprogramme, 1918–1930* (Leipzig 1931).

leaders, and the differences between Weimar and Bonn in this regard are similarly striking. Basically, the same kind of distinction has to be drawn—namely, between leadership of a narrow sector of society and leadership of a nation. Immediately apparent and most significant is that although the two-party system of Bonn might not be producing leaders of a different quality,[50] it has enabled the leaders whom it has chosen to lead. Unlike the Weimar party system, it has given them an opportunity to develop their leadership potential. Bauer, Wirth, Mueller, Cuno, Luther, even Stresemann—no Weimar chancellor approached the nearly universal respect and adulation offered an Adenauer. And although the quality of the Weimar chancellors (except for Stresemann) was not exceptional, the party system in Weimar tended to exaggerate their very mediocrity. The party system in Bonn bolsters and reinforces the man. Leadership in Weimar was leadership in spite of the party system, while in Bonn it is leadership because of and through the party system.

VII

That parties can form coalitions of broad general interests in society has a very important impact upon the way they function. What was it that kept them from combining interests in Weimar? Surely this is the question, rather than why did they not *want* to embrace a diversity of social interests. It was to their own advantage to have as much and as wide an appeal as possible if they wished to influence governmental policies. (We assume that at least some of them did seek to exercise power.) Why was it labor *or* management, labor *or* farmer, Catholic *or* Protestant, Bavarian *or* Prussian, and not a union of these elements? To exercise power in Bonn, parties have had to offer themselves as coalitions which cut across differing strata of the social structure, and they have succeeded in doing so. Why not the parties in Weimar?

One possible answer, it seems to me, lies in the political orientations of the voters. People in Weimar viewed their interests as being mutually exclusive and irreconcilable. They would not have voted for a party leader or a party that had tried to form as broad a social coalition as possible. In fact, they did not. The DNVP sincerely tried to broadcast an image of itself as a party of people;[51] many parties did. Witness, for example, how many times one encounters the word *Volk* in the names of Weimar parties. The DNVP, however, made only nominal gains in the long run; and made them, in large part, because it was the successor to the earlier party of Junkers. In the popular image it remained a party of East Elbian landowners and large industry. Much the same could be said for the other major parties in Weimar—the Social Democrats and the German Populists, the Centrists and the German Democrats.

Voters in Weimar perceived their interests as worker or bourgeois, businessman or farmer, Catholic or Protestant, as being mutually exclusive. They tended to look

---

[50]For a comparison of elites, including political elites, in Weimar with those in Bonn, see Lewis J. Edinger, "Post-totalitarian Leadership: Elites in the German Federal Republic," *American Political Science Review*, LIV (March 1960).

[51]Liebe, 13.

upon the political process, in turn, as a game of winner take all. How could one party (like the CDU today) possibly pretend to represent such diverse and disparate interests? On the contrary, because it was not limited by cross-membership or competing and conflicting interests, each party could and was expected to make bold specific demands. As a voter in such a system, however, to choose the party most specifically geared to one's particular status and outlook was to run the danger of wasting one's vote. And in fact hundreds of thousands of votes were lost at every election by the failures of parties to attain the necessary 60,000 vote minimum;[52] still, the number of parties never decreased. "Better nothing at all than a compromise solution" is what these voters were in effect saying.

What holds for individuals holds *a fortiori* for associations in Weimar. In the first place, interest groups everywhere are organized around specific interests. Within them there is, relative to the compromise that has to occur among interests in society, little need for compromise. But second, there were in Weimar practically no attitude groups of a kind similar to our broadly based civic associations where people with potentially different interests could meet. As much as possible, conflicting interests had been organizationally segregated at all levels; neither the individual as an individual nor as a member of a group had to learn to reconcile major differences. Any political compromise that occurred had to occur in the Reichstag, but no one, least of all the *party* deputies themselves, had been prepared by society for the task they faced.

This process of segregation within society culminated in the democratic parties' attempts to encompass all areas of an individual's life. By 1933 affiliation with a particular party tended to absorb the whole man in all his capacities.[53] It was within his party citadel that he and his family carried on many of their activities, and it was from behind a party framework that he looked out upon the world. From the party organizations for young people,[54] to the variety of party activities designed for the wife and mother, to his own participation in the party's non-political organizations, he was first, last, and always committed to a particular party.[55]

[52]In 1924, for example, the number of votes so lost was 710,000 and in 1928 it was 1,320,000 (James K. Pollock, Jr., "The German Elections of 1928," *American Political Science Review*, XXII [August 1928], 700).

[53]This process was only a part, but nevertheless a significant part, of the tendency throughout the Republic for more and more spheres of life and more and more people to be politicized—art, theater, drama, the very old, and the very young. No activity and no person was immune from politicalization. See among others Roger H. Wells, *German Cities* (Princeton 1932), 93: "That city government is far more politicalized now than before the Revolution is generally admitted. . . ." Also Felix Bertaux, *A Panorama of German Literature* (New York 1935), 216, 277.

[54]"In almost every school class there are, five years after the revolution, even political parties among the children!" (Wilhelm Flitner, "Der Krieg und die Jugend," in Otto Baumgarten, *et al., Geistige und sittliche Wirkungen des Krieges in Deutschland* [Stuttgart 1927], 304).

[55]The DNVP, for example, had a *Bismarckbund* for youths of 18 and above; in each of its local organizations it had special committees to deal with men and women in different walks of life; and for the men, it had its own *Kampfbund* after 1930. Until then, the party had relied upon the closely akin *Stahlhelm*. Also very important were the party newspapers and motion pictures that were at the disposal of the DNVP through the Hugenberg combine. Hugenburg became the party leader in 1928.

The formation of party battle leagues (*Kampfbuende*) was the last stage in this process.[56] In organizing the *Reichsbanner Schwarz-Rot-Gold* in 1924, the parties of the Weimar Coalition were seeking to resist the increasing popularity of associaions of the Free Corps type such as the (*schwarz-weiss-rot*) *Stahlhelm*. But in the *Reichsbanner*, as in the *Kampfstaffeln* of the DNVP, the *Rote Frontkaempferbund* of the Communists, and the *Sturmabteilung* of the Nazis, the goal was one and the same: political quarantine, complete integration of the individual under the party banner, and his isolation as much as possible from other "contaminating" influences. By the end of the Republic, the parties—strictly speaking—were no longer parties; they had become movements. Society itself had been divided into a multiplicity of self-contained, non-communicating armed hierarchies. The attempt to reach rational settlements in the Reichstag had failed; differences would now be settled in the streets.

In Bonn this sense of urgency, of life and death, surrounding party and pressure-group activities is fortunately missing. And it is missing, above all, because the government has been able to legislate. The integration of millions of refugees (more than one of every four West Germans today is a refugee) is perhaps the Republic's most outstanding domestic achievement, all the more so when one tries to imagine what such a burden would have done to Weimar. The success in that area, however, is only part of an extensive program of social welfare all of which rests upon the remarkable economic recovery. There is conflict and competition among pressure groups in Bonn—a great deal of it. But in contrast to Weimar, "win a little here, lose a little there" seems to be a more characteristic attitude. A greater margin of error has been permissible.

VIII

If the differences between a two-party and a highly fragmented multiparty system are so great, the final and most important question we can ask about party and pressure-group politics is why a system tends toward one pattern or the other.[57] One possible condition for a multiparty system has already been suggested—namely, exclusive and irreconcilable voter orientations. Another way of saying much the same thing is that such a system is marked by an absence of political consensus. It was, however, not merely an absence of agreement on fundamentals that characterized Weimar Germany; it was much more intense disharmony and disagreement.[58] Certainly no political institution—most definitely not parliament—had the universal

---

[56] See Ernst H. Posse, *Die politischen Kampfbuende Deutschlands* (2nd ed., Berlin 1931); Robert G. L. Waite, *Vanguard of Nazism* (Cambridge, Mass., 1952).

[57] Although it is not of major significance, one cannot discount completely the impact of the electoral laws upon the party system. Obviously, electoral laws have made it more difficult for minor parties to compete in Bonn. But as we saw above, the 60,000 vote minimum did not discourage minor parties in Weimar. For a description of the electoral laws in Bonn, see Kitzinger, ch. 2. For a discussion of the impact of electoral laws in general upon a party system, see Harry Eckstein and David E. Apter, eds., *Comparative Politics* (Glencoe, Ill., 1963), Part IV.

[58] Neumann speaks of the lack of a "*Wir-Bewusstsein*" and "insufficient social homogeneity" in explanation of the party crisis (p. 106).

support of the German people. To the forces on the right, the Reichstag epitomized a system of *Kuhhandel* politics—i.e., of politics in which the interests of the state are bartered back and forth like cattle.[59] On the other hand, the President, particularly Hindenburg, and the bureaucracy were distrusted as being partial to Junkerdom.

The first German Republic followed immediately upon the most rapid period of industrialization that any nation had ever experienced, and the processes of industrialization continued throughout the Republic. Intense rationalization of business and industry, for example, characterized the period between 1924 and 1929.[60] As a consequence of industrialization, all Germany's social patterns were in rapid flux; and, to make matters worse, there were the turmoil and losses of war and catastrophic inflation. Confusion and disagreement regarding social and political norms were bound to be widespread.[61] Moreover, Germany had shifted overnight from the most absolutist Western regimes to the most radically democratic. The Weimar Constitution was hailed in its day as the most nearly perfect of democratic constitutions. It featured democratic devices from many states of the Western world, including the United States, Switzerland, Belgium, and Great Britain. Formally it was the most direct and popular democracy one could imagine in a state the size of Germany. Its very radicality, however, against a background of centuries of absolutism meant that large numbers of people would be alienated from it.

If a fragmented party system is characterized by extreme disunity, one might assume that Germany has experienced a decrease in political disunity since the fall of the first Republic. The evidence is necessarily indirect, but it comes from several sources. Most important is the overall decrease in social differences.[62] The middle class has shown a marked increase in size since Weimar, and class lines are no longer as clearly distinguishable. Otto Kirchheimer, for instance, has suggested that "Workers in general have ceased to consider themselves a specific class with a specific mission."[63] Class differences continue to exist; to surmount them, however, is no longer as difficult as it was in Weimar. West Germany, that is to say, has displayed a high rate of social mobility.[64]

Not only have differences among classes diminished and relaxed, but also the differences within the family units of these classes. "The social events of the war and postwar periods," writes the German sociologist, Helmut Schelsky, "have called

---

[59] Carl Schmitt, *Staat, Bewegung, Volk* (Hamburg 1933), 23–27.
[60] Robert A. Brady, *The Rationalization Movement in German Industry* (Berkeley 1933).
[61] See Hermann Hass, *Sitte und Kultur im Nachkriegsdeutschland* (Hamburg 1932), 141; Georg Steinhausen, *Deutsche Geistes und Kulturgeschichte* (Halle 1931), esp. 11.
[62] "A rise in the social scale at one end and a fall at the other have produced one relatively equal and uniform social class [in West Germany]" (Helmut Schelsky, "Elements of Social Stability," *German Social Science Digest* [Hamburg 1955], 115). See also Ralf Dahrendorf, "Recent Changes in the Class Structure of European Societies," *Daedalus*, XLIII (Winter 1964).
[63] "West German Trade Unions," *World Politics*, VIII (July 1956), 507.
[64] Morris Janowitz, "Social Stratification and Mobility in West Germany," *American Journal of Sociology*, LXIV (1958), 10–11. He concludes that "The weight of the evidence rests on the side of the conclusion that the consequences of social stratification and social mobility are now operating to decrease traditional class-consciousness and to increase social consensus concerning internal matters."

forth extraordinarily important changes in the situation and structure of the German family."[65] In general, the changes he refers to have been toward a partnership with equality of rights within the family.[66] Moreover, the framework of government in Bonn is not—as was Weimar's—radically democratic; by that token, it is more consonant both with German political traditions and family authority patterns.

If there is indeed less political discord in Bonn, the style of politics might also be expected to reflect that change. Voters and politicians alike are probably less ideological and dogmatic, more pragmatic and concrete in their demands. "Germans tend to be satisfied with the performance of their government," Almond and Verba found, "but to lack a more general attachment to the system on the symbolic level. Theirs is a highly pragmatic—probably overpragmatic—orientation to the political system. . . ."[67] In the same vein, Schelsky called the postwar generation *Die skeptische Generation* because German youth are so skeptical of ideologies and so heavily oriented toward results. Extremist parties, right and left, have fared very poorly in Bonn at the hands of both the voters and the courts. The Social Democrats have rid themselves of their Marxist ideology[68] and their open hostility to the Church. There has been a marked decline in pitched political battles, political murders, and political activity in the streets; party battle leagues are unknown. Numerically, Protestants and Catholics are more evenly matched. The restrained wage demands of the German worker in the immediate postwar world and the absence of political strikes are further indices of the magnitude of the change since Weimar.[69] Bonn, in fact, has been a paragon of orderly and moderate political activity. And if general political system affect is low, as Almond and Verba note, agreement on the goals of the system is overwhelming.

One could hardly expect a high degree of system identification in Bonn, considering, first, its brief history; second, its indefinite nature until the framework for a united Germany is settled; and, third, its origins during a period of occupation and uncertainty about the future of Germany. But in the short run, general system affect would not seem to be as important for party and pressure group as the decrease in disunity. Despite the absence of political affect, political parties in Bonn have been able to form coalitions of diverse social groups simply because the differences among groups are no longer so deep and divisive. Moreover, that the demands from the system are so concrete suggests that the parties can combine and reconcile them with other competing interests—something the parties of Weimar were unable to do with ideologies.

Much has been written about the "miracle" of German recovery from World War II. For a relatively brief period in Weimar—for the five years between 1924 and 1929, to be exact—the German nation also enjoyed the fruits of an economic "miracle." In those five short years Germany managed to recover from a condition

[65] Helmut Schelsky, *Die skeptische Generation* (Duesseldorf 1957), 127.
[66] *Ibid.*, 148.
[67] *The Civic Culture* (Princeton 1963), 429.
[68] See the *Grundsatzprogramm der Sozialdemokratischen Partei Deutschlands* (Bonn 1959), esp. 13–14. This is the so-called Godesburger Program.
[69] Henry C. Wallich, *Mainsprings of the German Revival* (New Haven 1955), 11–12.

which experts believed would cripple the country for half a century. Yet the differences between the two periods of recovery are as night and day. For the purposes at hand, the most important difference lies in the changed political orientations. Whether or not one happened to be a Marxist in Weimar, Marxism defined the framework for analysis. The political or economic process was looked upon as a struggle between workers and management in regard to mutually exclusive and necessarily hostile class interests, and it was a struggle to the end.[70] Relatively speaking, in Bonn it has not been a question of the improvement of conditions for the workers or of more profits for the entrepreneurs, but of mutual gain and satisfaction through increased productivity. Collective bargaining and codetermination have all but replaced armed struggles, political strikes, and lockouts.[71]

Why has there been such a widespread decline in social, political, and economic differences, in ideological differences in the broadest sense? What brought about these dramatic changes? Most important, I believe, was the reign of the Third Reich itself. National Socialism came into power as a social revolution,[72] and it instituted policies that brought far-reaching changes in the German social structure. It sought to destroy all traditional social differences and to reorient a totally coordinated society around the *common* "ties of blood." The rise with Nazism of a new class of leaders predominantly from the lower middle class,[73] the release under totalitarianism of the individual from all traditional ties, and his subsequent incorporation into the new mass movement all led to a lessening of social extremes. Burckhardt talks of a citizenry's being equalized under dictatorship; how much more equal men can be made under totalitarianism! But in addition one cannot discount the impact of the war: loss of the territory east of the Elbe which was the seat of Junkerdom; exodus from the east and extensive internal migration; devastation, turmoil, and poverty in postwar Germany—all these also operated to lessen the divisions and differences within the society.

One relatively large segment of the West German population still feels alienated by increasing industrialization and the course of postwar politics: the German farmer. He is particularly uneasy about the pressure from Common Market developments. Alone among major economic groups in Bonn today, the farmer continues to retain some vestiges of the political radicalism of Weimar. His marches on Bonn recall the activities of the so-called "Green Front"[74] after 1928—except that today he marches with a tractor. There also exists a striking disparity between the farmers' high percentage of voter turnout, on the one hand, and their general lack of political awareness, on the other hand, as compared with all other economic and social groups.[75] According to Reigrotzki, "The independent farmers stand behind every other group in their intellectual participation in politics not only with

[70] See Bracher, chap. 8.
[71] Theodor Geiger speaks of this as the "institutionalization of class antagonism" (*Die Klassengesellschaft im Schmeltztiegel* [Cologne 1949], 184).
[72] Hermann Rauschning, *The Revolution of Nihilism* (New York 1939), esp. 87-88.
[73] Daniel Lerner, *The Nazi Elite* (Stanford 1951).
[74] Erwin Topf, *Die Gruene Front* (Berlin 1933).
[75] Reigrotzki, 103.

respect to the depth of their political awareness, but also with respect to their preparedness to make a political decision."[76] The independent farmers, he suggests, appear to be part of the group with the "highest opinion rigidity."[77] All observers agree that the farmers are among the most powerful groups in Bonn today, if not *the* most powerful.[78] The future of the Bonn Republic will depend upon its ability to enter this problem area boldly and with imagination.

[76]*Ibid.*, 242.
[77]*Ibid.*, 105.
[78]In addition to Pritzkoleit cited above, see Eschenburg, 65–66.

CHAPTER 18

The New Germanies:
Restoration, Revolution, Reconstruction

RALF DAHRENDORF

It is often taken for granted that the German history of the past hundred years is a sequence of discontinuities. To look for consistency in modern Germany history, let alone a guiding principle, seems hopeless in view of the extreme changes in her political make-up. But it is always risky to confuse the letter of a Constitution with "legal reality," to confuse political institutions with the social structure on which they are based. For, on closer inspection, the changes in German *society* in the course of the last century—as against Germany's political institutions—do indeed reveal a consistent and recognisable principle. The history of German society since the mid-19th century can be understood as a painful—sometimes slow, sometimes more rapid—but persistent movement towards modernity and away from the traditional authoritarian structures of pre-Enlightenment and pre-industrial days. Kant's famous definition of the Enlightenment—"the German's escape from his self-caused backwardness"—might not be a bad description of German social development over the last century, were it not that a neat formula of this kind extenuates the suffering, the violence, and the terror that were involved.

To-day German social historians tend to agree with Thorstein Veblen's thesis (put forward as early as 1915) that modern German society must be understood in terms of the peculiar form the process of industrialisation assumed in Germany. Whereas in Britain and France the industrial revolution was accompanied, or even preceded, by a bourgeois revolution, Germany's famous "revolution from within"

Reprinted from Ralf Dahrendorf, "The New Germanies: Restoration, Revolution, Reconstruction," ENCOUNTER, *XXII, No. 4 (April, 1964), 50–52, 54–58, by permission of the author.*

(as Heine was the first to call it)—*i. e.*, German Idealism—proved a poor substitute for the power-claims successfully advanced elsewhere by a self-confident bourgeoisie. Germany's rapid industrialization—in three waves during the 1850s, 1870s, and 1890s—did not lead to the social and political hegemony of a new enterpreneurial class. Instead, an older ruling class—largely Prussian, consisting of aristocratic civil servants, officers, diplomats and landowners—strengthened its position. There were economic reasons for this—*e.g.* the development of agricultural prices—but the causes of this surprising development were, above all, social. Members of the new entrepreneurial class were less interested in being represented in Parliament than in receiving the honorary title of *Leutnant* on their 60th birthdays—in being able to document, with a *"von"* before their name, that they had found recognition in the eyes of the old upper class. It is typical of this development that the science of Political Economy was transformed, in Germany, into something called *Nationalökonomie*, and that German liberals became known as National Liberals. If by "capitalism" we understand the social and political predominance of the private entrepreneur, industrial Germany in the late 19th century was not capitalist, but presented rather the paradox of a fuedal-industrial society.

This odd combination of modernity in economic affairs and backwardness in social affairs—unique at the time—had important political effects. Representative government was the indispensable instrument for a bourgeoisie advancing a claim to power. Only by equal representation could it hope to make its voice heard. Equal representation, and the counterplay of government and opposition, were therefore the formal expression of its will-to-power. But the German bourgeoisie did not advance a claim to political power. Rather, it permitted the authoritarian state to survive industrialisation: a state resting on the assumption that certain individuals, by virtue of very special insight, and guided by the "well-understood interests" of their subjects, are called upon to make all the political decisions. Not without reason has this state been described as "paternalistic": as in the Wilhelminian family, the authority of the father, harsh in punishment but genuinely concerned with the welfare of his subjects, was all-pervading.

Still, whatever the merits of a sketch of this kind, it is bound to mislead. It misconstrues reality, partly by over-simplifying it, but even more because it reduces a dynamic historical process to a static picture. In point of fact, this strange combination of an industrial economy with feudal values and authoritarian structures did not go unchallenged even in Imperial Germany. But the process of dismantling this combination and replacing "traditional" by "rational" values, authoritarian by modern institutions, remained extremely slow. The attempt, in 1918–19, to catch up with the bourgeois revolution after the completion of industrialisation (and to combine it with the anticipated proletarian revolution) failed because the old ruling class and its values, though damaged, had been by no means destroyed by the "revolution of 1918." As soon as the Weimar Republic found itself in trouble—whether in the field of foreign affairs, economic policy, or internal order—there arose a call for the "proven" authoritarian leaders of the past. The Hindenburg elections, government by emergency decree, the coalition of Hitler's first cabinet—all bear witness to this tendency. It is a commonplace that Hitler, too, was carried to power on the waves of this prevailing anti-parliamentarianism.

But it was, and is, an error to identify anti-parliamentarianism with National Socialism. Hitler and his party did not represent the Prusso-German tradition of the authoritarian welfare state. National Socialism marks the entry into German history of "modern politics"—in its ugliest, totalitarian form. In their impact on German society, one might say, the Nazis were the executors of the German revolution that never took place. As so often, the horrors of revolution grew with belatedness. Still, National Socialism completed for Germany the break with those traditional authoritarian values, institutions, and leaders which had characterised German society throughout the period of industrialisation. The most obvious evidence of the revolutionary impact of the Nazi régime is the near-total destruction of the old Prusso-German upper class—through the rise of a party élite of "new men"; through the disappearance of Prussia as a political and administrative entity; through the death of almost an entire generation of young officers from old aristocratic families in the campaigns of the first years of the war; through the systematic liquidation of the old German upper class after the revolt of July 20th, 1944; and finally through the destruction of the economic basis of the old élite as a result of the division of Germany. All these events can be seen as stepping-stones in the painful journey of German society out of a traditional, authoritarian past into modern political reality.

The contradiction, then, between a modern industrial economy and a pre-industrial—or, at least, pre-capitalist—structure of society, dominated some seven decades of German history.

Not until 1945 was the contradiction fully resolved (if one can say this without cynicism), in the *tabula rasa* situation of a country whose economic capacity had fallen back to an almost pre-industrial state, at precisely the moment its traditional economic and social structure had lost its basis. Thus, German social development in the post-war era was fraught with several alternative possibilities—the only possibility that was ruled out was a reconstruction of that authoritarian tradition which, even in the Weimar Republic, was still the strength as well as the weakness of the country. One often hears to-day, in East and West Germany, references to "authoritarian" traits in politics and society. But if the term is used at all precisely, it can refer only to fast-vanishing remnants of a lost past. German social development in the postwar era is in no sense a "restoration": it is a road to new destinations.

For subsequent generations, the most important change may well seem one which has little to do with these social developments—I mean, of course, the division of Germany. And, beyond that, the fact that while both halves have turned away from the older traditions, they have turned in radically different directions. In a sense, of course, a unitary German society within set territorial boundaries is itself less than a century old. But the divergences in the recent development of East and West Germany are starker even than were the divergences between the German states before the creation of the Empire in 1871. Despite the official hostility, one cannot overlook a certain inner convergence of these German states in the 18th and 19th centuries. But to-day, official hostility is reflected in a real divergence of social development. Before 1871, there were several German states—but one German society. To-day we may still insist that there are not two German states. But even the most enthusiastic advocates of reunification would find it difficult to deny the existence of two German societies.

In 1945, both halves of Germany faced much the same challenge. The economy was in desperate straits; no able and self-confident group of leaders was in sight; all traditional rules of social and political behavior had lost their validity. Yet, from the first day of reconstruction, the response to the challenge differed radically in the two halves of Germany. There were, to be sure, differences between the Western zones of occupation as well—some of which are noticeable, even to-day, in the appearance of cities, in the peculiar features of institutions such as the press or local government. But these differences are as nothing compared to the profound contrast between the Eastern and the Western Zones. What started out as the influence of the occupying powers soon merged with more or less indigenous forces and traditions, as the Western Zones grew into the German Federal Republic, the Eastern Zone into the German Democratic Republic. Whatever one may think of the legitimacy of the two governments—and although no democrat will hesitate to choose the Federal Republic —it becomes ever harder to deny the reality of these two political entities and their social bases.

These are historical events of some consequence, concerning even those not directly involved in them. The question is: whether the texture of German society before 1945 has proved strong enough to impose its unifying tradition on a divided Germany, or whether external force can so determine the course of history that a society grown together over many decades can be torn apart within fifteen years. Nor can we shirk the question whether German reunification is not becoming an increasingly remote hope as the two halves stabilise their differences. Whatever the answers to these burning (and rarely discussed) questions, the fact remains that since 1945 the two halves of Germany have grown apart socially as well as politically. For both, Germany's authoritarian tradition came to an end with the National Socialist revolution. In the East, the resulting vacuum was filled by totalitarian institutions and the new social structure supporting them. Whereas the West, with its new social and political institutions, launched the first successful experiment in liberal representative government in German history.

Even before the Berlin Wall, it was not enough to judge East German developments merely from the fluctuating refugee statistics. Of course, these figures did register changes in political pressure, and pointed to the basic lack of legitimacy of the current political order. Nevertheless, the fact is that most of those whose home was in East Germany are still living there, and would still be living there even if they had a chance to leave. Obviously, people's unreadiness to leave does not necessarily imply recognition of a régime. But, at least in its consequences, the absence of active protest is a kind of acceptance. And in this sense—not to speak of the many whose career depends on the present political set-up in the East—the political and social development of East Germany since the war has a more than ephemeral reality.

The most important single feature of East German social development has been the emergence, or rather the systematic creation, of a new and homogeneous upper class. In amusing contrast with their ruling Marxist ideology, the Soviet occupation forces and their German satellites have reconstructed East German society from its "superstructure" rather than from its "real basis." For, however loath one is to explain historical events in conspiratorial terms, there can be little doubt that this process was the result of deliberate planning. As early as May, 1945, the Soviet authorities began to build up a new political and economic élite. The leaders of the four original parties (Communists—KPD, Social Democrats—SPD, Christian Democrats—CDU; Liberal Democrats—LDP) differed considerably in social background and political orientation. In the beginning there was, moreover, an administrative élite of experts; and many sections of society were not immediately affected by the changes at all. But within a year, in the spring of 1946, the second stage in the formation of a closed political élite had begun with the systematic exclusion of "difficult" groups and invididuals. There were early indications of this trend in the so-called "Anti-fascist Front of Democratic Unity." There was the compulsory fusion of the Communist and Social Democratic parties into the Socialist Unity Party (SED) in 1946. A little later came the systematic subversion of the so-called bourgeois parties (CDU and LDP), as well as the creation of a number of satellite parties of the SED (the National Democratic Party, the Peasants' Union, etc.). Since about 1948, East Germany has possessed an increasingly homogeneous political class, fundamentally united in outlook, despite the surface appearance of organisational variety.

A similar process—if with a different intention—was very nearly brought about by the Nazis during the twelve years of their rule. What is new in East Germany, and perhaps of lasting importance, is the next step in the creation of a ruling class. Since 1948, if not earlier, there has been a clearly recognisable attempt to combine all the élite of society—politicians, the military, lawyers, professors, economic leaders, and artists (and, in the end, church leaders as well)—into a uniform "new class." This has not proved equally easy in all cases; even to-day the process is by no means complete. But we have to recognize that the creation of a homogeneous ruling class in East Germany has been largely realised. Nationalisation of all private firms has turned managers into state functionaries. In the reconstruction of the Army, the officer corps proper had almost no say (though here, too, certain traditions of the German army are likely to impede its complete *Gleichschaltung*). The abolition of most of the university faculties of law, and the invention of the "People's Judge," meant a break with one of the central institutions of German tradition (though the common historical background of the legal systems remains a unifying factor even to-day). The complete switch-over of education to a system of "polytechnical education," and the privileged position of real and so-called "workers and peasants," as well as the general indoctrination with Marxism-Leninism, has already made East and West German degrees and diplomas almost impossible to compare (though in some fields a common scholarly tradition remains alive). The only structure which has resisted many attempts at political domination, and has still not been completely subdued, is the Lutheran church.

It is perhaps necessary to emphasise that these trends are not superficial changes that can be reversed in a year or two. The man who has become a People's Judge by virtue of a diploma acquired from the new Academy of Administration; the Manager of the large State enterprise; the General of the People's Army; the Dean of the Workers' and Peasants' Faculty; these are not simply imposed office-holders, but new social dignitaries whose very existence is bound up with the new society. It is probably safe to assume that one-tenth of the population of East Germany is personally tied to the régime in this way. And it is clear that a homogeneous upper class of this kind, interchangeable in its membership, perpetually refreshed but not modified by planned social mobility, is a powerful basis for totalitarian political institutions and contributes to their stability.

There is much else to support the thesis that East German totalitarianism has acquired a social basis. All analysts of National Socialism have noted the connection between totalitarian institutions and efforts to cater for, and eventually win over, the youth of a country. Those who have reason to be worried about their legitimacy need to concern themselves, above all, with youth. For youth is not only "the future," it is (as we know from student risings in many countries, and specifically from the Hungarian revolution of 1956) also a potential threat. There are many indications of this concern of East Germany's leaders with youth—a concern which often takes the form of giving young people highly responsible positions in order to tie them to the existing order. Despite the high number of young people among the refugees from East Germany, not all of these measures are likely to have been successful.

Regarding this stabilisation of totalitarian political institutions, I should mention a further point. The ideology of Marxism-Leninism in itself is hardly more than a toy for certain party theoreticians. There is, however, a notion, widespread not only in East Germany but also among refugees in the West, according to which "the whole" is more important than its parts, and the welfare of "society" and "the state" takes precedence over the welfare of the individual. People who have turned their backs on the DDR in disgust and despair often remark after a few weeks in the West, "there is one point in favour of the East—there 'the whole' still means something! In the West everybody just follows his own wishes. . . ." Precisely because such remarks—unintentionally, to be sure—cast doubt on the very basis of Western freedom, they show the remarkable impact of certain value-patterns in the Communist East: value-patterns which are more in keeping with totalitarian institutions than many of their adherents realise. For the totalitarian state rests much of its authority on this permanent emphasis on the precedence of "the whole" —which in effect means the new ruling class—over the individual.

It is useless to indulge in speculation about how long the present régime in East Germany can survive. For years there have been indications of internal instability. The new ruling class may be homogeneous; but it is also very mediocre. Occasional waves of terror may succeed in maintaining totalitarian institutions; but they do not enhance the legitimacy of the régime. In any case, belief in the precedence of "the whole" loses much of its strength if Big Brother controls every aspect of life too noticeably. And yet, more has happened East of the Elbe since 1945 than the imposition of a rule of terror based on Soviet bayonets. To-day, East German political institutions have a social substructure which contributes to their maintenance. Any liberalisation of East Germany would require not only the abolition of the régime, but also a fresh social revolution. The fact that there have been profound changes in East German society provides its rulers with their only reasonably reliable guarantee of stability. But it also presents Germans with an intractable problem where reunification is concerned.

A man from Mars, standing for half an hour in the city centres of Leipzig and Frankfurt-on-Main, would hardly guess that both cities were part of the same country less than twenty years before. In fact, of course, the most striking difference between the two cities—the grey austerity of Leipzig, and the conspicuous wealth of Frankfurt—is a significant pointer to West German development since the war. The most frequently repeated cliché about West Germany—the "economic miracle"—is indeed, properly understood, her most surprising and most characteristic feature.

In the first place, the mere extent of West Germany's economic development since the war is astonishing. Germany's economy (much like that of Britain and the United States) struggled desperately throughout the inter-war period to return to the level of 1913. That level has been left behind. Indices of production are not always satisfactory measures of industrial development, but in this case the total index does hint at the truth.

If we take the total industrial production of Germany in 1913 as 100, production had sunk to 38 by 1919. The temporary recovery up to 1929 (103) was soon undone during the Depression (1932: 60). It was not until 1939 that the level of 1913 was reached and surpassed (126). In so far as it is possible to produce an index for 1945–46, 40 per cent of 1913 seems too high rather than too low an estimate. West German industrial production to-day, on the other hand, with an index of considerably above 300 (1963), and still growing in volume, bears witness to an unprecedented economic explosion for the post-war years.

Germany has shared with other Western countries, of course, this transition to mass production and affluence. More important, in our context, is a second feature, an aspect of the economic miracle which is peculiar to Germany and which, in terms of German history, probably constitutes the real German miracle. I have referred to the peculiarities of industrialisation in Germany. Though largely based on private capital, industrialisation in Germany was "national-liberal," that is, national-conservative in character. But in 1945—or, more precisely, in 1948—after the cruel destruction of the old Prusso-German ruling group, Germany had the unique opportunity of repeating the process of industrialisation from its beginnings. And then something totally unexpected happened. This second industrialisation took place in a strictly liberal-capitalist spirit, so that (if this does not sound paradoxical) the economy and society of post-war Germany provide the first example of true capitalist organisation in all German history.

It is hard to recall, after 15 years, just how surprising this development was. Even now, one is tempted to say that the liberal experiment of 1948-9 went against all economic rationality. On the eve of the currency reform, laymen and experts were convinced that gradual, limited liberalisation was the only path away from rationing, economic controls, and planning to a more prosperous society. Yet they were quite wrong. Joseph Schumpeter had predicted for the economy—and Karl Mannheim for the whole of society—the inevitability of increasing state intervention. No wonder people in all political parties were convinced, in 1949, of the necessity of far-reaching economic planning. To-day, we are faced with the remarkable fact that public property, in the second half of the 20th century, is being turned back to private hands. There is no doubt that entrepreneurial success in post-war Germany was, to some extent, a result of the abstinence of the state, especially in tax matters—and is therefore not entirely due to the spontaneous initiative of the new middle class. Still, the fact remains that vigorous private initiative—or, what is merely another way of saying the same thing: an unbounded individual desire for gain—was the motive force of the economic recovery of Western Germany. Of course, there can no longer be, in our time, a "pure" capitalism, such as existed in the early period of English and American industrialisation. But, as far as Germany is concerned, she has never before come as close to capitalist patterns of social organization as in the years since 1948.

Many features of contemporary German society show the effects of this economic miracle. Like East Germany, West Germany had to develop a new class of leaders after the disappearance of the old upper class. It would be an exaggeration to say that this task had already been completed. Obviously, an unplanned social development proceeds somewhat more slowly than one which is conceived and controlled in detail. Here, one must agree with the recent comment of the Bonn correspondent of *The Times* that there are no real signs of the emergence of "Society" with a capital S in Bonn. But many holders of leading political and economic positions in the Federal Republic, and to a lesser extent in other spheres of society, *do* differ considerably from their predecessors. The most pronounced change is the much greater influence of economic leadership groups. The proportion of cabinet ministers and members of parliament who, by family or occupation, have connections with industry has risen very considerably. In the economic sphere itself, the urge to exert political pressure has grown stronger, now that even medium-scale entrepreneurs no longer have anybody "above them" to tell them what to do or not to do. Indeed, perhaps the very absence of "Society" is a sign of this change. A tendency towards *nouveau riche* behavior—a liking for medieval choir-stools in the dining room, for conspicuous consumption, a horrific lack of taste in matters of art and literature—is one of the unmistakable characteristics of this new society in West Germany.

These developments chiefly concern the top of the social structure. But a corresponding change in values can be observed throughout West German society. As against the heroic past, with its emphasis on *Gemeinschaft* and hard work, the whole of German society is to-day (to quote a phrase of Schumpeter's about capitalism) "bathed in an economic light." So striking is this new illumination that the foreign visitor, in search of the much-praised and much-abused "German national character" of the past, finds hardly a trace of it. The Germans of to-day are neither particularly industrious nor particularly enamoured of the military—neither particularly subservient nor particularly romantic. Probably the most crucial change in the social values of West Germans is the setting up of individual success and private pleasure as the twin guide-lines of behavior. What scornful critics denounce as "materialism" and "exaggerated individualism" is in fact a necessary correlate of any quasi-capitalist social order—the desire to increase individual happiness, coupled with a readiness to protest against any external intervention in one's own pattern of life (especially if it comes from the state).

The significance of these new developments is that they have reversed the historic trend of German society towards a monolithic type of social organisation. When a sample of people in the Federal Republic was asked recently which social groups they thought were "too powerful," opinion was almost equally divided between Industry, the Church, and the Military. This seems to indicate that there are several groups competing for influence, and that no single group is capable of assigning all others their place within the whole. The concept of "pluralism" has lost almost all its meaning by frequent and indiscriminate use. But if one means by it the existence of competing groups of more or less equal rank, West German society can be accurately described as "pluralistic." Pluralism is, of course, always precarious. At any given time, one or other competitor in a pluralistic system tends to have a slight advantage over the others. But what matters, and what is new in West Germany's development, is that no one social group holds the reins of power permanently in its hands. Society, therefore, remains sufficiently flexible to give everybody a chance to see his claims represented.

It is here that the political significance of the changes in the West German social structure becomes apparent. Now that quasi-capitalist patterns of social organisation have emerged, there is, for the first time in German history, hope that German society will provide a stable foundation for representative government. Representative institutions remain a dead letter if rigidities of the social structure effectively prevent any chance of a non-violent change. This was one of the main problems of the Weimar Republic. But the social changes since the war have undone many of these rigidities. An entrepreneurial ruling class is more likely to fight off state tutelage than a ruling class of civil servants, army officers, and landlords. The man who has grown accustomed to determining his own course of action, and who makes personal happiness the criterion of his actions, is not so easily persuaded to accept authoritarian or totalitarian rule. Where there is a plurality of groups competing for domination, each one has an interest in the maintenance of conditions of free and fair conflict. In this sense, the conclusion seems justified that the structural changes in West German society since the war hold out a better chance for representative government than ever before in German history.

It would be pleasant to conclude this analysis of some changes in the German social structure on an optimistic note. But this favourable prognosis for German democracy should be qualified by adding that it is intended in relative terms—that is, by comparison with earlier phases of German history. And since not even the most ardent admirer of Germany could describe our country as a natural breeding-ground of democracy, this conclusion may not amount to very much. Indeed, a number of reservations and threats to the development I have depicted are too clearly in evidence.

In the first place, the very combination of the quantitative and the qualitative economic miracle—the affluent society and liberal elements in the new Germany—raises serious questions. Are not these changes a mere consequence of economic prosperity? Are not the changes in West German society as superficial as those in East Germany? Might not a new economic crisis have consequences like those of the Depression of 1929? Although objections of this order are widespread (both inside and outside Germany), their basis in fact is by no means as certain as is often believed. Thus, the causal connection between the economic crisis and Hilter's rise to power is at best tenuous—especially in view of the fact that those most directly hit by the crisis, the unemployed, did *not* turn to the Nazis in significant numbers. Still, this cannot invalidate the objection that the real test for the stability of representative institutions in West Germany is yet to come, and that less favourable economic developments would be one of the conditions of this test.

Again, it is impossible to be dogmatic as to whether those elements in the German social structure, which in the past were such effective obstacles to democracy, have really been destroyed as effectively as my interpretation suggests. There are clearly many remnants of the past in German society: "compromised" individuals in prominent positions in politics, the army, the legal system and elsewhere. There are also extremist quasi-political groupings, especially in the wake of the refugee organizations—though these have mostly shrunk to mere cadres of functionaries. Perhaps more important, there are unchanged patterns in important fields such as education. No doubt, all these traces of Germany's authoritarian past are merely leftovers, and therefore less important than those who see West Germany only in their light so often assume. But one cannot deny that resistance to representative democratic government is present in many a dark corner of German society.

Conscious anti-democratic tendencies apart, the most profound resistance to the changes I have described is found in an attitude of mind the connection of which with political democracy may not be immediately evident. As against what Talmon has called "totalitarian democracy," representative government involves recognition of the necessary interplay of incompatible interests and ideas. Conflict, the antagonism of government and opposition—but also of interest groupings in all spheres of society—is the lifeblood of democracy. Those who reject conflict as such, therefore, go some way towards rejecting representative institutions. Yet this attitude is very common in Germany. In lieu of political parties fighting one another for power, people would prefer the seeming harmony of an all-party coalition. There is (at least among non-participants) a widespread desire to settle industrial strife by state intervention or by some utopian once-and-for-all solution. Similar reactions are to be found within the educational and legal communities, and within many private organisations. But nowhere is this *penchant* for settling problems "for good" by one decisive—and, in effect, authoritarian—decision so evident as in people's attitudes to social conflict.

I have said that pluralism—the full competition of large groupings and institutions—often rests on a precarious equilibrium. That is why it is difficult to allay fears that one of the competing institutions—industry, the church, the military—might become so powerful that it would use its predominance to destroy the competitive system itself. On the other hand, an attempt to analyse these fears more precisely ("the Federal Republic is a clerical state," "Big Business decides everything") tends to reveal that they are debating points rather than sober descriptions of social reality.

There is, however, one threat to the development of more liberal patterns in German society which might lead to the abolition of social pluralism and which ought not to be underestimated. This arises from the fact that social changes in East as well as West Germany have taken place in the context of a world conflict between East and West. It appears to be a general law that human groups react to external pressure by increased internal coherence. In the East-West conflict, each society finds itself in such a position of pressure from without. For totalitarian countries, this does not constitute any particular threat, since they are in any case based on a structure of total mobilisation. But external pressure may lead the liberal societies of the West to restrict internal liberties in the name of resistance to totalitarian pressure. There are some indications that both the United States and West Germany will be exposed to this danger if they refuse to recognise new political developments which are in fact reducing the pressure from without. Here is a new threat to liberal political and social structures, and one difficult to counter. The paradoxical possibility that democracy can be destroyed while it is being protected deserves the attention of all who are concerned for the survival of representative institutions.

There can be no doubt, of course, that the Federal Republic is superior to the German Democratic Republic in terms of the legitimacy of its political order. In West Germany, it is true, there could be a more active acceptance of the new political institutions—but then legitimacy is perhaps always a passive phenomenon of the absence of revolt. Nevertheless, it would be irresponsible of me to describe the changes in West German society as final, and representative institutions as secure.

The two halves of Germany reacted in very different ways to the *tabula rasa* of 1945. With some exaggeration, one might say that the one half has chosen the totalitarian, the other the representative face of the Janus-head of modern politics as its model. This ambiguous fact points to the chance of a reunification in the foreseeable future. It points equally to the possibility—is it hope or danger?—of an internal, perhaps violent, breakdown of the régime in the East, as well as the threat of a renewed defeat of representative democracy in the West.

CHAPTER 19

Three Constitutional Courts: A Comparison

TAYLOR COLE

Two years ago, when an astute critic made a half-century appraisal of comparative politics in the United States, he reminded us that the American Political Science Association was founded in 1903 as an outgrowth of moves to establish a National Conference on Comparative Legislation.[1] During the more than half-century that followed, the writings in comparative government and politics have reflected the influences which have made themselves felt in the discipline as a whole. The attention given by Charles E. Merriam after World War I to "informal government," "underlying processes and relations," and "social bases of political cohesion" is fully appreciated now by those who are projecting comparative studies of political socialization.[2] In the 1930s, Carl J. Friedrich's writings pointed up the need for more adequate conceptualization when combined with appropriate appreciation of empirical research.[3] Mention should also be made of the earlier works of Herman Finer.[4] In their respective ways, albeit in varying degrees, all of these writers recognized the need for an increased emphasis upon the informal and extra-legal

---

[1] Sigmund Neumann, "Comparative Politics: A Half-Century Appraisal," *Journal of Politics*, Vol. 19, pp. 369–90 (1957).
[2] Note particularly his *Making of Citizens* (Chicago, 1931), with the subtitle, "A Comparative Study of Methods of Civic Training," in which Merriam sought to summarize and provide a central interpretation for eight country studies in a series on civic training.
[3] See the introductory chapter of his *Constitutional Government and Politics* (Boston, 1937), subsequently published in revisions under the title of *Constitutional Government and Democracy*.
[4] Esp., *Theory and Practice of Modern Government* (2 vols., London, 1932).

Reprinted from Taylor Cole, "Three Constitutional Courts: A Comparison," THE AMERICAN POLITICAL SCIENCE REVIEW, LIII, No. 4 (December, 1959), 963–84, by permission of the author and the publisher.

factors affecting the political process, and for more concern with generalization and theory.

Prior to World War II, there had been a growing belief in some quarters that much of the work in comparative government was a mere parochial accumulation of facts about Western institutions.[5] The needs experienced during the War called for far more systematic classification and interpretation of existing data, as well as for the use of new sources of information. Developments of the postwar period—the cold war, the anti-colonial movements, and the growth of nationalism in various parts of the globe—helped place a premium on policy-oriented research.

In this setting, and cognizant of the efforts in the social science disciplines to seek for a greater comparability of research findings, the Committee on Comparative Politics of the Social Science Research Council was established in 1954. On the basis of some common orientation and with a bias for functional analysis, this committee has encouraged a large number of studies focused on political groups in Western and non-Western countries.[6] These efforts are necessarily experimental. They have provoked some questions as to whether the interest group approach will lead to better understanding of political institutions and behaviors than would other approaches.[7] Notwithstanding, this "pioneering" represents one of the most promising and provocative group research efforts in comparative politics today.

Work in comparative administration has also developed rapidly since World War II under the auspices of governments, universities, and professional societies. Moreover, serious attempts are being made to move beyond the "action research" which has held the center of the stage in the past.[8] New American journals have appeared, such as the *American Journal of Comparative Law,* and *Comparative Studies in Society and History.* Both foreign and domestic institutes and centers are directing greater attention to research in comparative politics. This is also true of the international associations, including the International Political Science Association. *The Study of Comparative Government and Politics,* edited by Gunnar Heckscher,[9] and *Interest Groups on Four Continents,* edited by Henry W. Ehrmann,[10] embody

---

[5] E.g., Roy C. Macridis, *The Study of Comparative Government* (New York, 1955).

[6] Over 100 articles, and formal and informal papers, have resulted from the work of the Committee. For an explanation of its evolving rationale, see Gabriel A. Almond, "A Comparative Study of Interest Groups and the Political Process," this REVIEW, Vol. 52, pp. 270–82 (1958), and Lucian W. Pye, "Political Modernization and Research on the Political Socialization Process" (mimeo, July, 1959). The major collective and interpretive effort of this Committee to date is the forthcoming volume, *The Politics of the Underdeveloped Areas,* which deals with the characteristics and classification of the political systems and the process of political development in the new countries of Africa, South America, South Asia, and the Middle East, by Gabriel A. Almond, James S. Coleman, Lucian W. Pye, George O. Blanksten, Dankwart A. Rustow, and Myron Wiener.

[7] Joseph LaPalombara, "The Utility and Limitations of Interest Group Theory in Non-American Field Situations" (mimeographed ms. of paper to appear in a forthcoming issue of the *Journal of Politics*), p. 6.

[8] Note the current program of the American Society for Public Administration. On the literature, see, for example, Robert V. Presthus, "Behavior and Bureaucracy in Many Cultures," *Public Administration Review,* Vol. 19, pp. 25–35 (1959).

[9] (London, 1957).

[10] (Pittsburgh, 1958).

some of the reflections of foreign and American political scientists on the scope, methodology, objectives, and trends in comparative politics.

As Charles S. Hyneman recently reminded us in his *The Study of Politics*,[11] there are many paths for the laborer to follow in this vineyard. Each of us will doubtless be guided in that direction to which his interest, training, and experience point.

This evening, we have set for ourselves a modest assignment in an area which has received too little attention and which is deserving of more study in the future. That will be to examine, comparatively, the Constitutional Courts of three Western European countries—those of West Germany, Italy, and Austria. We shall discuss the reasons for their creation, some of their most significant decisions, and the general position which they occupy in their own political systems.

I

Constitutional Courts have been established in the three countries in accordance with the provisions of their respective Constitutions, as implemented by the necessary legislation. These Constitutions are the Austrian one of 1920 (as amended in 1925 and 1929) which was reinstituted during the uncertain postwar period in 1945, the Italian Constitution of 1948, and the West German Basic Law of 1949. Though the Austrian Constitution presents a special case, all three may be classed as of post-World War II vintage. The Constitutions of West Germany and Italy were the product of negative revolutions, reflecting a deep distaste for the "dismal past." As characterized by our Chairman of this evening, "the political theory of the new Constitutions which are democratic in the traditional Western sense . . . revolves . . . around four major focal points which distinguish them from their predecessors: (1) reaffirmation of human rights, *but* (2) efforts to restrict these rights in such a way as to make them unavailable to the enemies of constitutional democracy, (3) stress upon social goals and their implementation through socialization, *but* (4) efforts to circumscribe the goals and their implementation in such a way as to prevent the reemergence of totalitarian methods and dictatorship."[12] To achieve these goals, the specially provided Constitutional Courts were to play an important part.

In seeking the explanations for the adoption of these special Courts, we are reminded that judicial review in continental Europe, as in the United States, had its roots in the higher law background and conceptions of ancient and medieval times. The precedents for special courts to protect the fundamentality of the Constitution can be traced at least as far back as the 18th century, when written constitutions came into being,[13] with the proposals of Abbé Siéyès in the 1790s for the creation of a constitutional jury. The work of a succession of distinguished advocates of

---

[11] (Urbana, 1959).
[12] C. J. Friedrich, "The Political Theory of the New Democratic Constitutions," *Review of Politics*, Vol. 12, pp. 217–18 (1950).
[13] David Deener, "Judicial Review in Modern Constitutional Systems," this REVIEW, Vol. 46, pp. 1079–83 (1952).

judicial review, who refused to accept some of the implications and influences stemming from the French Revolution, followed at later periods.

That this heritage and this special advocacy were alone inadequate to account for the later creation of Constitutional Courts is evident from a glance at the history of Western Europe. It was not until postmortems on World War I that the Austrian Constitutional Court came into being. And there were particular considerations after World War II which gave an impetus to the establishment of special courts in West Germany and Italy.

The most obvious influence was that of national precedent. In the case of West Germany, there were precedents which could be traced from the constitutional proposals of the National Assembly of 1848 down to the history of the National Supreme Court (*Reichsgericht*), and of the High Court (*Staatsgerichtshof*) of the Weimar period. The High Court had jurisdiction over the settlement of disputes between states (*Länder*) and between states and the *Reich*, as well as over impeachment cases. The Supreme Court passed upon the compatibility of state laws with federal laws, and it reviewed the constitutional validity not only of state, but also in several instances of federal legislation. But there were various limitations which operated to restrict the scope and effectiveness of the activities of these Courts. In Austria, traces of the Austrian Constitutional Court can be found in constitutional developments of the period between 1848–1851, and especially in the establishment in 1867 of the Court of the Empire (*Reichsgericht*). As time went on, this Court exercised jurisdiction over the claims of the provinces (*Länder*) against the Empire and *vice versa*; it dealt with conflicts of competence between judicial and administrative authorities at both the provincial and national level, and with complaints of citizens over the violation of constitutionally guaranteed political rights after other remedies had been exhausted. As for Italy, though there were certain pre-1922 procedures and institutions which pointed toward judicial review, these were of limited significance. The first noteworthy Italian precedent was provided by the Sicilian High Court, created by the Regional Statute of May 15, 1946.

Foreign example can also be stressed. Certain of the practices and procedures in Switzerland, particularly the use of the constitutional complaint, were given serious attention during the drafting of the Bonn Basic Law. The exercise of judicial review by the United States Supreme Court has received continued attention in European countries. As one Italian professor has observed, the "impact of *Marbury* vs. *Madison* was felt in Italy almost a century and a half after the decision."[14] To these considerations may be added indirect pressures which were brought to bear by the occupying powers after 1945, especially by the United States in Germany. However, evidence that direct Allied pressure was responsible for the final action taken is lacking in all three instances.[15]

[14] Giuseppino Treves, "Judicial Review of Legislation in Italy," *Journal of Public Law*, Vol. 7, p. 345 (1958).

[15] Even in Germany, there is considerable evidence for the view expressed by Rudolf Katz, Vice President of the Federal Constitutional Court, that there was no necessary causal relationship between the original Allied demands and the final German action. See comments

But it was definitely the reaction to excesses of the Fascist and Nazi regimes which was the most important factor in the decisions finally taken in Austria to restore her Constitution of 1920, as amended in 1925 and 1929, with its provision for a Constitutional Court; and in Italy and Germany, to establish new Courts. There was remarkable unanimity among most of the democratic parties in all three countries to grant the power of judicial review to some type of court. This same reaction helps explain the incorporation of elaborate bills of rights, to protect the individual, and of federalistic arrangements which, while borrowing from the past, were directed against the centralization of the Fascist period. Judicial review, in some hands, was widely accepted as necessary to safeguard these guaranteed liberties and arrangements. Disagreements existed over the type of court, its organization and composition, and over the method of selecting the judges. The answers were provided by the special Constitutional Courts. In short, external influences and pressures combined with domestic concerns to explain the final decisions which were taken.

The most controversial of these decisions, as evidenced in the debates in the Constituent Assembly in Austria in 1919–20, and in the German Parliamentary Council and the Italian Constituent Assembly between 1946–49, turned on the manner of selecting the judges. These discussions were concerned with the degree of independence to be accorded the Court from the political departments, especially from the parliament. In all three instances, compromises were effected, which provided for a method of selection differing from that used in choosing the judges of the highest regular court and which allowed for some participation by both houses of parliament in the selection process. Today, in Germany the 20 judges of the two Senates, or "twin courts," are selected by parliament, one-half of them by the *Bundestag* and one-half by the *Bundesrat*; six of them are chosen from the judiciary for life and the remainder for 8-year terms. In Italy, one-third of the 15 judges are chosen by the magistracy of the three highest Courts (Cassation, Council of State, and Accounts), one-third by the two houses of parliament sitting together, and one-third by the President—all eventually to hold office for 12-year terms. In Austria, the President, the Vice President, and 6 of the additional 12 judges, as well as 3 substitutes, are appointed by the President from nominees of the Federal Government; the remaining 6 judges, and 3 substitutes, are appointed in part on the recommendation of the lower house, the National Council, and in part on the recommendation of the upper house, the Federal Council. Unlike a majority of the German justices and all of the Italian justices, these are appointed for life. No instances of the use of the removal power over members by the Courts[16] during the post-World War II period have been recorded, though there have been resignations. In the three countries, provisions are made for the appointment to the Courts of

---

and literature cited in the author's "The West German Federal Constitutional Court: An Evaluation after Six Years," *Journal of Politics*, Vol. 20, pp. 283–84 (1958), and "The *Bundesverfassungsgericht*, 1956–1958: An American Appraisal," *Jahrbuch des Öffentlichen Rechts*, Vol. 8, pp. 29–47 (1959).

[16]In West Germany, by the Federal President upon the request and with the consent of the Court; in Italy and Austria, by the Courts acting directly.

practicing judges and high administrative officials, as well as professors of law. Indeed, the substantial percentage of professors on all of the Courts[17] reflects the long-established practice in continental European Countries to look toward the universities in making high judicial appointments.

Partisan considerations have played their part in the selections, though it has not always been possible to document the extent to which such factors have been controlling. The many early criticisms of the German procedure of selection have been based in part on the charge that political affiliation was more important than professional attainment, but it should be noted that all of the German judges except two have been elected unanimously, and re-election has been customary. The operation of *Proporz*, that is, the proportional allocation of administrative and other posts on the basis of the strength of the two major coalition parties, was discussed in 1957, in connection with the appointment of a new president of the Austrian Constitutional Court.[18] Italy had the greatest difficulties in securing the enabling legislation necessary to implement the constitutional provisions regarding the Court[19] and in selecting the judges after the implementing legislation had been finally passed in 1953. The requirement that 5 of the judges must be selected by a three-fifths majority of the two houses of parliament, where approximately 40 per cent of the seats were held by the left-wing parties which demanded representation, presented particular difficulties and provided the background for much of the maneuvering which occurred.[20]

But, withal, there has been only limited criticism of the judges after their selection; on the contrary, there has been general commendation, though the anonymity attached to the method of making decisions and the absence of dissenting opinions may offer protection from public criticism of the partisan and incompetent judge. In short, the judges have become "judicialized" rather than "politicized" with the passage of time.

The jurisdiction of the West German Constitutional Court is the most extensive and that of the Italian Court the most limited of the three Courts. With variations as

---

[17]Of the present 20 justices on the German Federal Constitutional Court, seven are professors (who retain their professional status on a part-time basis); and of the 15 justices in Italy at the end of 1958, some 10 held the title of professor. The President and Vice President, as well as other members of the Austrian Constitutional Court today, are professors in Vienna and other universities. The Constitutional Council of the Fifth French Republic (which can hardly be designated as a "constitutional court") contains no professor of law for special reasons. See Stanley H. Hoffmann, "The French Constitution of 1958: I. The Final Text and Its Prospects," this REVIEW, Vol. 53, p. 341, n. 37 (1959).

[18]See *Berichte und Informationen*, Dec. 6, 1957, and *Die Wochen-Presse*, Dec. 29, 1956 and Jan. 19, 1957; also Herbert P. Secher, "Coalition Government: The Case of the Second Austrian Republic," this REVIEW, Vol. 52, p. 799 (1958).

[19]One of the several reasons for this delay was the hesitancy of parliament to set up a body which would restrict parliament's powers. See John Clarke Adams and Paolo Barile, "The Italian Constitutional Court in Its First Two Years of Activity," *Buffalo Law Review*, Vol. 7, pp. 250–265 (1957–58). This difficulty has reminded these two authors of the legendary story of Bertoldo, who was sentenced to be hanged and then was entrusted with the responsibility of finding an appropriate tree. In Bertoldo's case there were explainable delays.

[20]Note the account in David G. Farrelly, "The Italian Constitutional Court," *Italian Quarterly*, Vol. 1, pp. 53–56 (1957).

to scope and application, the Courts in all three countries have the power to review the constitutionality of federal and state, or provincial and regional, legislation.[21] They pass upon disputes involving "conflicts of competence" between the central governments and the states, provinces, or regions, as well as between these latter political units. They also can decide jurisdictional disputes between "organs" of government at the national level in West Germany and Italy, and between the courts, or courts and administrative authorities, in Austria. They can try impeachments or accusations against certain officials at the national level in West Germany and Italy, or against federal and provincial officials in Austria. Both the Austrian and West German Courts have some jurisdiction in cases involving disputed elections and international law. In addition, each Court possesses some special competences which are unique to it. For example, the West German Court may pass upon the constitutionality of political parties and the forfeiture of basic rights. Advisory opinions were authorized by legislation in Germany until the repeal of the empowering provisions in 1956.

But a mere mention of the competences of the Courts will tell little, without a recognition that their functioning depends heavily upon the nature of the social structures within which they operate. These societies have been referred to as fragmented ones in which there is sharp competition between political cultures. The extent of political involvement by the citizen and the development of institutional pluralism vary in the three countries; the degree of consensus, on procedural if not substantive matters, is lower in all instances than that to be found, for example, in Britain.[22] The legal backgrounds of the three countries, with their differing ingredients of Roman and Germanic law, affect the position of the judges. And there are many other considerations which have a bearing on the role of the Courts. The federalism of West Germany and Austria, and the "attenuated federalism" of Italy, merit particular attention. The nature of the party system (with the trend toward the two-party system in West Germany, government by party cartel "with built-in opposition" in Austria,[23] and shifting coalitions based upon a mass party in the center of Italy) affects the legislative product of the parliaments which is subject to review by the Courts. We must of necessity leave these matters with only passing mention, though with a full appreciation of their significance in appraising the work of the Constitutional Courts.

II

In examining the work of the Courts, some attention may be directed to selected decisions dealing with (1) equality before the law, (2) federalism, (3) delegation

[21] All three differentiate between "incidental" proceedings, arising out of a pending trial, and "principal" proceedings, *i.e.,* those instituted by a governmental organ.

[22] Note the provocative comparisons in Herbert J. Spiro, *Government by Constitution* (New York, 1959), ch. 22.

[23] An expression used by Otto Kirchheimer, "The Waning of Opposition in Parliamentary Regimes," *Social Research*, Vol. 24, pp. 127–56 (1957); *cf.* Charles A. Gulick, "Austria's Socialists in the Trend toward a Two-Party System: An Interpretation of Postwar Elections," *Western Political Quarterly*, Vol. 11, pp. 539–62 (1958).

of legislative powers, and (4) legislation and public service relationships dating from the Fascist and National Socialist periods.

The Constitutions of each of the three countries contain an almost identical guarantee that "all persons shall be equal before the law."[24] In addition, they include certain other clauses which are to make more specific the general guarantees. The differences among these reflect the varied historical circumstances under which these constitutional provisions had their origins.

The significance of the equality before the law guarantees must be viewed in the light of the accessibility of the Courts in question. West Germany, to speak generally, provides a more liberal access to the Court than either Austria or Italy. For the individual, two avenues are open to the German Court, the most widely used being the constitutional complaint.[25] Under this arrangement, any person can question before the Court a law, an act having the force of law, or an administrative decision and order, which violates his constitutional guarantees, including equal protection before the law. He likewise may, during proceedings in regular courts, secure the judicial review of legislation, which allegedly infringes his rights, though the courts themselves must determine whether the access to the Constitutional Court is justified. While the Austrian system also provides two somewhat comparable modes of access, they are more narrowly construed than in Germany. In Italy, the access to the Constitutional Court is still more limited, as there is only the one procedure of judicial review. The institution of constitutional complaint, as it exists in Germany and, in a modified form in Austria,[26] is unknown in Italy.

The West German Court, in the cases before it, has applied the general principle that equal protection prohibits differential treatment of that which is essentially equal but "it does not prevent that which is essentially unequal from being treated by the legislature differentially in proportion to its inequality."[27] There must be a reasonableness of classification in all instances, even outside the range of the specific prohibition of discrimination on the grounds of sex, race, descent, language, place of birth, or religious belief. At the same time, the Court—applying the guarantee of equal rights to men and women—has recognized permissible distinctions involving the biological and functional differences of the sexes. Thus, on May 10, 1957[28] the Court rejected two constitutional complaints by male plaintiffs who alleged that the provisions of the Criminal Code under which they had been committed violated

---

[24]West Germany: Art. 3; Austria: Art 7; Italy: Art. 3. In the following discussion, I am heavily indebted to Mr. W. R. Dallmayr for his assistance.

[25]Though a number of nuisance and facetious complaints are submitted, the Court in West Germany has based an increasing percentage of its important decisions on selections from the 4,800 complaints which had been made prior to December, 1958. An illustration of the facetious complaint was one which contended that the refusal of police to extend the time during which "bars" might be kept open violated the constitutionally guaranteed right of freedom of assembly.

[26]Against individual decrees and acts of the administration, but not against laws, ordinances, or court rulings.

[27]1 *Entscheidungen des Bundesverfassungsgerichts* (hereafter cited as *B.V.G.E.*) 52; cf. Gerhard Leibholz, *Die Gleichheit vor dem Gezetz* (2d ed., Berlin and Munich, 1959), pp. 1-12.

[28]6 *B.V.G.E.* 389.

Article 3 of the Basic Law, in that these provisions provided no punishment for women convicted of comparable offenses. In a highly publicized decision announced on July 29, 1959,[29] the Court held that certain provisions of the Civil Code[30] violated Article 3, Sections 2 and 3 (as well as Article 6), of the Basic Law in that they denied the equal status of man and wife with regard to their children. Several housewives had brought constitutional complaints before the Court alleging that these statutory provisions accorded to the father the right of legal representation and certain other rights with respect to the child, and thereby discriminated against the mother. The Court found no biological or functional differences between man and wife which justified the statutory differentiations.[31]

This was the first major instance involving family relations where the Court has based its decision specifically upon the provisions of the Basic Law concerning equality of the sexes. The Austrian Court had in the previous year held the provisions of Section 26, paragraph 3, of the Income Tax Law of 1953, providing for joint taxation of man and wife, to violate Article 7 of the Constitution in differentiating between the sexes for tax purposes.[32] This case, and others involving the granting of concessions and the treatment of public employees, illustrate the extension of the applicability of the equality before the law provisions to new aspects of social relationships.

Aside from the fact that access of the individual to the Courts is limited in Italy, there are other considerations which explain why the Italian Court seems to have accepted the most restrictive view on equal protection. In review cases, it allows the widest discretion to a legislative finding of facts. Said the Court in a decision in 1957: "the evaluation of the relevance of the diversity of situations in which the individuals subject to the legal regulations find themselves cannot but be reserved to the discretion of the legislature, as long as the limits specified in the first paragraph of Article 3 are observed";[33] however, the "principle of equality is violated when the legislators subject to an indiscriminate discipline situations which they consider themselves and declare to be different."[34]

---

[29]1BvR 205/58.

[30]Secs. 1628 and 1629, paragraph 1. These provisions had not been altered by the Equal Protection Law of June 18, 1957. See J. Leyser, " 'Equality of the Spouses' Under the New German Law," *American Journal of Comparative Law*, Vol. 7, pp. 276–87 (1958).

[31]We cannot avoid quoting from an editorial in an American newspaper which commented on this decision: "Thus from Karlsruhe comes the news that father no longer has the last word. It is triumph for the species. Of course, at this point, the German wife has only acquired a sort of deadlock. There is no last word. Give her time, however, and we may be sure that she not only will have deprived mere man of the last word but will, as has her American counterpart, have appropriated it, herself." *Durham Sun*, July 5, 1959, p. 3.

[32]Decision of March 29, 1958; G 1, 2, 3, 5, 29, 30/58. The Federal Constitutional Court in West Germany had in 1957 invalidated somewhat comparable provisions of the Income Tax Law of 1951, but, while expressing some doubts as to the compatability of these statutory provisions with Art. 3 of the Basic Law, the West German Court had based its decision on the grounds of violation of Art. 6, paragraph 1, of the Basic Law. 6 B.V.G.E. 55.

[33]*Raccolta Ufficiale delle Sentenze e Ordinanze della Corte Costituzionale* (hereafter cited as *R.U.*), No. 3, Vol. 2, 1957, p. 21, at 27.

[34]*R.U.*, No. 53, Vol. 6, p. 68 (1958); also cited in Treves, "Judicial Review of Legislation in Italy," *loc. cit.*, p. 351.

It has been particularly in cases involving equality before the law that the West German Federal Constitutional Court has given some evidence of its recognition of a higher law above the positive law, that is, of a superior and unwritten constitutional law. Though there were earlier statements by the Court to which natural law adherents might point, the Court perhaps gave its clearest expression of the acceptance of a "hierarchy of norms within the Basic Law" and of certain natural law "guidelines" in a decision of December 18, 1953,[35] involving the equality of the rights of men and women. There the Court did acknowledge the possibility in "extreme cases" of conflicts between the positive law of the Basic Law and of the higher law.[36] The Court was here reflecting something of the natural law revival in post-World War II Germany, which had resulted in part from a reaction against the earlier positivist justifications for the Nazi regime. Since 1953, it appears that the Court has been deliberately more careful in its references. It has tended more to stress the "basic principles" of the Constitution as expressed in the specific provisions of the Basic Law, and it is being cautious in providing continuing opportunity to reopen the debates on "unconstitutional constitutional norms."[37]

Of the three Courts, the Italian Court has been the most careful to confine its reasoning narrowly to the provisions of the Constitution and to avoid overt reference to value judgments based on natural law in its decisions. Certainly, the Italian Court has insisted that its jurisdiction is limited to examination of the compatibility of laws and of acts having the force of law with the Constitution, and that it is not competent to pass upon the constitutionality of constitutional norms. The position of the Austrian Court appears to be different from the other two: it has not rejected completely the review of constitutional norms, as has the Italian Court, nor has it claimed the right to subject constitutional provisions to review in the light of higher or natural law precepts. In its much discussed decision on provincial citizenship on December 12, 1952,[38] the Court recognized that it could not review the substance of constitutional provisions in the light of higher or supra-positive ideas "since, in general, any standard for such an examination is missing."[39] It has, however, insisted upon its power to decide whether a proposed amendment involves a "total

---

[35] 3 B.V.G.E. 225.

[36] Such words and phrases as "supra-positive basic norms," "natural justice," "fundamental postulates of justice," "norm of objective ethics," etc., have been used in cases. *Cf.* Heinrich Rommen, "Natural Law in Decisions of the Federal Supreme Court and of the Constitutional Courts in Germany," *Natural Law Forum*, Vol. 4, pp. 1–25 (1959); Gottfried Dietze, "Unconstitutional Constitutional Norms? Constitutional Development in Postwar Germany," *Virginia Law Review*, Vol. 42, pp. 1–22 (1956).

[37] *Cf.* the author's "The West German Federal Constitutional Court: An Evaluation after Six Years," *loc. cit.*, pp. 300–304, and the literature there cited.

[38] *Sammlung der Erkenntnisse und wichtigsten Beschluesse des VGH* (hereafter cited as *Slg.*), No. 2455.

[39] For pertinent comments by the three Presidents of the Court during the period since 1946, note Ludwig Adamovich, "Probleme der Verfassungsgerichtsbarkeit," *Juristische Blätter*, Vol. 72, p. 73 (1950), and "Die Verfassungsmässige Funktion des Richters," *Österreichische Juristenzeitung*, Vol. 9, p. 410 (1954); Gustav Zigeuner, "Zehn Jahre Verfassungsgerichtshof in der Zweiten Republik," *Juristische Blätter*, Vol. 78, pp. 631–32 (1956); and the somewhat more natural-law oriented position of the present President, Walter Antoniolli, "Gleichheit vor dem Gesetz," *ibid.*, Vol. 78, pp. 611 ff. (1956).

revision of the Constitution" and hence is subject to a popular referendum.[40] In this instance, the Court must go beyond the formal requirements of enactment to a consideration of those basic constitutional principles whose alteration would involve "total revision."[41] Thus, in all three countries, the quest of the judges for foundations on which to base some of their decisions regarding individual rights and, specifically, the application of the equality before the law provisions of the Constitutions, continues.

The restraints which are imposed by a federalistic system and by federalistic arrangements upon the exercise of arbitrary power at the center were recognized by many of the framers of the Bonn Basic Law and the Italian Constitution. They also helped influence the sequence of events in Austria in 1919–20 and, again, during 1945–46. Since only four of the regions in Italy have as yet been created, the relationships in that state can be called only pseudo-federalistic. Nevertheless, the regions which have been established are guaranteed a significant degree of autonomy which can be altered only by constitutional amendment.

It is easy to overstress the centralizing trends in West Germany, unless there is adequate appreciation of the functioning of the *Bundesrat* and of the Federal Constitutional Court. Two of the most significant decisions of the Federal Constitutional Court, in particular, have evidenced its efforts to draw the lines between the competence of the Federal and the state governments. In the highly controversial Concordat case, decided on March 26, 1957,[42] while the Court recognized that the Concordat of 1933 was still a binding treaty, it did sustain the school legislation of Lower Saxony as falling under its reserved powers.

Several events in the spring and early summer of 1958 provided the setting for the much publicized atomic rearmament referenda cases involving the states of Hamburg, Bremen, and Hesse and decided on July 30, 1958.[43] In the play of party politics, the Social Democratic Party had sought and failed to secure the passage by the *Bundestag* of an act to provide for a national referendum on atomic rearmament. It resorted to other tactics to secure what it termed "consultative plebiscites." The states of Hamburg and Bremen, both with legislative bodies containing Social Democratic majorities, passed legislation for holding referenda at the state level. At the request of the Federal Minister of Interior, the Federal Constitutional Court issued on May 27, 1958, restraining orders to prevent the implementation of state laws pending a final decision by the Court as to their constitutionality. The Court, in following certain selected arguments of the Federal government in its joint decision, found the acts of Hamburg and Bremen to be unconstitutional. They represented attempts to provide for the participation of the citizens "in an area within the

---

[40] See Constitution, Arts. 44 and 140.
[41] At this point, as Professor Felix Ermacora has said, "the Constitutional Court is . . . the guardian of the Constitution and also the guarantor of the implementation of the requirements of direct democracy." "Die Bedeutung der Überprüfung von Bundesverfassungsgesetzen durch den Österreichischen Verfassungsgerichtshof," *Juristische Blätter*, Vol. 75, p. 539 (1953).
[42] 6 B.V.G.E. 309.
[43] 2 BvF 3/58 and 2 BvF 6/58. See also 2 BvG 1/58 of July 30, 1958.

exclusive jurisdiction of the Federal Government." In addition, "instructions" through referenda from the people of the state to "representatives" were violative of the Basic Law. The Court, in its brief decision, was particularly parsimonious in its discussion of Article 28 of the Basic Law, which recognizes the right of the states to deal with their own constitutional organization as long as they meet "republican, democratic, and social rule of law" requirements. But, recognizing the restricted grounds on which it based its decisions, the Court did give evidence that it would impose limits on the efforts of the states to explore at the behest of a political party uncharted jurisdictional areas under the federal system. These, and other recent decisions,[44] indicate some of the efforts of the Federal Constitutional Court to draw the lines between the competences of the federal and state governments.

The Constitutional Court in Austria has been faced with equally complex problems. After World War II, this Court has taken again as a basis for some of its decisions the theory of freezing of the distribution of competences (*Versteinerungstheorie*) at a given time in the constitutional development during the First Republic. The date chosen was that of the effectiveness of the first constitutional amendment on October 1, 1925.[45] Nevertheless, certain general trends in the decisions of the Court may be noted. During the first years after 1946, the Court's decisions were seemingly directed toward the protection of the modest sphere of reserved powers of the provinces.[46] However, in later years, the Court has more frequently decided in favor of the Federation. Thus, by a decision in 1951, the Constitutional Court upheld the second Nationalization Law of March 26, 1947, which recognized the power of the Federation to nationalize electricity and power plants.[47] Again in 1952, the Court sustained the law of 1949 providing for the equalization of economic burdens as falling within federal jurisdiction.[48] In a suit brought in 1953, while several provisions of the law establishing the Federal Chamber of Commerce were invalidated, the essential contentions of the plaintiff government of Vienna were rejected.[49] In 1954, the Court recognized that the

---

[44] Note the decision of June 16, 1959, in which the Court held a Federal Law concerning the Payment of Compensation Claims of 1956 to be incompatible with Article 120 of the Basic Law in that it required the states to bear expenditures which represented obligations of the Federal Government (2 BvF 5/56); and the decision of July 14, 1959, in which the Court held that the 1957 Federal Law for the Establishment of a Foundation called "Prussian Cultural Property" and the Transfer of Assets of the former *Land* Prussia was not in violation of Article 135 of the Basic Law. 2 BvF 1/58.

[45] *Slg.*, Nos. 2217 (1951), 2319 (1952), 2546 (1953), and 2721 (1954).

[46] Note *Slg.*, No. 2087 (1951), where the Court criticized the Federation for using the powers granted in Art. 12, Sec. 1 (under which the *Bund* lays down the "basic principles" and the province retains the power of execution) in such a way as to infringe upon the competences of the province by providing detailed regulation of the subject matter in question. For earlier post-war cases, see Paul L. Baeck, "Postwar Judicial Review of Legislative Acts: Austria," *Tulane Law Review*, Vol. 26, pp. 76–77 (1951–52).

[47] *Slg.*, No. 2092 (1951).

[48] *Slg.*, No. 2264 (1952). See the criticisms of this decision in Hans Spanner, "Die Prüfung von Gezetzen und Verordnungen durch den Verfassungsgerichtshof in der Zeit von 1950–1952," *Österreichische Zeitschrift für Öffentliches Recht*, Vol. 6, pp. 181–82 (1954).

[49] *Slg.*, No. 2500 (1953). See also H. P. Secher, "Representative Democracy or Chamber State" (mimeographed paper delivered at the 1959 Annual Meeting of the Midwest Conference of Political Scientists), pp. 9–10.

control of radio fell entirely within the jurisdiction of the Federation;[50] in 1956, the first Nationalization Law of 1946 was sustained.[51] These decisions must, of course, be compared with those which have favored the provinces.[52] But they must also be read in the light of the realities of the coalition government and of the changing international status of Austria, which have served to encourage federal legislation tending to narrow progressively the area of provincial autonomy.

Though the constitutional provisions have been only partially implemented, regionalism in Italy has provided more than its share of legal controversy. Indeed, more than half of the 381 cases "disposed of" by the Italian Constitutional Court prior to March 31, 1957 involved disputes between the central government and the regions.[53] The most controversial questions have involved the relations between the Sicilian High Court, which was created in accordance with Articles 24–30 of the Special Statute for Sicily on May 15, 1946, and was authorized to pass upon the constitutionality of laws enacted by the Sicilian legislature and the compatibility of national laws with the Regional Statute.[54] After the Italian Constitutional Court began to function in 1956, the problem of the relationship of the two Courts arose. In a decision of February 27, 1957[55] the Constitutional Court refused to recognize the possibility of a coexisting and competing jurisdiction with the High Court, at least over subjects within the competence of the Constitutional Court. But there are still unsettled questions involving the relationships between the two, as recent decisions of the Constitutional Court bear witness.[56] Indeed, the Italian Constitutional Court has evidenced a cautious approach in its efforts to demarcate the autonomous sphere of the sensitive regions.[57]

Thus, the West German Court is looked upon more as a protector of the reserved powers of the states than is the Austrian Court. The decisions of the Federal Constitutional Court appear to be of the greater significance in the total political picture, but its competences are broader. Its decisions have commanded more attention, but it is the newer creation. There has been less concern generated by the decisions of the Austrian Court, possibly because there is little evidence that its decisions have threatened major parts of the legislative program of the coalition government. The jurisdictional controversies between the central government and the regions in Italy, while occupying much of the Court's attention, necessarily have limited application.

[50] *Slg.*, No. 2721 (1954).
[51] *Slg.*, No. 3118 (1956).
[52] For references to certain of these cases, including ones involving hunting, real estate transactions, area planning, etc., see Felix Ermacora, "Die Entwicklung des Österreichischen Verfassungsrechts seit dem Jahre 1951," *Jahrbuch des Öffentlichen Rechts*, Vol. 6, p. 339 (1957), and *Der Verfassungsgerichtshof* (Vienna, 1957), pp. 145–46. Note, particularly, a decision of June 28, 1958 (G 32/58) in which the Court declared unconstitutional a federal law of 1957 levying import duties on certain products.
[53] Adams and Barile, *loc. cit.*, p. 258.
[54] The Sicilian Statute was converted by the Constituent Assembly, under pressure of time, into a Constitutional Law of Feb. 26, 1948, No. 2.
[55] *R.U.*, No. 38, Vol. 2, p. 375.
[56] *E.g.*, a decision of January 24, 1958. *R.U.*, No. 7, Vol. 5, p. 61 (1958).
[57] Note the discussion by P. Biscaretti di Ruffia, "The First Two Years of Functioning of the Italian Constitutional Court," *Il Politico*, Vol. 23, pp. 477 ff. (1958).

The concern over the dangers of unlimited delegation of legislative powers was reflected in the attempts of the Constitution makers to place constitutional restraints upon such delegation, as, for example, in Article 80 of the Basic Law in West Germany. This Article, empowering legislative bodies to authorize the Federal Government, a minister, or a state government to issue decrees implementing legislation, requires that the "content, purpose, and scope" of the statutory basis be specific. In 1956,[58] and again in 1958,[59] the Court has found provisions of legislation to be lacking in clarity as to "content, purpose, and scope" insofar as they authorized certain implementing decrees. But, said the Court in 1958, in a case involving designated paragraphs in the Price Law of 1948,[60] it is not necessary that "content, purpose, and scope" be expressly stated in the statutory basis; it suffices if they can be deduced from the whole statute, its styling, its meaning in context, its history. "This can be done in the present case."

In Italy, Article 76 of the Constitution provides that "the exercise of the legislative function cannot be delegated to the Government unless directive principles and criteria have been determined and only for a limited time and for definite purposes." There the Court has recognized that the determination of cases involving the unconstitutional delegation of legislative powers is one of its most important tasks.[61] The Court has invalidated, as being in effect "unconfined and vagrant," a law which left to the administrative authorities the determination of contributions (and of the persons required to contribute) to the tourist offices.[62] While it did not do so, the Court might well have borrowed from the language used by Justice Cordozo in 1935 in dissenting in the *Panama Refining Co.* and in concurring in the *Schechter* cases.[63]

The complicated history of restraints on legislative delegation, and of the legality of "law-amending ordinances" in Austria, defies brief summarization. But, according to numerous decisions of the Court, Article 18 of the Constitution permits the legislature to authorize the issuance only of implementing and not of "law-amending" ordinances; in order to justify the implementation, the statutory basis must prescribe the essential limits within which the intended regulations will be confined and the purposes toward which they will be directed.[64] The Court has not hesitated to strike down statutory provisions which have violated these requirements. In short,

---

[58] 5 B.V.G.E. 71.
[59] 7 B.V.G.E. 282.
[60] 8 B.V.G.E. 274.
[61] Note the comments of President Azzariti at the beginning of the second year of activity of the Court, in *R.U.*, Vol. 3, pp. 13–14 (1957). The Italian Court has held that both the law of delegation and the authorized act are subject to its review. *R.U.*, No. 3, Vol. 2, p. 21 (1957). *Cf.* Giovanni Cassandra, "The Constitutional Court of Italy," *American Journal of Comparative Law*, Vol. 8, pp. 4–5, n. 8 (1959); Gaetano Sciascia, "Die Rechtsprechung des Verfassungsgerichtshofs der Italienischen Republik," *Jahrbuch des Öffentlichen Rechts*, Vol. 6, pp. 7–9 (1957).
[62] *R.U.*, No. 47, Vol. 2, p. 507 (1957).
[63] 293 U.S. 388 (1935) and 295 U.S. 495 (1935); Adams and Barile, *loc. cit.*, p. 259.
[64] *E.g., Slg.*, Nos. 2109 (1951), 2276 (1952), 2462 (1953), and 2664 (1954).

in differing ways but with rather similar results, the Constitutional Courts of the three countries have been concerned with the application of constitutional provisions designed to prevent the legislature from leaving ill-defined and broad discretion in the hands of the administrator.

The Courts have been called upon to pass on the constitutionality of post-World War II legislation which dealt with public officials and military personnel in service during the Nazi regime in Germany, and of legislation enacted during the Fascist and Nazi regimes in Italy and Austria and affecting the private rights of individuals.

In West Germany, Article 131 of the Bonn Basic Law provides that the legal relationship of persons, including refugees and expellees, who were in the public service on May 8, 1945 and had been excluded from the public service on other than civil service or salary grounds, and who had not received positions comparable to their previous posts, was to be regulated by law. Such a law was passed on May 11, 1951. The Court rejected on December 17, 1953,[65] constitutional complaints of certain public officials who alleged that various constitutional rights had been violated by the law. The Court pointed out that, while under international law the state had retained its identity after 1945, the public service relationship had fundamentally changed during the Nazi regime. Consequently, the legislature could, in the exercise of its discretion, determine the status of the plaintiffs without allowing them any grounds for constitutional complaint against the law based upon their previous public service relationships. Similarly, the Court held that the *Wehrmacht* ceased to exist with the unconditional surrender of German military forces in 1945, and rejected the constitutional complaints entered by various officers, officials, and members of the former *Wehrmacht* directed against the Law of May 11, 1951.[66] In particular, the Court in a constitutional complaint of a former official of the *Gestapo*, took the opportunity to answer various criticisms of its previous "131" decisions and presented a lengthy and devastating analysis of the nature of the public service relationship during the Third Reich.[67] The tenor of the Court decisions, in passing upon the rights of officialdom of a previous totalitarian regime, has been consistent with its application of Article 21 of the Basic Law, under which the Nazi-oriented Socialist Reich Party was dissolved and its assets confiscated in 1952,[68] and the successor case, decided in 1956,[69] in which the Communist Party of West Germany was subjected to the same penalties.

In Italy, a large percentage of the "civil liberties cases," have involved the constitutionality of legislation which was enacted during the Fascist period (including the Criminal Code, the Code of Criminal Procedure, and the Police Law of 1931). Indeed, roughly one-third of the first forty decisions of the Court involved the constitutionality of criminal laws and regulations, most of them of Fascist

[65] 3 B.V.G.E. 58.
[66] 3 B.V.G.E. 288.
[67] 6 B.V.G.E. 132.
[68] 2 B.V.G.E. 1.
[69] 5 B.V.G.E. 85.

vintage.⁷⁰ To take a few examples: in 1956, Article 157 of the Police Law of 1931 providing for repatriation to the community of origin by administrative decree was held to be incompatible with the Articles of the Constitution guaranteeing the inviolability of personal liberty and freedom of travel;⁷¹ in 1957, a section of the Police Law of 1931 requiring notification in case of religious ceremonies outside of churches, irrespective of the place where held, was considered inconsistent with the Constitution;⁷² in the following year, the Court invalidated the provisions of a Law of 1942, which left to administrative officials the discretion to authorize the opening of private schools.⁷³ "Such a system," preserved "even after the collapse of the regime which established it," said the Court, "is incompatible with the meaning which the Republican Constitution attributes to the freedom of the school."

Under the provisional Constitution of Austria of 1945, two constitutional transitional laws were passed, the one to nullify constitutional provisions of the period after 1933, and the other, to deal with the period after 1938.⁷⁴ The latter of these transitional laws provided that "all laws and ordinances . . . passed after March 13, 1938 which are incompatible with the existence of a free and independent Austria or with the principles of true democracy, or which contradict the legal conceptions of the Austrian people or reflect typical National-Socialist ideas, are abrogated." This provision which, at least after 1953, the Court has held to be applicable without any governmental ordinance designating the laws or ordinances to be abrogated, has provided the basis for several decisions of the Austrian Court.⁷⁵

There is no need to mention the several illustrations which might be cited. Suffice it to say that in Austria, as in Italy, the Court has been continuously concerned with an examination of legislation, or legal norms, dating from the previous regimes and has invalidated many of them which have been violative of the Constitution. In so acting, the Courts in these two countries have perhaps offered some instigation to parliaments which have been slow to revise legislation still bearing some of the Fascist and National Socialist substance as well as imprint.

If we have dwelt at some length on certain selected decisions of these Courts, it

⁷⁰Since the Constitution was silent on this point, there was doubt as to whether the Court has the power to pass upon the constitutionality of "anterior legislation," but the Court in its first decision laid all questions at rest as to its jurisdiction. *R.U.*, No. 1, Vol. 1. p. 25, (1956). See David G. Farrelly and Stanley H. Chan, "Italy's Constitutional Court: Procedural Aspects," *American Journal of Comparative Law*, Vol. 6, p. 326 (1957).

⁷¹*R.U.*, No. 2, Vol. 1, p. 41 (1956).

⁷²*R.U.*, No. 45, Vol. 2, p. 491 (1957); cf. *R.U.*, Nos. 13 and 14, Vol. 5, pp. 101–107 (1958).

⁷³*R.U.*, No. 36, Vol. 5, p. 231 (1958).

⁷⁴*StGBl.*, Nos. 4 and 6.

⁷⁵For example, in 1953, in the decision which declared governmental proclamation unnecessary, the Court abrogated a National Socialist Law of November 5, 1935 on Exchanges, Vocational Guidance, and Procurement of Apprentices, which had been extended to Austria after the Nazi *Anschluss*, as reflecting "typical National Socialist ideas" and as "being incompatible with . . . true democracy." *Slg.*, No. 2620 (1953). The plaintiff had, moreover, been deprived of certain rights guaranteed under Article 12 of the Basic Law of 1867 and under Article 83, paragraph 2, of the Constitution.

has been to indicate the ways in which the Constitutions are being interpreted by the judges of the Constitutional Courts. They have clearly pointed out some of the effective constitutional limits beyond which the legislator and the administrator cannot go in their actions affecting individual rights.

III

Any conclusion regarding the role of the Constitutional Courts must be highly tentative and subject to much more critical examination. The record of the West German Constitutional Court has occasioned more comment than that of either Italy or Austria, possibly because of the breadth of its jurisdiction, its daring during the formative years, and the controversial character of some of its decisions.

In the relation of the Constitutional Courts to other governmental organs at the national level, there have been crisis periods in each of the countries. The German crisis occurred during 1952–53, when the consideration of the European Defence Community Treaties eventuated in what one critic called a "period of judicial frustration."[76] However, despite the critical position taken at that time in certain official quarters, the Adenauer Government has looked with increasing sympathy upon the Court in recent years. The *Bundesrat* has furnished more friendly support for the Court than has the governmental coalition in the *Bundestag*. The Social Democratic Party, as the opposition party, and the governments of certain of the states, as the weaker elements in the federal system, have viewed the Court as the protector of the rights of minorities. Although there have been various proposals coming from several circles for the reform of the Court, such minor changes in composition, organization, and jurisdiction as were made by legislation in 1956 and 1959, have emanated from the Court itself.

The crises in the brief history of the Italian Court were those which took place during the long period of delay after 1948, before implementing legislation could be enacted, and after 1953, before the judges were finally appointed. The assortment of internal and external problems faced by the Court, culminating in the final resignation of its first President, De Nicola, in 1957, were brought sharply to public and parliamentary attention. However, the reticence of the Court to go behind a legislative finding of facts in Italy and the limited exercise of the power to invalidate statutes enacted since 1948 have kept parliamentary criticism at a minimum.

In Italy, dissatisfied groups and organizations have on occasion attacked decisions of the Court. For example, the Communists have objected to certain ones respecting land reform legislation; the Church, to others involving the application of constitutional provisions regarding freedom of worship. Some opposition to the Court has also come from the lower bureaucracy. But the really violent opposition has emanated from the regions, especially from Sicily. These reactions, when coupled with the lethargy of the Italian populace toward the Constitution, have combined to create a negative image of the Court which is gradually being erased.

[76] Karl Löwenstein, "The Bonn Constitution and the European Defence Community Treaties, A Study in Judicial Frustration," *Yale Law Journal*, Vol. 64, pp. 805–39 (1955).

In Austria, neither of the major political parties nor any important pressure groups have made the Constitutional Court a target for continuing criticism. There have been past occasions, as in 1956–57, when partisan differences almost involved the Court, but these were exceptional instances. The relationship between the Constitutional Court and the other highest courts has provoked some controversy,[77] and there has been continuing academic discussion of the right of access to and the jurisdiction of the Court. Those who favor an expansion of its jurisdiction sometimes look toward West Germany; those who favor a more restricted status, may point toward Italy. But the recent constitutional law and legislation of 1958 dealing with the Court have resulted in only slight changes in its jurisdiction and organization. In Austria, as in West Germany and Italy, there has been general acceptance of the Court, though without either generous enthusiasm or violent criticism.

There have been problems of implementation of decisions. Some have been considered in West Germany, in connection with the atomic rearmament referenda cases, and others with decisions requiring parliamentary action.[78] The failures on the part of the parliament and the bureaucracy in Italy, to accept his strictures as to implementation, help explain De Nicola's threatened resignation in 1956 as President of the Court. But the record does not show any situation comparable to the effective nullification of a Court's decision, as occurred in the United States during President Jackson's administration following the Cherokee Indian cases.[79] There have been more warnings to the Federal Constitutional Court of Germany to exercise "intellectual humility," and "self restraint" in not pushing its jurisdictional bounds beyond the limits of the feasible and the practicable, than there have been in Italy and Austria, where the more limited jurisdiction of the Courts and the greater hesitancy to question legislative enactments have been evidenced. Its record indicates that the West German Court is seeking to follow this advice, and is sensitive to the charges of "judicial legislation," but it apparently has been unable to extricate itself from involvement with what the United States Supreme Court would call political questions.

The Constitutional Courts in Europe are in part the products of reaction against a gloomy past, as previously mentioned. Some of their activities have been devoted to a liquidation of this heritage and to a prevention of its repetition. But, today, the Courts are increasingly faced with the new issues which have developed during the post-World War II period. These new issues, as well as the old ones, have continued to invove the application of the pertinent constitutional provisions regarding equality before the law, federalism, and the delegation of legislative powers.

The idea that courts, or some judicial body, should serve as the final guardian of the constitution had its roots and origins in Europe. It has seen its widest acceptance and expansion in the United States. In turn, American application and judicial

---

[77] See *Juristische Blätter*, Vol. 79, pp. 263–65, and 287–89 (1957).
[78] 8 *B.V.G.E.* 1, of June 11, 1958.
[79] *Cherokee Nation v. Georgia*, 5 Peters 1 (1831), and *Worcester v. Georgia*, 6 Peters 515 (1832).

experience have helped undergird the European precedents and theoretical support for the formation of special judicial bodies to guarantee the fundamentality of their constitutions.

Today, there are those who believe that the significance of judicial review in the United States is diminishing and that our Supreme Court can no longer serve as an effective protector of individual liberties and minority rights against legislative majorities and executive discretion. Is it possible, asked one thoughtful observer, that we may borrow in the future from the experience of these European Constitutional Courts rather than contribute to it—that there will be another period in the "give-and-take between the new and the old worlds?"[80]

However, it is still too early in their history to speculate about the future of these Constitutional Courts. During the past decade, they have not faced that type of crisis which economic adversity, the messianic leader, or foreign military experiment might provide. Until such a time there will be uncertainty as to the degree to which constitutional democracy today reflects an active faith, and the extent to which it is the formal expression resulting from Allied political pressure, a prosperous economy, and anti-totalitarian resentment.[81] Only then will we know how deeply rooted are the constitutions for which these Courts serve today as interpreters and guarantors.

[80] Gottfried Dietze, "America and Europe—Decline and Emergence of Judicial Review," *Virginia Law Review*, Vol. 44, p. 1272 (1958).
[81] *Cf.* Leonard Krieger, *The German Idea of Freedom* (Boston, 1957), p. 468.

CHAPTER 20

Disaffection and Participation in Western Democracies: The Role of Political Oppositions

GIUSEPPE DI PALMA

Research on American political behavior consistently shows that people who hold a disaffected and pessimistic view of politics are less likely to participate in politics. Thus, political inactivity is common among people who are cynical and suspicious

This is a revised version of a paper presented at the 1968 Annual Meeting of the American Political Science Association, Washington–Hilton Hotel, Washington, D.C., Sept. 2–7. Research leading to the paper was made possible by a grant from the Institute of International Studies at Berkeley. Technical assistance and computer time were provided by the Survey Research Center and the Computer Center at Berkeley. Printed by permission of the author.

about politics, who put little trust in politicians, who show feelings of anomie and alienation from the political system, who find politics distant from and inconsequential to their lives, who believe that politics does not improve their lot, and who have little confidence in themselves and in their ability to influence political events. It is argued that these people do not participate because they find it difficult to understand politics and its operations, or believe that they cannot manage the political environment, or think that the system is not sensitive to the aspirations and demands of average voters like themselves.

There is probably no other area of mass political behavior which has been so heavily researched and where the findings approach more closely the empirical standing of a law.[1] There are still, however, a few questions worth exploring concerning the explanations of these findings and the degree to which they may hold in countries other than the United States.

One interpretation of their findings relies on the dynamics of belief formation and functioning, and suggests that attitudes of disaffection may lead to low rates of political activity under most conditions and in most cultures. Studies of belief formation and change often show that persons who hold a derogatory view of a given object do not necessarily act either to change their view into a more positive one or to change the nature of the object. This is especially true if the person attaches little importance to his particular view, as is often the case with political beliefs, if he finds little support in his environment because, for instance, few people share his belief or few can be organized.[2] Under these conditions a person who decides to act in order to change either the nature of the object or his view of it may find his decision too stressful in the short range and little rewarding in the long range. On the other hand, if he holds his belief, even if it is negative and derogatory, this, per se, may reward him because it maintains his cognitive consonance and balanced functioning.[3] Thus withdrawal, rather than further involvement, may appear the easiest way to preserve the economy of the person's belief system.

In the present case it can be argued that initial feelings of political disaffection, instead of motivating people to participate so as to shape the political system more to their liking, will induce them to sever their relations with politics and to make politics a more remote and therefore less painful experience in their lives. If

---

[1] The research literature is very large. Two relevant summaries and discussions can be found in Lester W. Milbrath, *Political Participation* (Chicago: Rand McNally & Co., 1965) and Herbert McClosky, "Political Participation," in *International Encyclopedia of the Social Sciences*, Vol. XII (New York: The Macmillan Company and The Free Press, 1968), 252–65.

[2] For a review and interpretation of the literature on the conditions leading to attitude change and on the strategies and outcomes of change, see Daniel Katz, "The Functional Approach to the Study of Attitudes," *Public Opinion Quarterly*, XXIV (1960), 163–204. The whole Summer 1960 issue of the journal is devoted to attitude change. On the saliency of political attitudes in the American public see Philip E. Converse, "The Nature of Belief Systems in Mass Publics." in David E. Apter (ed.), *Ideology and Discontent* (New York: The Free Press, 1964), 206–61.

[3] The classical statement on this point is M. Brewster Smith, Jerome S. Bruner, and Robert W. White, *Opinions and Personality* (New York: John Wiley & Sons, Inc., 1956), esp. chaps. 3 and 10. See also Katz, *op. cit.*

disaffected people participate at all, participation is of a symbolic, protest, and consummatory nature, and involves isolated and sporadic acts—voting, striking, protesting a belief or a grievance. The very performance of any one of these acts may exhaust the further desire to participate.

Another interpretation which supplements the first concerns the nature of the American political system and, in particular, the functioning of its parties in office and in opposition. This interpretation asserts that in American society political disaffection has never been organized politically owing to its existence only among small minorities, to the relative lack of large scale or serious social cleavages from which it can draw nourishment, and, above all, to the operations of a party system bent on compromise and insensitive to the issue of disaffection. Of these factors it is the party system and the role of political oppositions that will serve as the focuses of this paper.

At least four characteristics of the American party system, all well known to students of American politics, help to explain why disaffected people find it difficult to participate:[4]

1. Whatever differences in political interests and ideals exist in American society, these are not strongly related to social and demographic characteristics of the population or to party preference. Sociodemographic groups are as heterogeneous in their political attitudes as political parties are in their social composition and in the attitudes of their followers.

2. Even though activists and leaders of the two major parties (in contrast with party followers) differ significantly in many political beliefs and in some aspects of their ideologies, they share a strong attachment to a common core of substantive and procedural tenets of American democracy.[5] These so-called "rules of the game" refer to fair play, respect for procedural rights, protection of the rights of others, constitutional government, majority rule, minority rights, freedom of thought and organization, freedom of opposition, equal protection of the law and due process, reciprocity, tolerance of political diversity, willingness to compromise. This combination of issue cleavage and procedural consensus means acceptance of the reality of conflict, but it also means an even deeper conviction that the most satisfactory resolution of conflict requires a set of rules which do not prejudge the outcome of the contest and which guarantee the contestants reasonably equal opportunities to prevail.[6] These rules are embodied in most political institutions and structures and

[4] The best and most recent treatment of American oppositions and the political organization of dissent can be found in Robert A. Dahl, "The American Oppositions: Affirmation and Denial," in Robert A. Dahl (ed.), *Political Oppositions in Western Democracies* (New Haven: Yale University Press, 1966), chap. 2. The chapter contains extensive references to the relevant literature and thoroughly develops the four points above.

[5] The most significant evidence on both points is offered by Herbert McClosky, "Consensus and Ideology in American Politics," *American Political Science Review*, LVIII (1964), 361–82; and Herbert McClosky et al., "Issue Conflict and Consensus among Party Leaders and Followers," *American Political Science Review*, LIV (1960), 406–27.

[6] On the prerequisites of effective conflict regulation in Western societies and its effects on the expression and outcome of conflict, see Ralf Dahrendorf, *Class and Class conflict in Industrial Society* (Palo Alto, Calif.: Stanford University Press, 1959), esp. chap. 6.

make compromise, bargaining, and incrementalism stable features of American party politics.[7]

3. The presence of issue conflict between the two parties does not prevent them from inventing bipartisan solutions to many issues. Since both parties are loose alliances of diverse interests operating in a nonparliamentary system, this favors issue-specific bipartisan coalitions in which alignment is affected by constituency interests more than by party affiliation. It has also been suggested that bipartisan channels are used as a last resort when very divisive issues are involved, such as those having to do with civil rights, education, regional cleavages, religion, and foreign policy. Strictly partisan and majoritarian solutions of such issues are very much resented by the minority, may render cleavages irreconcilable, and hence threaten the viability and survival of the rules of the game. This makes the eventual search for bipartisan solutions that much more likely, although by no means easier.[8]

4. Partially as a consequence of the above factors, American oppositions are not stably and easily identifiable with one or the other party. Although the leaders and activists of the two parties differ in beliefs and behavior, party discipline is not very strong and their behavior is not exclusively determined by their party affiliation. Oppositions either work through both parties and within the common rules of the game, or, if they present radical alternatives outside the two parties, these alternative movements are small, fragmented, and often ineffective. In either case, oppositions are unstable, dispersed, incohesive, diverse in strategies, and have a high turnover in personnel.

One can easily understand why this type of party system may make it very difficult for disaffected people to participate. These people typically reject the basic rules of the game which are the root and branch of the American system. They may have political needs and aspirations of their own, but cannot recognize them in American politics as it is defined and conducted by most of its actors. Lacking major political organizations which appeal to their disaffection and which offer opportunities for identification, disaffected people are confined to the margins of American politics; they either abandon the political arena or engage in ineffective and sporadic action charged with paranoia and overt hostility.

According to Robert Dahl, "In the conventional liberal-democratic theory of politics, political dissenters are a valuable source of enlightenment. . . . In the United States, however, it is doubtful whether the process has worked quite this way. Frustration and alienation seem to encourage paranoid interpretations of political life, emotionalism and styles of thought that do not produce debate and discussion

---

[7]The concept of incrementalism as a policy-making strategy in American politics is formally developed in David Braybrooke and Charles E. Lindblom, *A Strategy of Decision* (Glencoe: The Free Press, 1963).

[8]Peter Drucker has recently written that the acceptance of conflict on economic issues, the transformation of many issues in economic ones, and bipartisanship on issues which cannot be solved in economic terms, are among the political strategies which have been successfully used as a unifying force in American politics. He questions, however, the continuing success of these strategies for the future, when great but unforeseeable political innovations are needed and expected. See Peter F. Drucker, "On the 'Economic Basis' of American Politics," *The Public Interest*, X (Winter, 1968), 30–42.

but hostility and rejection."[9] Dahl contrasts the United States with West European countries such as France, Germany, and Italy, where ideological controversy has been very intense and where political dissent has achieved strength and influence through mass political organizations.[10] It could be suggested—although Dahl does not make this argument—that if political cleavage on ideological bases is greater in Europe, this may encourage participation by the disaffected.

The purpose of this paper is to introduce evidence supporting the proposition that disaffection is actually as much a source of political inactivity in Europe as it is in the United States. Although there are significant differences between the American and European party systems, there are also similarities, for they are all systems operating within advanced Western societies. I shall argue that it is these similarities in parties and party systems which make a European who is disaffected from politics just as unlikely to participate as an American. To lend support to this proposition I will compare the United States with Germany, Italy, and Great Britain.

HYPOTHESES

Disaffected people may become politically active only if they can find in their society objects and institutions with which they can identify and which can replace identification with the political system. These institutions must be capable of offering the disaffected new goals and norms of conduct which radically challenge the goals and norms of the political system.[11] Such institutions, however, are hardly more typical of European countries than they are of the United States.

Radical forces challenging the very form of the political system and offering a real source of counteridentification tend to emerge mostly during the period of nation-building. Secession, regional independence, revolution, or a complete constitutional overhauling of the policy have often been the goals of political oppositions precisely at a time when the issues of the form of government and the distribution of power make new political systems rather undefined and unstable entities, hard to identify with and unable to cope with a radical challenge. In the United States and Western Europe today, however, issues of proper allocation of welfare resources and proper development of human opportunities have become the central concerns for competi-

---

[9] Dahl, "The American Oppositions" pp. 67–68 (emphasis supplied).
[10] *Ibid.* p. 65.
[11] A parallel may be drawn here with Merton's treatment of values–means conflict in his theory of anomie. See Robert K. Merton, *Social Theory and Social Structure* (Glencoe: The Free Press, 1957), chap. 4. According to Merton, "rebellion" against the goals and institutional means of society demands not only their rejection but also their substitution with new goals and institutionalized means. "Retreatism," on the other hand, results when rejection of old goals and means is not followed by the adoption of new ones. Although couched in the language of social typologies, it is possible to see in Merton's argument the suggestion that withdrawal, rather than rebellion, will ensue if institutional support for one's disaffection is not found. A similar line is developed by Everett Hagen in his discussion of how innovational character develops in groups which are deprived and discriminated against by dominant groups. Innovation develops after a period of withdrawal only when certain conditions are met. One is that the dominant group becomes a source of negative identification and a sense of one's group worth and goals is developed. See Everett E. Hagen, *On the Theory of Social Change* (Homewood, Ill.: Dorsey Press, Inc., 1962), esp. chaps. 9, 10, 11.

tive politics.¹² This change in the object of political competition with respect to the period of nation-building occurs as the political system becomes more clearly defined in its constitutional form, in its institutions, and in its relation to internal forces. Today the main task of political oppositions rarely involves offering total and sweeping alternatives to the basic framework of the political system. Nor do the typical issues of modern politics rest as heavily on absolute value considerations as did the issues involved in the creation and development of new nations.¹³ Thus the competing forces which operate within the political system are not so much sources of counteridentification as part and parcel of the system itself. They are institutions which mediate and make possible one's identification with the larger system.

These considerations apply to European countries as well as to the United States, and to moderate parties as well as to so-called "radical" opposition parties, although, in this sense, the use of the word *radical* is somewhat of a misnomer today. Except for the English Conservative party, all the major European parties I will consider were born as parties of protest. This is obviously the case with the Socialist parties, but it is the case with the Christian Democratic parties as well.¹⁴ Both types of parties emerged at the turn of the century or during World War I as a challenge both to the governing liberal and nationalist parties and to the kind of bourgeois society they were building. To label such parties today as parties of protest, however, does not at all clarify the nature of their current policies, the style of their protest, and the limits of their challenge to the established constitutional systems. While some of these parties might in the past have represented a total and revolutionary alternative to the system, most of them have more recently assumed national or local governmental responsibilities and share them with other political forces, as have the Christian Democratic parties of Germany and Italy which have been in power for about twenty years. The same can be said of the Socialist parties of England and Germany.

It would seem that exceptions occur only in the cases of the Communist and Socialist parties of Italy, but even in these cases the contrast with the experience of other European parties is by no means as striking as one may believe. Both parties have long held political power in Italian society through their control of innumerable local governments and their large share in cooperative and union movements. Since 1964 the Socialist Party has been the second partner in a government coalition with the Christian Democratic and other moderate parties. Even though the Socialists have lost substantial electoral strength because of their decision to join a moderate government, and even though the Communists, by remaining in opposition, have been the chief beneficiaries of this loss, the Communist party has not necessarily

---

¹²See Reinhard Bendix, "Social Stratification and the Political Community," *European Journal of Sociology*, I (1960), 181–210.

¹³For a treatment of instrumental and transcendental ("consummatory") values and their role in political modernization, see David E. Apter, *The Politics of Modernization* (Chicago: University of Chicago Press, 1965), chaps. 1 and 3.

¹⁴The early development of Christian Democracy in Western Europe is treated in Michael P. Fogarty, *Christian Democracy in Western Europe, 1820–1953* (Notre Dame, Ind.: University of Notre Dame Press, 1957). See especially Part II and chap. 20.

remained steadfastly attached to unreconstructed opposition. The party has long since found little comfort in its role as permanent opposition; the decision of the Socialists to join the government has led the Communist leadership to a serious re-evaluation of the function of Communism in Italian society.[15] If Communists and Socialists ever had subversive aspirations, these are in many ways only vestiges of the past. Even in the case of the Communist Party, its long-standing professed allegiance to the constitutional order is no mere electoral façade but an attitude which, with all its contradictions and instrumentalism, has deeply affected the party's very style of operation.

In sum, the European parties we are dealing with are large and established political forces whose stakes in the political order are many and long-standing, although issues and conflicts do exist between parties and conflict does touch aspects of the constitutional order. For some countries conflict is indeed very severe and threatening, as recent disturbances in Europe indicate. However, there exist certain fundamentals in the constitutional order which regulate conflict and which major parties never challenge. These have to do with the legitimacy of the political system and its institutional form, support for competitive politics, and recognition of civil, political, and social rights as they are understood in Western nations. While this seems obvious, some analyses of European party politics tend to overlook these fundamentals in their attempts to differentiate and classify parties and build ideal types of parties and party conflict.[16]

These developments which European and American parties share have a number of clear effects on the place of the disaffected in mass politics. Insofar as major European parties today aim at political power through electoral and constitutional processes, an appeal to a voter who is disaffected from these processes would be against their purpose. Minor parties whose only vocation is the propagation of opinions and ideas may afford the risk of unpopularity; however, this is a luxury which major parties can ill afford, for the strength of their appeal remains based on group factors and on long-standing traditional allegiances. If this argument is correct, we should find empirical support for it in the attitudes and participation of party supporters in the United States and Western Europe. We should find, for example, that in Europe, as well as in the United States, sociodemographic factors rather than political disaffection significantly determine preference between major parties. European supporters of so-called "radical" parties should be no more disaffected than supporters of moderate parties. Furthermore, since disaffection may

[15] Sidney G. Tarrow, *Peasant Communism in Southern Italy* (New Haven: Yale University Press, 1967), contains one of the most recent and most insightful statements on the reform traditions of the Italian Communist Party. Even Joseph LaPalombara, who has taken strong issue with the thesis of the party's "deideologization," does not deny that the party has long been reform oriented. See Joseph LaPalombara, "Decline of Ideology: a Dissent and an Interpretation," *American Political Science Review*, LX (1966), 5–16.

[16] For an extreme example of this tendency, see Giovanni Sartori, "European Political Parties: The Case of Polarized Pluralism," in Joseph LaPalombara and Myron Weiner (eds.), *Political Parties and Political Development* (Princeton, N.J.: Princeton University Press, 1966), 137–76. Sartori's depiction of the Italian party system and of party conflict is far removed from mine.

be related to sociodemographic factors which, in turn, affect party preference, any relation between party preference and disaffection should be explained by the different social bases of parties. Finally, if major radical parties do not offer disaffected people stable and significant sources of counteridentification, disaffection should always present political participation, irrespective of country and of the kind of party one supports.

DATA AND MEASURES

The data for the United States, Germany, England, and Italy come from four sample surveys conducted in 1959 by Gabriel Almond and Sidney Verba.[17] In each country a multistage probability sample of approximately 1,000 adults was interviewed.

Mainly people have been examined who sympathize with the major parties: the Conservative and Labor parties in England, the Christian Democratic (CDU) and Social Democratic (SPD) parties in Germany, the Christian Democratic (DC) and the Communist (PCI) and Socialist (PSI)[18] parties in Italy, the Republican and Democratic parties in the United States. Minor parties and the parties of the "right" cannot be considered because too few respondents express sympathy with them. Parties of the "left" have been considered as the opposition and the possible repositories of disaffection because of their traditions of radicalism and protest. It is quite irrelevant for our purposes which party was in power at the moment of the survey.

One scale of political participation and several indices and scales tapping attitudes toward politics and the political system have been developed for this study. Interitem analysis was used to build the scales from a common pool of interview items. The internal reliability of each scale was calculated through a modified version of the Spearman-Brown split-half reliability formula.[19] All scales have at least .80 as a reliability coefficient. They are all scored in the direction suggested by their names and they all have a minimum score of zero.

Political Participation This 21-point multidimensional scale is made up of items concerning rather simple forms of participation, such as voting, seeking political information, discussing politics, belonging to a political club or organization, being informed about political leaders and institutions.

---

[17]Since 1967 the Italian Socialist party has again merged with the Italian Social Democratic party. At the time the data used in this study were collected, the former party had only recently broken its 1934 "pact of united action" with the Communist party over which the social Democrats had eventually seceded from the Socialists in 1947. The name of the new party is now United Socialist party.

[18]Gabriel A. Almond and Sidney Verba, *The Civic Culture* (Princeton, N.J.: Princeton University Press, 1963), appendices A and B.

[19]J. P. Guilford, *Psychometric Methods* (New York: McGraw-Hill Book Company, 1954), chap. 14.

$$\text{reliability} = \frac{nr}{1+(n-1)r}$$

n = number of items in the measure
r = mean interitem correlation coefficient for the matrix of items in the measure

Political Efficacy This 19-point scale has been built from statements of agreement or disagreement with items such as the following: "Some people say that politics and government are so complicated that the average man cannot really understand what is going on." "People like me don't have any say about what the government does." Measures with the same or similar names and similar items are found in many political survey studies.[20] Political Efficacy measures how the political actor evaluates his skills and resources, how he perceives the role which ordinary citizens are allowed to play in politics, and what he feels about the accessibility of the system to free and open communication.

System Proximity This 7-point scale taps the citizen's perception of the relative proximity or remoteness of the political system as well as the extent to which politics affect his life. The following item typifies the scale: "Thinking about your national government, about how much effect do you think its activities, the laws passed and so on, have on your day-to-day life? Do they have a great effect, some effect, or none?"

System Commitment This 13-point scale includes questions concerning the obligations which the respondents feel toward their government. The answers are used to rank respondents along a continuum. At the bottom of the continuum are placed people who feel no obligation and commitment to their system. They are followed by people who recognize only passive or legally sanctioned duties. At the top of the continuum are placed people who see their commitment as requiring more active and sophisticated types of performance, involving an obligation to use their civic and political rights fully.

System Satisfaction The System Satisfaction 8-point scale measures whether people evaluate the performance of the political system as detrimental or beneficial to its citizens, whether the operation of the government improves or worsens conditions in the country.

Given its content, the System Satisfaction measure seems to tap a dimension somewhat different from the other three attitude scales. Political Efficacy, System Proximity, and System Commitment seem to measure political disaffection: whether or not an individual rejects the political system, is alienated from it, or is capable of establishing any profitable relation with it. System Satisfaction, on the other hand, seems to imply an overall political evaluation, a simple statement of like or dislike for the system's output. A person can be displeased with how the government operates and yet feel committed to it, find politics highly relevant to his life, and feel that he is politically competent. System Satisfaction, used together with the other scales, will show to what extent basic disaffection (i.e., feelings tapped by the first three scales) rather than simple dissatisfaction (i.e., feelings tapped only by the last scale) endangers political participation.

[20] Angus Campbell, Gerald Gurin, and Warren E. Miller, *The Voter Decides* (Evanston, Ill.: Row, Peterson and Co., 1954), appendix A. Robert A. Dahl, *Who Governs?* (New Haven: Yale University Press, 1961). pp. 286–293. Elizabeth Douvan and Alan M. Walker, "The Sense of Effectiveness in Public Affairs," *Psychological Monographs*, LXX, 22 (1956). Heinz Eulau and Peter Schneider, "Dimensions of Political Involvement," *Public Opinion Quarterly*, XX (1956), 128–42.

DATA ANALYSIS

Sociodemographic Factors and Patry Preference While scoial group factors traditionally affect party preference, no single group factor should normally be a very strong predictor of it for various and well-known reasons. First, several group factors involved in party preference cross-cut, so that only rarely do people possess all of the group characteristics that pressure one toward a particular party preference. Second, the general increase in the affluence of Western societies contributes to the blurring of social and group barriers and differences, and creates a large middle-class stratum whose status is uncertain and whose partisan choices are less easily predicted. Third, parties themselves actively seek supporters among groups other than those which have been traditionally committed to them in order to enlarge their electoral following. This is obvious in two-party and modified two-party systems like England, Germany, and the United States. But there is also an incentive for parties to attempt expansion in any party system when party leaders believe that by appealing to new groups the party will not lose its traditional electorate. When there are no other strong parties to which a party's traditional electorate is likely to defect, this condition is met. Thus, both the Christian Democratic party and the extreme left in Italian politics can appeal to groups other than their traditional electorate because there are no substantial forces capable of challenging the Christian Democrats from the right or the Communists from the left.

The suggestion that group factors are not strong predictors of party preference and that the influence of these factors may actually be on the decline is not new and has been presented repeatedly.[21] Although basically correct, it can be overstated. Traditional group allegiances may remain in operation even after the "objective" reasons for such an allegiance have disappeared, possibly because of the vested interest that groups have in a party when they remain attached to it for a long time, or because of the stability of human habits and identifications.[22] The decline in the

---

[21] For several analyses of the social bases of party preference, see Mattei Dogan, "Les Clivages Politiques de la Classe Ouvrière," in Léo Hamon (ed.), *Les Nouveaux Comportments Politiques de la Classe Ouvrière* (Paris: Presses Universitaires de France, 1962), 101–27, and "La Stratificazione dei Suffragi," in Alebtro Spreafico and Joseph LaPalombara (eds.), *Elezioni e Comportamento Politico in Italia* (Milano: Edizioni di Comunitá, 1963), 407–74. Seymour M. Lipset, "Party Systems and the Representation of Social Groups," *European Journal of Sociology*, I (1960), 3–38. Seymour M. Lipset, et al., "The Psychology of Voting: an Analysis of Political Behavior," in Gardner Lindzey (ed.), *Handbook of Social Psychology*, Vol. II, (Cambridge, Mass.: Addison-Wesley Publishing Co., Inc., 1954), 1124–1175. Ralf Dahrendorf, "Recent Changes in the Class Structure of European Societies," in Stephen R. Graubard (ed.), *A New Europe?* (Boston: Houghton Mifflin Company, 1964), 291–336. Juan Linz, "The Social Bases of German Politics" (unpublished Ph.D. dissertation, Columbia University, 1958). Otto Kirchheimer, "Germany: the Vanishing Opposition," in Dahl, *Political Oppositions*, chap. 7.

[22] On this point see V. O. Key, Jr. and Frank Munger, "Social Determinism and Electoral Decision: the Case of Indiana," in Eugene Burdick and Arthur J. Brodbeck (eds.), *American Voting Behavior* (Glencoe: The Free Press, 1959), chap. 15; and Raymond E. Wolfinger, "The Development and Persistence of Ethnic Voting," *American Political Science Review*, LIX (1965), 896–908. See also Richard Hamilton for evidence that increased affluence is accompanied by changes in social structure which may favor an increase in leftist votes. Rich-

importance of group factors may actually be taking place at a slower rate than we think[23] and, in fact, there is yet only uncertain evidence that such a decline is taking place either in the United States or in Europe.[24] Finally, because some group factors such as religious and regional cleavages may decrease in importance, the relative importance of other group factors like occupation or socioeconomic status may increase.[25] Despite all these considerations, it nevertheless remains true that today's parties are rarely able to build their own strength on any specific group in society, and that no group of major size, except for the Negroes in the United States, gives its overwhelming allegiance to a single party.

Table 1 demonstrates that single social group factors often have an impact on party preference, although such impact is rarely very strong. The social factors considered include education, family income, personal occupation (if employed) or occupation of the head of the house, sex, and religiosity (as measured through reported church attendance).[26] The table reports the effect of single group factors on party preference before and after statistical adjustments for third factors. The adjusted percentages are obtained through a version of partial regression analysis applied to categorical variables, and they estimate the genuine effect of each group factor on party preference after the effect of all other group factors is accounted for.[27] Thus, for example, by comparing adjusted with unadjusted percentages, it is clear that the overwhelming support which women in Italy and Germany give the Christian Democratic party is entirely due to their other group-characteristics and, in particular, to their deep religious attachments. The role of social group factors appears even clearer in Table 2, which reports Morgan's beta coefficients for the relation between group factors and party preference. Morgan's betas are analogous to partial beta coefficients for continuous variables, and indicate the amount of change in

---

ard F. Hamilton, *Affluence and the French Worker in the Fourth Republic* (Princeton, N.J.: Princeton University Press, 1967).

[23] Alessandro Pizzorno, "The Individualistic Mobilization of Europe," in Graubard, *A New Europe?*, pp. 265–90.

[24] See Seymour M. Lipset, "The Changing Class Structure and Contemporary European Politics," in Graubard, *op. cit.*, pp. 337–69. Also Robert A. Alford, *Party and Society* (Chicago: Rand McNally & Co., 1963).

[25] *Ibid.*, chaps. 11 and 12, esp. pp. 326–36.

[26] The exact nature of religiosity as a group factor remains uncertain. We are not investigating what other characteristics accompany it, to what extent religious or nonreligious people represent a homogeneous and distinctive group, and what cultural factors mark them. Equally unexplored are the reasons why religiosity should be linked to party preference, and why this should be true in some countries but not in others. See, on some of these points and on religious and secular-radical appeals to the working–class electorate, Charles Y. Glock and Rodney Stark, *Religion and Society in Tension* (Chicago: Rand McNally & Co., 1965). Our purpose is not to explore these points, but simply to emphasize that religious people in some countries have traditionally shown certain party preferences.

[27] The test has been developed by Alan Wilson. See Alan Wilson, "Analysis of Multiple Cross–Clasifications in Cross–Sectional designs," revision of a paper presented to the American Association for Public Opinion Research (Excelsior Springs, Missouri, May, 1964). The version of the test used here assumes additivity, which is a fair assumption. A feature of the test allows a comparison between actual and estimated party preference for each cell of a full cross-tabulation of all group factors. No significant differences emerge between actual and estimated percentages, except when cells contain very few cases.

Table 1. SYMPATHIZERS FOR TWO MAJOR PARTIES PREFERRING THE LEFT PARTY BY SOCIODEMOGRAPHIC FACTORS AND WITHIN NATION—UNADJUSTED AND ADJUSTED PERCENTAGES[1]

		ENGLAND			GERMANY			ITALY[2]			USA		
		(N)	UNAD.	AD.	(N)	UNAD.	AD.	(N)	UNAD.	AD.	(N)	UNAD.	AD.
Sex:	Male	(342)	54%	53%	(200)	53%	50%	(127)	32%	24%	(340)	59%	59%
	Female	(321)	52	52	(197)	42	45	(104)	18	27	(363)	62	63
Church Attendance:	Weekly	(132)	52	54	(130)	22	23	(153)	6	6	(333)	62	62
	Less	(531)	53	52	(267)	60	59	(78)	64	64	(370)	60	59
Education:	Elementary	(402)	60	55	(342)	52	50	(151)	28	25	(229)	68	66
	More	(261)	42	50	(55)	22	30	(80)	20	27	(474)	58	59
Occupation:	Manual	(441)	66	66	(251)	57	53	(130)	32	29	(412)	68	68
	Non-manual	(222)	26	28	(146)	32	38	(101)	17	21	(291)	51	51
Income:	Low	(334)	61	55	(108)	47	45				(180)	59	54
	High	(329)	44	51	(289)	48	48				(523)	62	63

[1] Percentages are adjusted for all sociodemographic factors except the one which is analyzed. All adjustment factors are dichotomized. The dividing point for yearly income is £650 for England, DM 4,200 for Germany, $3,000 for the USA. Income is not used in Italy because the number of respondents reporting income is too low for adjustment. At any rate, even before adjustment, low income in Italy does not lead to preference for the left (24% of low-income and 25% of high-income respondents prefer the left).

[2] In Italy Communists and Socialists have been merged to represent the left party.

Table 2. SOCIODEMOGRAPHIC FACTORS AND PARTY PREFERENCE—MORGAN BETAS[1]

	SEX	CHURCH ATTENDANCE	EDUCATION	OCCUPATION	INCOME
England	...	...	...	.35	...
Germany		.34	.15	.15	...
Italy		.63	...	.10	...
USA	...	...	...	.17	...

[1]Morgan betas below .10 are not reported. All factors are dichotomized as in Table 1.

party preference due to a standardized change in one group factor after the other group factors have been controlled.[28] Tests of significance are not reported either in these tables or those to be presented later since the sampling procedures followed in the four countries make all conventional tests too generous.

From the analysis three group factors of genuine importance for party preference emerge: education, occupation, and religiosity. Only occupation is important in the United States and England, particularly so in England, a country where cultural and regional cohesiveness and lack of substantial ethnic and religious cleavages leave occupational status as one of the very few significant group factors.[29] Education and religiosity are the key factors in Germany. Religiosity, much more than any other factor, explains party preference in Italy.[30]

Disaffection and Party Preference

The arguments and evidence given so far suggest that there is little or no relation between political disaffection and preference for the parties of the left, and that any relation found is an artifact of the somewhat different social bases of the parties. Hence, group factors rather than disaffection determine party choice.

It is possible that while disaffection is not what motivates party preference, certain parties unintentionally obtain a share of support from the disaffected because they traditionally enjoy the allegiance of social groups in which disaffection tends to be higher. However, even if this is true, it does not necessarily mean that disaffection

[28]On Morgan betas, see James W. Morgan, *et al.*, *Income and Welfare in the United States* (New York: McGraw-Hill Book Company, 1962), appendix E.

[29]The findings concerning the effects of occupation in England are very close to the survey findings reported by the English Gallup for all post-war elections. See Henry Durant, "Voting Behaviour in Britain, 1945–1964," in Richard Rose (ed.), *Studies in British Politics* (London: Macmillan & Co. Ltd., 1966), pp. 122–28.

[30]The fact that religiosity is the main genuine factor in the voter's choice between Communism and Christian Democracy is well known to the Communist party, which has made the so-called "dialogue with the Catholics" one of the cornerstones of its electoral politics. Many authors, however, continue to emphasize the importance of socioeconomic factors. See, among others, Dogan, "Les Clivages Politiques," and most recently Giordano Sivini, "Gli Iscritti alla Democrazia Cristiana e al Partito Comunista Italiano," *Rassegna di Sociologia*, III (1967), 429–70. The great but unrecognized importance of religiosity as a factor in party alignments in many European countries has been recently stressed by Converse. See Philip E. Converse, "Some Priority Variables in Comparative Electoral Research," *Occasional Paper #3* (Glasgow: University of Strathclyde). Powell's study of politics in an Austrian community reports that religiosity has an impact on party preference that far exceeds that of social status factors. See G. Bingham Powell, Jr., *Fragmentation and Political Hostility in an Austrian Community* (unpublished Ph.D. dissertation, Stanford University, 1968).

will be greater among supporters of the left than it is among supporters of the center.

First, although leftist parties differ somewhat from center parties in the social background of their supporters, each party is still rather heterogeneous in its social bases. Indeed, only in Italy, and only in the case of religiosity, have we found a social group factor which explains practically all variation in party preference. Otherwise all parties in all countries receive support from people of all social backgrounds.

Second, statistical evidence from our survey data (not presented here for reasons of space) indicates that social group factors, in their turn, are only moderately related to political disaffection.[31] In none of the countries considered is disaffection the exclusive property of any of the groups which are important for party preference. Furthermore, while low socioeconomic status leads to disaffection and to support for the left, being a woman or being religious leads, at least in some countries, to disaffection and to support for the center. Hence some parties of the center also receive their share of support from the disaffected.

Political disaffection is tapped by our measures of Political Efficacy, System Proximity, and System Commitment. By way of contrast, with disaffection I will consider how dissatisfaction with system performance—measured by System Satisfaction—is related to party preference. Owing to the more directly politico-ideological overtones of the System Satisfaction measure, we cannot exclude the possibility of some relation between dissatisfaction and preference for the parties of the left. Indeed, what is involved in the measure is not exactly deep alienation from the political system as such, but a simpler statement that its operation and its policies do not improve living conditions.

Table 3 tends to confirm our expectations. The entries in the table are the mean attitude scores among the sympathizers of the center and the left parties before and after group factors have been controlled. The group factors are those considered in the previous tables, and the adjusted means are obtained through the same version of partial regression analysis used before. The evidence from England, Germany, and the United States is the clearest. At times, as in the case of England, disaffection and dissatisfaction are somewhat greater among sympathizers of the left parties, but the differences are not strong and tend to disappear when group factors are controlled. Here dissatisfaction does not behave differently from disaffection.

The case of Italy is less clearcut even after accounting for social group differences between parties. Italians who sympathize with the left not only show less satisfaction with the system, but also feel that the system is less close to their lives. This may mean that Italians in general, and not only Communists and Socialists, are unable or unwilling to evaluate their government independently of the political parties that happen to be in power and of their personal party preferences. (One should remember that, with the exception of the Fascist period, Italians have known only governments of the center.) Or it may simply mean that resentment toward the

[31] This and other findings which are reported but not printed in this paper appear in my forthcoming book on mass politics and participation in Western democracies.

Table 3. MEAN POLITICAL ATTITUDES BY PARTY PREFERENCE (TWO MAJOR PARTIES) AND BY NATION. UNADJUSTED AND ADJUSTED FOR GROUP FACTORS[1]

	MEAN POLITICAL EFFICACY					
	UNADJUSTED				ADJUSTED	
	CENTER	(N)	LEFT	(N)	CENTER	LEFT
England	12.0	(313)	10.9	(350)	11.8	11.2
Germany	11.8	(208)	11.9	(189)	11.8	11.9
Italy	10.3	(172)	11.0	(59)	10.3	11.2
USA	12.7	(275)	12.1	(428)	12.4	12.3

	MEAN SYSTEM PROXIMITY					
	UNADJUSTED				ADJUSTED	
	CENTER	(N)	LEFT	(N)	CENTER	LEFT
England	4.1	(313)	4.0	(350)	4.0	4.1
Germany	4.1	(208)	4.2	(189)	4.0	4.3
Italy	4.2	(172)	3.1	(59)	4.2	2.8
USA	4.7	(275)	4.4	(428)	4.6	4.5

	MEAN SYSTEM COMMITMENT					
	UNADJUSTED				ADJUSTED	
	CENTER	(N)	LEFT	(N)	CENTER	LEFT
England	7.4	(313)	6.3	(350)	7.2	6.5
Germany	8.1	(208)	7.7	(189)	8.0	7.8
Italy	5.5	(172)	5.4	(59)	5.6	5.3
USA	9.1	(275)	8.7	(428)	8.9	8.9

	MEAN SYSTEM SATISFACTION					
	UNADJUSTED				ADJUSTED	
	CENTER	(N)	LEFT	(N)	CENTER	LEFT
England	7.0	(313)	6.6	(350)	6.9	6.7
Germany	7.1	(208)	6.8	(189)	7.1	6.9
Italy	6.9	(172)	4.6	(59)	6.9	4.6
USA	7.3	(275)	7.1	(428)	7.3	7.2

[1]Group factors relate both to party preference and to political attitudes. They include education, occupation, income, sex, and church attendance. In Italy income is not used because of the high number of nonrespondents. Factors are dichotomized as in Table 1.

system in Italy has, in fact, a leftist tradition and is still partially expressed through support for the parties of the left. The first possibility is actually more dangerous for the system than the second. If many Italians cannot hold a concept of the system as a stable entity endowed with the characteristics and functions which set it apart from any party in power, popular attachment to the system is very unstable and precarious. One wonders what might happen to the attitudes of supporters of the center if the government were to move rapidly to the left. While remoteness and dissatisfaction are linked with preference for the left, System Commitment is not. In the case of Political Efficacy we actually find a reversal, and people who sympathize with the left feel more powerful and politically competent, even after group factors are accounted for.

There may be a technical reason for some of the Italian findings. Other data analysis indicates that those Italian respondents who acknowledge a sympathy for the left are only a few out of a larger number of respondents who have hidden their leftist leanings. Indeed, the electoral strength of the Communist and Socialist parties is much greater than what emerges from the interviews. Thus the respondents who readily declare their sympathy for the left may be a special group with peculiar characteristics of commitment to their party, and of personal strength and articulateness. Because of this they may also be more effective politically and more outspoken about their opposition than the average leftist.

The findings on party preference and political attitudes are strengthened by another test whose results are not presented here. In this test I have controlled for degree of party identification by dividing the respondents into those who do not identify with any party, those who report simple leanings toward a party, those who declare themselves supporters of a party, and those who are members of a party or partisan club. Disaffection uniformly decreases as party identification increases, irrespective of party and country. The greatest disaffection occurs among people who do not sympathize with any party. Dissatisfaction, on the other hand, is not affected by the level of party identification, except among Italian Communists and Socialists, where dissatisfaction is highest among party members and lowest among leaners.

The findings on party identification signify that all the major parties tend to recruit their partisans and activists among the people who are least disaffected from their polity and to socialize them in their positive attitudes. In other words, the results of party recruitment and socialization are such that they bring partisans of different parties closer together in terms of some basic political values. Also, the similarity in basic values becomes greater as the involvement and identification with parties increases.

Disaffection and Participation—the Role of Parties

When the pattern of recruitment and socialization of the major American and European parties leaves disaffected people at the margins of party life and encourages the development of more positive political attitudes, then those people who remain disaffected will find it increasingly difficult to participate in politics, no matter what party they support and what country they come from.

Table 4 reports the average political participation in the four countries before and after adjusting for cross-national differences in the four measures of attitudes toward politics. The first column indicates clear differences in the participation of the four countries, especially between Italy and the other countries. The second column reveals that these differences are to a large extent due to corresponding national differences in political attitudes. Further evidence also proves that the strength of the association

Table 4. MEAN POLITICAL PARTICIPATION UNADJUSTED AND ADJUSTED FOR FOUR POLITICAL ATTITUDE SCALES[1]

	(N)	MEANS UNADJUSTED	ADJUSTED
England	(963)	7.0	6.8
Germany	(955)	7.3	6.9
Italy	(995)	4.4	6.0
USA	(970)	7.9	6.8

[1] The political attitude scales are Political Efficacy, System Proximity, System Commitment, and System Satisfaction.

between attitude measure and participation is similar across countries and that the impact of the System Satisfaction on participation results from the former's association with the three measures of disaffection. In all countries the more disaffected people are, the less they participate; dissatisfaction, however, does not preclude participation.

The relation between attitudes and participation does not change when party preference is considered. Tables 5 to 8 report the average political participation scores separately for sympathizers of the center and for those of the left, as affected by their political attitudes. There is no indication that any party, not even a party of opposition, gives its disaffected supporters an opportunity to participate which is equal or better than that of supporters who hold a more trusting and effective relation with their polity.

The case of dissatisfaction (Table 8) deserves separate attention. In all the

Table 5. MEAN POLITICAL PARTICIPATION BY EFFICACY AMONG SYMPATHIZERS OF TWO MAJOR PARTIES

		(N)	CENTER MEAN	(N)	LEFT MEAN	(N)
England:	Low	(207)	5.7	(80)	4.4	(127)
Political	Middle	(274)	8.0	(121)	6.6	(153)
Efficacy	High	(278)	9.3	(165)	8.8	(113)
	Total	(759)	8.1	(366)	6.5	(393)
Germany:	Low	(140)	4.5	(89)	5.8	(51)
Political	Middle	(157)	7.5	(83)	8.5	(74)
Efficacy	High	(259)	10.0	(147)	10.3	(112)
	Total	(556)	7.8	(319)	8.8	(237)
Italy:	Low	(131)	3.4	(107)	4.5	(24)
Political	Middle	(76)	6.3	(59)	5.7	(17)
Efficacy	High	(95)	9.4	(69)	8.5	(26)
	Total	(302)	5.9	(235)	6.4	(67)
USA:	Low	(149)	4.9	(53)	4.5	(96)
Political	Middle	(255)	7.9	(194)	7.7	(161)
Efficacy	High	(311)	10.7	(106)	9.5	(205)
	Total	(769)	8.8	(307)	7.8	(462)

Table 6. MEAN POLITICAL PARTICIPATION BY SYSTEM PROXIMITY AMONG SYMPATHIZERS OF TWO MAJOR PARTIES

		(N)	CENTER MEAN	(N)	LEFT MEAN	(N)
England: System Proximity	Low	(227)	6.4	(96)	5.3	(131)
	Middle	(253)	7.9	(136)	6.3	(117)
	High	(279)	9.4	(134)	7.8	(145)
	Total	(759)	8.1	(366)	6.5	(393)
Germany: System Proximity	Low	(156)	6.3	(90)	7.1	(66)
	Middle	(146)	7.9	(95)	8.1	(51)
	High	(254)	8.7	(134)	10.0	(120)
	Total	(556)	7.8	(319)	8.8	(237)
Italy: System Proximity	Low	(104)	4.5	(68)	5.7	(36)
	Middle	(97)	5.3	(75)	6.8	(22)
	High	(101)	7.3	(92)	8.1	(9)
	Total	(302)	5.9	(235)	6.4	(67)
USA: System Proximity	Low	(126)	5.0	(42)	5.4	(84)
	Middle	(246)	8.4	(97)	7.1	(149)
	High	(397)	10.1	(168)	9.2	(229)
	Total	(769)	8.8	(307)	7.8	(462)

Table 7. MEAN POLITICAL PARTICIPATION BY SYSTEM COMMITMENT AMONG SYMPATHIZERS OF TWO MAJOR PARTIES

		(N)	CENTER MEAN	(N)	LEFT MEAN	(N)
England: System Commitment	Low	(225)	6.5	(87)	4.7	(138)
	Middle	(311)	8.2	(154)	7.0	(157)
	High	(223)	8.9	(125)	8.3	(98)
	Total	(759)	8.1	(366)	6.5	(393)
Germany: System Commitment	Low	(111)	5.4	(63)	6.2	(48)
	Middle	(209)	7.6	(125)	9.0	(84)
	High	(236)	9.2	(131)	9.7	(105)
	Total	(556)	7.8	(319)	8.8	(237)
Italy: System Commitment	Low	(138)	4.2	(107)	5.7	(31)
	Middle	(74)	6.2	(58)	6.6	(16)
	High	(90)	8.1	(70)	7.3	(20)
	Total	(302)	5.9	(235)	6.4	(67)
USA: System Commitment	Low	(175)	5.4	(60)	5.2	(115)
	Middle	(122)	8.2	(50)	8.5	(72)
	High	(472)	10.1	(197)	8.7	(275)
	Total	(769)	8.8	(307)	7.8	(462)

Table 8. MEAN POLITICAL PARTICIPATION BY SYSTEM SATISFACTION AMONG SYMPATHIZERS OF TWO MAJOR PARTIES

		(N)	CENTER MEAN	(N)	LEFT MEAN	(N)
England: System Satisfaction	Low	(139)	7.1	(60)	6.1	(79)
	Middle	(199)	7.9	(87)	6.5	(112)
	High	(421)	8.4	(219)	6.7	(202)
	Total	(759)	8.1	(366)	6.5	(393)
Germany: System Satisfaction	Low	(64)	5.0	(29)	7.3	(35)
	Middle	(197)	7.2	(107)	9.3	(90)
	High	(295)	8.6	(183)	8.8	(112)
	Total	(556)	7.8	(319)	8.8	(237)
Italy: System Satisfaction	Low	(77)	4.7	(43)	7.0	(34)
	Middle	(69)	4.5	(52)	6.6	(17)
	High	(156)	6.7	(140)	5.1	(16)
	Total	(302)	5.9	(235)	6.4	(67)
USA: System Satisfaction	Low	(67)	6.9	(22)	5.6	(45)
	Middle	(236)	8.3	(91)	8.1	(145)
	High	(466)	9.3	(194)	8.0	(272)
	Total	(769)	8.8	(307)	7.8	(462)

parties except the Italian left dissatisfaction behaves like disaffection: it is inversely related to participation. However, when dissatisfaction is adjusted for the three measures of disaffection, I have found that any relation between dissatisfaction and participation disappears. Again, sheer dissatisfaction with the operation of the government does not by itself prevent people from participating. In the Italian left participation is actually higher among dissatisfied people, even before controlling for disaffection. What prevents many Socialists and Communists who are dissatisfied from participating even more is their accompanying deeper disaffection. In a sense, these people find themselves in a cross-pressured situation, for their parties encourage dissatisfaction and penalize disaffection, two attitudes which unfortunately tend to be linked in the mind of the average voter.

CONCLUSIONS

This study has investigated followers' attitudes and behavior, more than linkages between followers' characteristics and party politics. I have speculated, however, on some of these linkages, and will conclude with some additional speculations.

Even if great political disaffection appears within a party or society, it is naive to expect that this would be automatically reflected in the style of politics of parties and leaders and in the role that disaffected people play. As Lijphart, Powell, and MacRae argue in their discussions of political cleavages and fragmentation in European societies, political leaders may actually act to bridge cleavages and to avoid some of their disruptive consequences.[32] This may increasingly be the case as cleavages become sharper and potentially more threatening, provided the leaders share a high stake in the preservation of the system and can rely on established mechanisms and institutions designed to mediate conflict. European societies and their institutions are differentiated enough that a degree of leeway can exist between mass behavior and the working of elites and institutions.

The same argument can be made with greater cogency when what is at stake is not only social and ideological cleavages in the mass public, but a more basic popular disaffection from politics which may threaten the very preservation of essential political processes and institutions. There are, to be sure, ways in which disaffection may filter into party politics. At times leaders may find it expedient to pay lip-service to it so as to maintain certain parts of their traditional electorate. Because of its long alliance with some marginal stratum of society, a party may recruit, wittingly or unwittingly, among the disaffected and thus accumulate a disproportionate share of them in its ranks. However, these aspects of mass politics have, at best, only indirect effects on the way the party operates as a national organization and allocates power and responsibility among its members. Disaffected people still appear to remain marginal within their parties.

A case in point is the Labor party in England. Our evidence clearly indicates that the Labor and the Conservative parties differ in their social composition and that Labor supporters show some aspects of disaffection not shown by the conservatives. Some authors have argued that the degree of factionalism which characterizes the internal party politics of British Labor—between the trade unions, local constituencies, and national leadership—reflects some elements of disaffection and resentment against the system.[33] Disaffection within the ranks of Labor, however, hardly finds expression in the official policies and the style of politics of the Labor party.

The case of Italy is even more illuminating. The Italian society, by comparison with other Western societies, contains deep and widespread disaffection.[34] Yet disaffection is not always a significant and major motivation in party alignment; at times the Communist and Socialist parties are electorally advantaged by it, yet disaffection even among Communists and Socialists does not lead to participation, and rarely motivates the appeals, strategy, and ideology of these parties. In sum, one crucial intervening variable between disaffection and its expression through politics and participation is the nature of the party's response to it and the nature of the party system. What matters is not only how the rank and file of parties typically feel toward the political system, but also, and especially, how the party leaders behave

---

[32] Arend Lijphart, *The Politics of Accommodation* (Berkeley: University of California Press, 1968). Powell, *Fragmentation and Political Hostility*. Duncan MacRae, Jr., *Parliament, Parties, and Society in France 1946–1958* (New York: St. Martin's Press, Inc., 1967).

[33] See Robert T. McKenzie, *British Political Parties* (New York: St. Martin's Press, Inc., 1955), Parts II and III; also Martin Harrison, *The Trade Unions and the Labour Party Since 1945* (London: George Allen and Unwin Ltd., 1960).

[34] The point is amply illustrated in Almond and Verba, *The Civic Culture*, and in Joseph LaPalombara, "Italy: Fragmentation, Isolation, and Alienation," in Lucien W. Pye and Sidney Verba (eds.), *Political Culture and Political Development* (Princeton, N.J.: Princeton University Press, 1965), pp. 282–329. It is further explored in my forthcoming book.

and what choices of action they offer their followers. Disaffection is, per se, a motive for withdrawal, and leads to action only if it finds leadership and organizational support.

There are some doubts whether this support has ever existed historically. Concerning the United States, Dahl's historical analysis emphasizes that disaffection was condemned to ineffectiveness from the very beginning of the American polity, owing to the relative lack of precise social cleavages, and, especially, to the major parties' failure to express them.[35] In a similar vein, Samuel Huntington suggests that since the United States began as a consensual society, with a strong egalitarian social inheritance and without an entrenched social class powerful enough to significantly resist change, mass participation spread early and automatically and did not have to follow class lines.[36] Further, the process was not a significant source of conflict over which government and parties organized and clashed. Thus participation in the United States was probably nourished from the beginning by a climate of consensus and attachment to the political system.

By contrast, in Europe—and especially in Continental Europe—disaffection and conflict were often the essence and the source of attachment to political oppositions. Popular participation was achieved through the operation of working-class political movements, was nurtured by a climate of political tension, was opposed by established traditional interests—and it developed in the long struggle for citizenship. In Germany and Italy meaningful universal suffrage was not granted until after World War I.[37] It is still open to question, however, how extensively this development of Continental mass politics did, in fact, motivate participation by the disaffected.

Popular movements sought to give political organization and direction to disaffection by offering the disaffected a new set of social goals—a total alternative to the contemporary society. Further, they helped the disaffected participate because they offered internal opportunities for equality, making equality a new means to bring about change. Popular oppositions, in other words, realized that participation by the marginal and disaffected can be achieved only if a sense of personal power, purpose, and commitment to a community is recreated, and if disaffection from the present society is replaced by identification with a future society and its present harbingers.

This type of subcultural participation, however, revealed itself to be difficult to achieve and, in the long range, unstable and difficult to maintain because, as Alessandro Pizzorno suggests, revolutionary goals and means were operative only as

---

[35] Dahl, *Political Oppositions*.

[36] Samuel P. Huntington, "Political Modernization: America vs. Europe," *World Politics*, XVIII (1966), esp. 405–7.

[37] Prussia had a system of plural voting by tax classes which practically barred all lower-class people from representation until 1919. In the German Reich universal manhood suffrage was recognized from 1870, but the popularly elected Reichstag was practically devoid of power until its fall. Italy did not achieve universal manhood suffrage until the acts of 1912 and 1919. For an account of the timing in the spread of mass suffrage, see Stein Rokkan, "The Comparative Study of Political Participation: Notes toward a Perspective on Current Research," in Austin Ranney (ed.), *Essays on the Behavioral Study of Politics* (Urbana: University of Illinois Press, 1962), esp. pp. 72–90.

popular movements formed, and were thereafter superseded by new organizational demands.[38] The movements, even while maintaining countervalues, were compelled to act in the larger society, to accept some of its values and processes, and to adopt an internal division of labor that reflected external demands. Hence movements that did not achieve their original objectives in a short period of time tended to revert to inequality as an organizational principle and to develop adaptive goals as their code of operation. Equality and the organization of the disaffected were replaced by a more selective strategy of political education designed for a limited number of party activists who eventually developed at least as much commitment to adaptive goals and organizational means as to ultimate revolutionary objectives.

In sum, as European popular oppositions have progressively become established mass organizations, they have limited their ability to reach the disaffected and to educate their potential followers—a limitation almost built in the operation of mass organizations. The irony here—an irony well understood by Roberto Michels in his treatment of social democracy at the turn of the century—is that the very success of the oppositions in training a hard core of politically educated and skillful party activists with a stake in the larger politics set a ceiling against internal participation which ultimately deprived party followers, especially the socially and psychologically marginal, of a further opportunity to participate in the larger society.[39]

This has been invariably true whether or not the changes which popular movements have undergone have been total and irreversible. Some movements, as Pizzorno suggests of Italy, have acquired a double, if transitional, personality; they accept the system's values and rules of the game at their leadership level, but remain isolated and disaffected subcultures among the rank and file.[40] Other movements have progressed somewhat more swiftly toward becoming at all levels a more integral part of the larger society. In either case the consequences are close; the traditional oppositions leave no significant role for the disaffected to play and offer only some of them an opportunity to change their sentiments.

If the traditional oppositions have long abandoned their original goals, is their room for new and significant forces of dissent in the politics of Western democracies? Both in Europe and the United States small root-and-branch movements of unreconstructed opposition, at times devoid of electoral ambitions, have been rising. These movements of opinion often contain libertarian characteristics appealing to restricted elite, intellectual, and youth groups. Their stated purposes are debate and the search for a new style of politics.

At times these groups are splinter parties or factions within established parties; they may also appeal to small centers of old working-class opposition. Such is the case of certain intellectual and labor union groups within the English Labor party, and of small "Chinese" factions inside and outside many European Communist

---

[38] Alessandro Pizzorno, "Introduzione allo Studio della Partecipazione Politica," *Quaderni di Sociologia*, XV (1966), esp. 256–61 and 273–77. My paraphrase of Pizzorno's argument in the above paragraph does not do justice to the richness of his reasoning.

[39] Roberto Michels, *Political Parties* (New York: Crowell Collier and Macmillan, Inc.,

parties. A more successful group with some electoral following is the Italian Socialist Party of Proletarian Unity, a coalition of Socialist intellectuals and cadres who recently abandoned the Socialist party, and who in many ways place themselves to the left of the Communist party.[41]

Often these groups are in no way related to parties and play no partisan role, although they may have ambitions of political power. They may be outside political competition and the political rules of the establishment, and therefore more directly represent intellectual and moral interests, as seen by the New Left and the peace movements in England and the United States, and the recent student movements in the United States and in many European countries.

Whether and how these forces may channel certain types of disaffection and become the nucleus for new alternatives and new modes of political conflict thus become significant issues. Nothing that has been written in the previous pages denies that conflict and ideology are very much alive in industrial societies. However, all major political forces, including traditional oppositions, tend to agree on a set of stringent rules regulating and managing conflict. One objective of some of the new oppositions is to challenge these rules because, from their viewpoint, such rules resist change, discourage debate and participation, and prevent consideration of new issues and new alternatives. Some social commentators see in the rise of new oppositions significant innovations and an anticipation of the types of conflict which will characterize the post-industrial societies. Others consider them a safety valve in a consensual society or, in some instances, reactionary expressions by historical irrelevants.[42] To discriminate the new from the old and the relevant from the irrelevant in these political forces will require future studies.

[41] In the national elections of May 19, 1968, the party won 4.5 percent of the votes, apparently at the expense of the Socialist party.
[42] See, on the last point, Zbigniew Brzezinski, "Revolution and Counterrevolution," *The New Republic*, CLXII n. 23 (June 1, 1968), 23–25.

Part Five

THE POLITICAL SYSTEM OF THE SOVIET UNION

INTRODUCTION

V. I. Lenin was both a scholar and a politician—though more the latter than the former. When he made predictions about a Communist classless society (Chapter 21), he was not only interested in comparing the stages of social and political evolution abstractly, but also in providing a blueprint for Communist Russia. Now, after more than half a century of Soviet rule, the Soviet Union has obviously diverged from Lenin's model. Milovan Djilas, the Yugoslav Communist leader who later became a critic of Communism, considers both the extent of the divergence and the causes of it from a comparative perspective in Chapter 22. Like Lenin, he distinguishes a number of different phases of Communist development, but his phases do not coincide with those described by Lenin.

Most analyses of the Soviet Union focus on the centrally important factor: the exercise and maintenance of power through the party and state apparatus. The remaining chapters in Part V are comparative treatments of different aspects of the Soviet Union by a historian, an economist, and a legal expert. In Chapter 23, Henry L. Roberts compares and contrasts Russia—not just the Soviet Union—with the West. As he points out, the many similarities and differences depend on the aspect subjected to comparison and the criteria applied to it. In Chapter 24, W. Donald Bowles discusses the important, but technically complex, question of Soviet economic development and the relevance of the Soviet economic model to the underdeveloped nations of the Third World. His analysis is illuminating not only for those interested in economics and in the underdeveloped areas, but also for students of politics, particularly Soviet politics. Bowles' comparative analysis is deliberately directed to the general reader; he emphasizes: "the use of technical jargon is minimized." In the final chapter Harold J. Berman discusses Soviet law. As the chapter title indicates, he compares Soviet Law and legal procedures and practices with American ones, but also with British and Continental European law—particularly that of France and Germany. In examining the Soviet political system, it is important to remember that law can be an alternative to and a safeguard against power, as well as an instrument through which power is exercised.

The reader is also referred to Chapter 3 by Carl J. Friedrich and Zbigniew Brzezinski and Chapter 4 by Gabriel A. Almond for discussions of the general characteristics of totalitarian political systems.

ABOUT THE AUTHORS

V. I. LENIN *was the leader of the Bolshevik Revolution of 1917 and the first ruler of the Soviet Union. Chapter 21 is an excerpt from his book* State and Revolution. *Another famous work by Lenin is* Imperialism: The Highest Stage of Capitalism.

MILOVAN DJILAS *was a leader of the Yugoslav Communist party and successively a minister, president of the Parliament, and vice president of Yugoslavia until 1954, when he was expelled from the party. He was imprisoned from 1956 to 1961 and again from 1962 to 1966. The* New Class *is only one of the books of this prolific author; other titles include* On New Roads to Socialism, Conversations with Stalin, *and* Land Without Justice.

HENRY L. ROBERTS *is professor of history at Columbia University, and was the director of Columbia's Russian Institute from 1956 to 1962. Among his works is the well-known book* Russia and America: Dangers and Prospects.

W. DONALD BOWLES, *professor of economics at American University, has written several other studies on the economy of the Soviet Union.*

HAROLD J. BERMAN, *professor of law at the Harvard Law School, has authored* Justice in the USSR. *He has also written a number of other books and articles primarily on various aspects of Soviet law.*

CHAPTER 21

The Nature of Communist Society

V. I. LENIN

TRANSITION FROM CAPITALISM TO COMMUNISM

Between capitalist and Communist society—Marx states—lies the period of the revolutionary transformation of the former into the latter. To this also corresponds a political transition period, in which the state can be no other than *the revolutionary dictatorship of the proletariat.**

This conclusion Marx bases on an analysis of the rôle played by the proletariat in modern capitalist society, on the data concerning the evolution of this society, and on the irreconcilability of the opposing interests of the proletariat and the bourgeoisie.

Critique of the Social-Democratic Programmes.—Ed.

Reprinted from V. I. Lenin, STATE AND REVOLUTION *(New York: International Publishers, 1932), pp. 71–85, by permission of INTERNATIONAL PUBLISHERS CO. INC. Copyright © 1932. Extract title has been provided by editor.*

Earlier the question was put thus: to attain its emancipation, the proletariat must overthrow the bourgeoisie, conquer political power and establish its own revolutionary dictatorship.

Now the question is put somewhat differently: the transition from capitalist society, developing towards Communism, towards a Communist society, is impossible without a "political transition period," and the state in this period can only be the revolutionary dictatorship of the proletariat.

What, then, is the relation of this dictatorship to democracy?

We have seen that the *Communist Manifesto* simply places side by side the two ideas: the "transformation of the proletariat into the ruling class" and the "establishment of democracy." On the basis of all that has been said above, one can define more exactly how democracy changes in the transition from capitalism to Communism.

In capitalist society, under the conditions most favourable to its development, we have more or less complete democracy in the democratic republic. But this democracy is always bound by the narrow framework of capitalist exploitation, and consequently always remains, in reality, a democracy for the minority, only for the possessing classes, only for the rich. Freedom in capitalist society always remains just about the same as it was in the ancient Greek republics: freedom for the slave-owners. The modern wage-slaves, owing to the conditions of capitalist exploitation, are so much crushed by want and poverty that "democracy is nothing to them," "politics is nothing to them"; that, in the ordinary peaceful course of events, the majority of the population is debarred from participating in social and political life.

The correctness of this statement is perhaps most clearly proved by Germany, just because in this state constitutional legality lasted and remained stable for a remarkably long time—for nearly half a century (1871-1914)—and because Social-Democracy in Germany during that time was able to achieve far more than in other countries in "utilising legality," and was able to organise into a political party a larger proportion of the working class than anywhere else in the world.

What, then, is this largest proportion of politically conscious and active wage-slaves that has so far been observed in capitalist society? One million members of the Social-Democratic Party—out of fifteen million wage-workers! Three million organised in trade unions— out of fifteen million!

Democracy for an insignificant minority, democracy for the rich—that is the democracy of capitalist society. If we look more closely into the mechanism of capitalist democracy, everywhere, both in the "petty"—so-called petty—details of the suffrage (residential qualification, exclusion of women, etc.), and in the technique of the representative institutions, in the actual obstacles to the right of assembly (public buildings are not for "beggars"!), in the purely capitalist organisation of the daily press, etc., etc.—on all sides we see restriction after restriction upon democracy. These restrictions, exceptions, exclusions, obstacles for the poor, seem slight, especially in the eyes of one who has himself never known want and has never been in close contact with the oppressed classes in their mass life (and nine-tenths, if not ninety-nine hundredths, of the bourgeois publicists and politicians are of this class), but in their sum total these restrictions exclude and squeeze out the poor from politics and from an active share in democracy.

Marx splendidly grasped this *essence* of capitalist democracy, when, in analysing the experience of the Commune, he said that the oppressed were allowed, once every few years, to decide which particular representatives of the oppressing class should be in parliament to represent and repress them!

But from this capitalist democracy—inevitably narrow, subtly rejecting the poor, and therefore hypocritical and false to the core—progress does not march onward, simply, smoothly and directly, to "greater and greater democracy," as the liberal professors and petty-bourgeois opportunists would have us believe. No, progress marches onward, *i.e.*, towards Communism, through the dictatorship of the proletariat; it cannot do otherwise, for there is no one else and no other way to *break the resistance* of the capitalist exploiters.

But the dictatorship of the proletariat—*i.e.*, the organisation of the vanguard of the oppressed as the ruling class for the purpose of crushing the oppressors—cannot produce merely an expansion of democracy. *Together* with an immense expansion of democracy which *for the first time* becomes democracy for the poor, democracy for the people, and not democracy for the rich folk, the dictatorship of the proletariat produces a series of restrictions of liberty in the case of the oppressors, the exploiters, the capitalists. We must crush them in order to free humanity from wage-slavery; their resistance must be broken by force; it is clear that where there is suppression there is also violence, there is no liberty, no democracy.

Engels expressed this splendidly in his letter to Bebel when he said, as the reader will remember, that "as long as the proletariat still *needs* the state, it needs it not in the interests of freedom, but for the purpose of crushing its antagonists; and as soon as it becomes possible to speak of freedom, then the state, as such, ceases to exist."

Democracy for the vast majority of the people, and suppression by force, *i.e.*, exclusion from democracy, of the exploiters and oppressors of the people—this is the modification of democracy during the *transition* from capitalism to Communism.

Only in Communist society, when the resistance of the capitalists has been completely broken, when the capitalists have disappeared, when there are no classes (*i.e.*, there is no difference between the members of society in their relation to the social means of production), *only then* "the state ceases to exist," and "*it becomes possible to speak of freedom.*" Only then a really full democracy, a democracy without any exceptions, will be possible and will be realised. And only then will democracy itself begin to *wither away* due to the simple fact that, freed from capitalist slavery, from the untold horrors, savagery, absurdities and infamies of capitalist exploitation, people will gradually *become accustomed* to the observance of the elementary rules of social life that have been known for centuries and repeated for thousands of years in all school books; they will become accustomed to observing them without force, without compulsion, without subordination, without the *special apparatus* for compulsion which is called the state.

The expression "the state *withers away*," is very well chosen, for it indicates both the gradual and the elemental nature of the process. Only habit can, and undoubtedly will, have such an effect; for we see around us millions of times, how readily people get accustomed to observe the necessary rules of life in common, if there is no exploitation, if there is nothing that causes indignation, that calls forth protest and revolt and has to be *suppressed*.

Thus, in capitalist society, we have a democracy that is curtailed, poor, false; a democracy only for the rich, for the minority. The dictatorship of the proletariat, the period of transition to Communism, will, for the first time, produce democracy for the people, for the majority, side by side with the necessary suppression of the minority—the exploiters. Communism alone is capable of giving a really complete democracy, and the more complete it is the more quickly will it become unnecessary and wither away of itself.

In other words: under capitalism we have a state in the proper sense of the word, that is, special machinery for the suppression of one class by another, and of the majority by the minority at that. Naturally, for the successful discharge of such a task as the systematic suppression by the exploiting minority of the exploited majority, the greatest ferocity and savagery of suppression are required, seas of blood are required, through which mankind is marching in slavery, serfdom, and wage-labour.

Again, during the *transition* from capitalism to Communism, suppression is *still* necessary; but it is the suppression of the minority of exploiters by the majority of exploited. A special apparatus, special machinery for suppression, the "state," is *still* necessary, but this is now a transitional state, no longer a state in the usual sense, for the suppression of the minority of exploiters, by the majority of the wage slaves *of yesterday*, is a matter comparatively so easy, simple and natural that it will cost far less bloodshed than the suppression of the risings of slaves, serfs or wage labourers, and will cost mankind far less. This is compatible with the diffusion of democracy among such an overwhelming majority of the population, that the need for *special machinery* of suppression will begin to disappear. The exploiters are, naturally, unable to suppress the people without a most complex machinery for performing this task; but *the people* can suppress the exploiters even with very simple "machinery," almost without any "machinery," without any special apparatus, by the simple *organisation of the armed masses* (such as the Soviets of Workers' and Soldiers' Deputies, we may remark, anticipating a little).

Finally, only Communism renders the state absolutely unnecessary, for there is *no one* to be suppressed—"no one" in the sense of a *class*, in the sense of a systematic struggle with a definite section of the population. We are not Utopians, and we do not in the least deny the possibility and inevitability of excesses on the part of *individual persons*, nor the need to suppress *such* excesses. But, in the first place, no special machinery, no special apparatus of repression is needed for this; this will be done by the armed people itself, as simply and as readily as any crowd of civilised people, even in modern society, parts a pair of combatants or does not allow a woman to be outraged. And, secondly, we know that the fundamental social cause of excesses which consist in violating the rules of social life is the exploitation of the masses, their want and their poverty. With the removal of this chief cause, excesses will inevitably begin to "*wither away.*" We do not know how quickly and in what succession, but we know that they will wither away. With their withering away, the state will also *wither away*.

Without going into Utopias, Marx defined more fully what can *now* be defined regarding this future, namely, the difference between the lower and higher phases (degrees, stages) of Communist society.

FIRST PHASE OF COMMUNIST SOCIETY

In the *Critique of the Gotha Programme*, Marx goes into some detail to disprove the Lassallean idea of the workers' receiving under Socialism the "undiminished" or "full product of their labour." Marx shows that out of the whole of the social labour of society, it is necessary to deduct a reserve fund, a fund for the expansion of production, for the replacement of worn-out machinery, and so on; then, also, out of the means of consumption must be deducted a fund for the expenses of management, for schools, hospitals, homes for the aged, and so on.

Instead of the hazy, obscure, general phrase of Lassalle's—"the full product of his labour for the worker"—Marx gives a sober estimate of exactly how a Socialist society will have to manage its affairs. Marx undertakes a *concrete* analysis of the conditions of life of a society in which there is no capitalism, and says:

> What we are dealing with here [analysing the programme of the party] is not a Communist society which has *developed* on its own foundations, but, on the contrary, one which is just *emerging* from capitalist society, and which therefore in all respects—economic, moral and intellectual—still bears the birthmarks of the old society from whose womb it sprung.*

And it is this Communist society—a society which has just come into the world out of the womb of capitalism, and, which, in all respects, bears the stamp of the old society—that Marx terms the "first," or lower, phase of Communist society.

The means of production are no longer the private property of individuals. The means of production belong to the whole of society. Every member of society, performing a certain part of socially-necessary work, receives a certificate from society to the effect that he has done such and such a quantity of work. According to this certificate, he receives from the public warehouses, where articles of consumption are stored, a corresponding quantity of products. Deducting that proportion of labour which goes to the public fund, every worker, therefore, receives from society as much as he has given it.

"Equality" seems to reign supreme.

But when Lassalle, having in view such a social order (generally called Socialism, but termed by Marx the first phase of Communism), speaks of this as "just distribution," and says that this is "the equal right of each to an equal product of labour," Lassalle is mistaken, and Marx exposes his error.

"Equal right," says Marx, we indeed have here; but it is *still* a "bourgeois right," which, like every right, *presupposes inequality*. Every right is an application of the *same* measure to *different* people who, in fact, are not the same and are not equal to

---

*Ibid.—Ed.

one another; this is why "equal right" is really a violation of equality, and an injustice. In effect, every man having done as much social labour as every other, receives an equal share of the social products (with the above-mentioned deductions).

But different people are not alike: one is strong, another is weak; one is married, the other is not; one has more children, another has less, and so on.

> ... With equal labour—Marx concludes—and therefore an equal share in the social consumption fund, one man in fact receives more than the other, one is richer than the other, and so forth. In order to avoid all these defects, rights, instead of being equal, must be unequal.*

The first phase of Communism, therefore, still cannot produce justice and equality; differences, and unjust differences, in wealth will still exist, but the *exploitation* of man by man will have become impossible, because it will be impossible to seize as private property the *means* of *production*, the factories, machines, land, and so on. In tearing down Lassalle's petty-bourgeois, confused phrase about "equality" and "justice" *in general*, Marx shows the *course of development* of Communist society, which is forced at first to destroy *only* the "injustice" that consists in the means of production having been seized by private individuals, and which *is not capable* of destroying at once the further injustice consisting in the distribution of the articles of consumption "according to work performed" (and not according to need).

The vulgar economists, including the bourgeois professors and also "our" Tugan-Baranovsky, constantly reproach the Socialists with forgetting the inequality of people and with "dreaming" of destroying this inequality. Such a reproach, as we see, only proves the extreme ignorance of the gentlemen propounding bourgeois ideology.

Marx not only takes into account with the greatest accuracy the inevitable inequality of men; he also takes into account the fact that the mere conversion of the means of production into the common property of the whole society ("Socialism" in the generally accepted sense of the word) *does not remove* the defects of distribution and the inequality of "bourgeois right" which *continue to rule* as long as the products are divided "according to work performed."

> But these defects—Marx continues—are unavoidable in the first phase of Communist society, when, after long travail, it first emerges from capitalist society. Justice can never rise superior to the economic conditions of society and the cultural development conditioned by them.**

And so, in the first phase of Communist society (generally called Socialism) "bourgeois right" is *not* abolished in its entirety, but only in part, only in proportion to the economic transformation so far attained, *i.e.*, only in respect of the means of production. "Bourgeois right" recognises them as the private property of separate individuals. Socialism converts them into common property. *To that extent*, and to that extent alone, does "bourgeois right" disappear.

*Ibid.—Ed.
**Ibid.—Ed.

However, it continues to exist as far as its other part is concerned; it remains in the capacity of regulator (determining factor) distributing the products and allotting labour among the members of society. "He who does not work, shall not eat"—this Socialist principle is *already* realised; "for an equal quantity of labour, an equal quantity of products"—this Socialist principle is also *already* realised. However, this is not yet Communism, and this does not abolish "bourgeois right," which gives to unequal individuals, in return for an unequal (in reality unequal) amount of work, an equal quantity of products.

This is a "defect," says Marx, but it is unavoidable during the first phase of Communism; for, if we are not to fall into Utopianism, we cannot imagine that, having overthrown capitalism, people will at once learn to work for society *without any standards of right*; indeed, the abolition of capitalism *does not immediately lay* the economic foundations for *such* a change.

And there is no other standard yet than that of "bourgeois right." To this extent, therefore, a form of state is still necessary, which, while maintaining public ownership of the means of production, would preserve the equality of labour and equality in the distribution of products.

The state is withering away in so far as there are no longer any capitalists, any classes, and, consequently, no *class* can be suppressed.

But the state has not yet altogether withered away, since there still remains the protection of "bourgeois right" which sanctifies actual inequality. For the complete extinction of the state, complete Communism is necessary.

HIGHER PHASE OF COMMUNIST SOCIETY

Marx continues:

> In a higher phase of Communist society, when the enslaving subordination of individuals in the division of labour has disappeared, and with it also the antagonism between mental and physical labour; when labour has become not only a means of living, but itself the first necessity of life; when, along with the all-around development of individuals, the productive forces too have grown, and all the springs of social wealth are flowing more freely—it is only at that stage that it will be possible to pass completely beyond the narrow horizon of bourgeois rights, and for society to inscribe on its banners: from each according to his ability; to each according to his needs!*

Only now can we appreciate the full correctness of Engels' remarks in which he mercilessly ridiculed all the absurdity of combining the words "freedom" and "state." While the state exists there is no freedom. When there is freedom, there will be no state.

The economic basis for the complete withering away of the state is that high stage of development of Communism when the antagonism between mental and physical labour disappears, that is to say, when one of the principal sources of modern *social* inequality disappears—a source, moreover, which it is impossible to remove immediately by the mere conversion of the means of production into public property, by the mere expropriation of the capitalists.

*Ibid.—Ed.

This expropriation will make a gigantic development of the productive forces *possible*. And seeing how incredibly even now, capitalism *retards* this development, how much progress could be made even on the basis of modern technique at the level it has reached, we have a right to say, with the fullest confidence, that the expropriation of the capitalists will inevitably result in a gigantic development of the productive forces of human society. But how rapidly this development will go forward, how soon it will reach the point of breaking away from the division of labour, of removing the antagonism between mental and physical labour, of transforming work into the "first necessity of life"—this we do not and *cannot* know.

Consequently, we have a right to speak solely of the inevitable withering away of the state, emphasising the protracted nature of this process and its dependence upon the rapidity of development of the *higher phase* of Communism; leaving quite open the question of lengths of time, or the concrete forms of withering away, since material for the solution of such questions is *not available*.

The state will be able to wither away completely when society has realised the rule: "From each according to his ability; to each according to his needs," *i.e.*, when people have become accustomed to observe the fundamental rules of social life, and their labour is so productive, that they voluntarily work *according to their ability*. "The narrow horizon of bourgeois rights," which compels one to calculate, with the hard-heartedness of a Shylock, whether he has not worked half an hour more than another, whether he is not getting less pay than another—this narrow horizon will then be left behind. There will then be no need for any exact calculation by society of the quantity of products to be distributed to each of its members; each will take freely "according to his needs."

From the bourgeois point of view, it is easy to declare such a social order "a pure Utopia," and to sneer at the Socialists for promising each the right to receive from society, without any control of the labour of the individual citizen, any quantity of truffles, automobiles, pianos, etc. Even now, most bourgeois "savants" deliver themselves of such sneers, thereby displaying at once their ignorance and their self-seeking defence of capitalism.

Ignorance—for it has never entered the head of any Socialist to "promise" that the highest phase of Communism will arrive; while the great Socialists, in *foreseeing* its arrival, presupposed both a productivity of labour unlike the present and a person not like the present man in the street, capable of spoiling, without reflection, like the seminary students in Pomyalovsky's book,* the stores of social wealth, and of demanding the impossible.

Until the "higher" phase of Communism arrives, the Socialists demand the *strictest* control, *by society and by the state*, of the quantity of labour and the quantity of consumption; only this control must *start* with the expropriation of the capitalists, with the control of the workers over the capitalists, and must be carried out, not by a state of bureaucrats, but by a state of *armed workers*.

Self-seeking defence of capitalism by the bourgeois ideologists (and their

---

*Pomyalovsky's *Seminary Sketches* depicted a group of student-ruffians who engaged in destroying things for the pleasure it gave them.—*Ed.*

hangers-on like Tsereteli, Chernov and Co.) consists in that they *substitute* disputes and discussions about the distant future for the essential imperative questions of present-day policy: the expropriation of the capitalists, the conversion of *all* citizens into workers and employees of *one* huge "syndicate"—the whole state—and the complete subordination of the whole of the work of this syndicate to the really democratic state of the *Soviets of Workers' and Soldiers' Deputies.*

In reality, when a learned professor, and following him some philistine, and following the latter Messrs. Tsereteli and Chernov, talk of the unreasonable Utopias, of the demagogic promises of the Bolsheviks, of the impossibility of "introducing" Socialism, it is the higher stage or phase of Communism which they have in mind, and which no one has ever promised, or even thought of "introducing," for the reason that, generally speaking, it cannot be "introduced."

And here we come to that question of the scientific difference between Socialism and Communism, upon which Engels touched in his above-quoted discussion on the incorrectness of the name "Social-Democrat." The political difference between the first, or lower, and the higher phase of Communism will in time, no doubt, be tremendous; but it would be ridiculous to emphasise it now, under capitalism, and only, perhaps, some isolated Anarchist could invest it with primary importance (if there are still some people among the Anarchists who have learned nothing from the Plekhanov-like conversion of the Kropotkins, the Graveses, the Cornelissens, and other "leading lights" of Anarchism to social-chauvinism or Anarcho-*Jusquaubout*-ism,* as Ge, one of the few Anarchists still preserving honour and conscience, has expressed it).

But the scientific difference between Socialism and Communism is clear. What is generally called Socialism was termed by Marx the "first" or lower phase of Communist society. In so far as the means of production become *public* property, the word "Communism" is also applicable here, providing we do not forget that it is *not* full Communism. The great significance of Marx's elucidations consists in this: that here, too, he consistently applies materialist dialectics, the doctrine of evolution, looking upon Communism as something which evolves *out of* capitalism. Instead of artificial, "elaborate," scholastic definitions and profitless disquisitions on the meaning of words (what Socialism is, what Communism is), Marx gives an analysis of what may be called stages in the economic ripeness of Communism.

In its first phase or first stage Communism *cannot* as yet be economically ripe and entirely free of all tradition and of all taint of capitalism. Hence, the interesting phenomenon of Communism retaining, in its first phase, "the narrow horizon of bourgeois rights." Bourgeois rights, with respect to distribution of articles of *consumption*, inevitably presupposes, of course, the existence of the *bourgeois state*, for rights are nothing without an apparatus capable of *enforcing* the observance of the rights.

Consequently, for a certain time not only bourgeois rights, but even the bourgeois state remains under Communism, without the bourgeoisie!

---

*Jusquaubout—combination of the French words meaning "until the end." Anarcho-*Jusqaubout*-ism—Anarcho-until-the-End-ism.—*Ed.*

This may look like a paradox, or simply a dialectical puzzle for which Marxism is often blamed by people who would not make the least effort to study its extraordinarily profound content.

But, as a matter of fact, the old surviving in the new confronts us in life at every step, in nature as well as in society. Marx did not smuggle a scrap of "bourgeois" rights into Communism of his own accord; he indicated what is economically and politically inevitable in a society issuing *from the womb* of capitalism.

Democracy is of great importance for the working class in its struggle for freedom against the capitalists. But democracy is by no means a limit one may not overstep; it is only one of the stages in the course of development from feudalism to captialism, and from capitalism to Communism.

Democracy means equality. The great significance of the struggle of the proletariat for equality, and the significance of equality as a slogan, are apparent, if, we correctly interpret it as meaning the abolition of *classes*. But democracy means only *formal* equality. Immediately after the attainment of equality for all members of society *in respect of* the ownership of the means of production, that is, of equality of labour and equality of wages, there will inevitably arise before humanity the question of going further from formal equality to real equality, *i.e.*, to realising the rule, "From each according to his ability; to each according to his needs." By what stages, by means of what practical measures humanity will proceed to this higher aim —this we do not and cannot know. But it is important to realise how infinitely mendacious is the usual bourgeois presentation of Socialism as something lifeless, petrified, fixed once for all, whereas in reality, it is *only* with Socialism that there will commence a rapid, genuine, real mass advance, in which first the *majority* and then the whole of the population will take part—an advance in all domains of social and individual life.

Democracy is a form of the state—one of its varieties. Consequently, like every state, it consists in organised, systematic application of force against human beings. This on the one hand. On the other hand, however, it signifies the formal recognition of the equality of all citizens, the equal right of all to determine the structure and administration of the state. This, in turn, is connected with the fact that, at a certain stage in the development of democracy, it first rallies the proletariat as a revolutionary class against capitalism, and gives it an opportunity to crush, to smash to bits, to wipe off the face of the earth the bourgeois state machinery—even its republican variety; the standing army, the police, and bureaucracy; then it substitutes for all this a *more* democratic, but still a state machinery in the shape of armed masses of workers, which becomes transformed into universal participation of the people in the militia.

Here "quantity turns into quality": *such* a degree of democracy is bound up with the abandonment of the framework of bourgeois society, and the beginning of its Socialist reconstruction. If *every one* really takes part in the administration of the state, capitalism cannot retain its hold. In its turn, capitalism, as it develops, itself creates *prerequisites* for "every one" *to be able* really to take part in the administration of the state. Among such prerequisites are: universal literacy, already realised in most of the advanced capitalist countries, then the "training and disciplining" of millions of workers by the huge, complex, and socialised apparatus of the post-office, the railways, the big factories, large-scale commerce, banking, etc., etc.

With such *economic* prerequisites it is perfectly possible, immediately, within twenty-four hours after the overthrow of the capitalists and bureaucrats, to replace them, in the control of production and distribution, in the business of *control* of labour and products, by the armed workers, by the whole people in arms. (The question of control and accounting must not be confused with the question of the scientifically educated staff of engineers, agronomists and so on. These gentlemen work today, obeying the capitalists; they will work even better tomorrow, obeying the armed workers.)

Accounting and control—these are the *chief* things necessary for the organising and correct functioning of the *first phase* of Communist society. *All* citizens are here transformed into hired employees of the state, which is made up of the armed workers. *All* citizens become employees and workers of *one* national state "syndicate." All that is required is that they should work equally, should regularly do their share of work, and should receive equal pay. The accounting and control necessary for this have been *simplified* by capitalism to the utmost, till they have become the extraordinarily simple operations of watching, recording and issuing receipts, within the reach of anybody who can read and write and knows the first four rules of arithmetic.*

When the *majority* of the people begin everywhere to keep such accounts and maintain such control over the capitalists (now converted into employees) and over the intellectual gentry, who still retain capitalist habits, this control will really become universal, general, national; and there will be no way of getting away from it, there will be "nowhere to go."

The whole of society will have become one office and one factory, with equal work and equal pay.

But this "factory" discipline, which the proletariat will extend to the whole of society after the defeat of the capitalists and the overthrow of the exploiters, is by no means our ideal, or our final aim. It is but a *foothold* necessary for the radical cleansing of society of all the hideousness and foulness of capitalist exploitation, *in order to advance further.*

From the moment when all members of society, or even only the overwhelming majority, have learned how to govern the state *themselves*, have taken this business into their own hands, have "established" control over the insignificant minority of capitalists, over the gentry with capitalist leanings, and the workers thoroughly demoralised by capitalism—from this moment the need for any government begins to disappear. The more complete the democracy, the nearer the moment when it begins to be unnecessary. The more democratic the "state" consisting of armed workers,

---

*When most of the functions of the state are reduced to this accounting and control by the workers themselves, then it ceases to be a "political state," and the "public functions will lose their political character and be transformed into simple administrative functions" (*cf.* above, Chap. IV, § 2 on Engels' polemic against the Anarchists).

which is "no longer a state in the proper sense of the word," the more rapidly does *every* state begin to wither away.

For when *all* have learned to manage, and independently are actually managing by themselves social production, keeping accounts, controlling the idlers, the gentlefolk, the swindlers and similar "guardians of capitalist traditions," then the escape from this national accounting and control will inevitably become so increasingly difficult, such a rare exception, and will probably be accompanied by such swift and severe punishment (for the armed workers are men of practical life, not sentimental intellectuals, and they will scarcely allow any one to trifle with them), that very soon the *necessity* of observing the simple, fundamental rules of every-day social life in common will have become a *habit*.

The door will then be wide open for the transition from the first phase of Communist society to its higher phase, and along with it to the complete withering away of the state.

CHAPTER 22

The Essence of Contemporary Communism

MILOVAN DJILAS

1.

None of the theories on the essence of contemporary Communism treats the matter exhaustively. Neither does this theory claim to do so. Contemporary Communism is the product of a series of historical, economic, political, ideological, national, and international causes. A categorical theory about its essence cannot be entirely accurate.

The essence of contemporary Communism could not even be perceived until, in the course of its development, it revealed itself to its very entrails. This moment came, and could only come, because Communism entered a particular phase of its development—that of its maturity. It then became possible to reveal the nature of its power, ownership, and ideology. In the time that Communism was developing and was predominantly an ideology, it was almost impossible to see through it completely.

Just as other truths are the work of many authors, countries, and movements, so it is with contemporary Communism. Communism has been revealed gradually, more

Reprinted from Milovan Djilas, THE NEW CLASS: AN ANALYSIS OF THE COMMUNIST SYSTEM *(New York: Frederick A. Praeger, Inc., 1957), Chap. 8, by permission of the publisher. Extract title provided by editor.*

or less parallel to its development; it cannot be looked upon as final, because it has not completed its development.

Most of the theories regarding Communism, however, have some truth in them. Each of them has usually grasped one aspect of Communism or one aspect of its essence.

There are two basic theses on the essence of contemporary Communism.

The first of them claims that contemporary Communism is a type of new religion. We have already seen that it is neither a religion nor a church, in spite of the fact that it contains elements of both.

The second thesis regards Communism as revolutionary socialism, that is, something which was born of modern industry, or capitalism, and of the proletariat and its needs. We have seen that this thesis also is only partially accurate: contemporary Communism began in well-developed countries as a socialist ideology and a reaction against the suffering of the working masses in the industrial revolution. But after having come into power in underdeveloped areas, it became something entirely different—an exploiting system opposed to most of the interests of the proletariat itself.

The thesis has also been advanced that contemporary Communism is only a contemporary form of despotism, produced by men as soon as they seize power. The nature of the modern economy, which in every case requires centralized administration, has made it possible for this despotism to be absolute. This thesis also has some truth in it: modern Communism is a modern despotism which cannot help but aspire toward totalitarianism. However, all types of modern despotism are not variants of Communism, nor are they totalitarian to the degree that Communism is.

Thus whatever thesis we examine, we find that each thesis explains one aspect of Communism, or a part of the truth, but not the entire truth.

Neither can my theory on the essence of Communism be accepted as complete. This is, anyway, the weakness of every definition, especially when such complex and living matters as social phenomena are being defined.

Nevertheless, it is possible to speak in the most abstract theoretical way about the essence of contemporary Communism, about what is most essential in it, and what permeates all its manifestations and inspires all of its activity. It is possible to penetrate deeper into this essence, to elucidate its various aspects; but the essence itself has already been exposed.

Communism, and likewise its essence, is continuously changing from one form to another. Without this change it cannot even exist. Consequently, these changes require continuous examination and a deeper study of the already obvious truth.

The essence of contemporary Communism is the product of particular conditions, historical and others. But as soon as Communism becomes strong, the essence itself becomes a factor and creates the conditions for its own continued existence. Consequently, it is evident that it is necessary to explain the essence separately according to the form and the conditions in which it appears and is operating at a given moment.

2.

The theory that contemporary Communism is a type of modern totalitarianism is not only the most widespread, but also the most accurate. However, an actual understanding of the term "modern totalitarianism" where Communism is being discussed is not so widespread.

Contemporary Communism is that type of totalitarianism which consists of three basic factors for controlling the people. The first is power; the second, ownership; the third, ideology. They are monopolized by the one and only political party, or—according to my previous explanation and terminology—by a new class; and, at present, by the oligarchy of that party or of that class. No totalitarian system in history, not even a contemporary one—with the exception of Communism—has succeeded in incorporating simultaneously all these factors for controlling the people to this degree.

When one examines and weighs these three factors, power is the one which has played and still continues to play the most important role in the development of Communism. One of the other factors may eventually prevail over power, but it is impossible to determine this on the basis of present conditions. I believe that power will remain the basic characteristic of Communism.

Communism first originated as an ideology, which contained in its seed Communism's totalitarian and monopolistic nature. It can certainly be said that ideas no longer play the main, predominant role in Communism's control of the people. Communism as an ideology has mainly run its course. It does not have many new things to reveal to the world. This could not be said for the other two factors, power and ownership.

It can be said: power, either physical, intellectual, or economic, plays a role in every struggle, even in every social human action. There is some truth in this. It can also be said: in every policy, power, or the struggle to acquire and keep it, is the basic problem and aim. There is some truth in this also. But contemporary Communism is not only such a power; it is something more. It is power of a particular type, a power which unites within itself the control of ideas, authority, and ownership, a power which has become an end in itself.

To date, Soviet Communism, the type which has existed the longest and which is the most developed, has passed through three phases. This is also more or less true of other types of Communism which have succeeded in coming to power (with the exception of the Chinese type, which is still predominantly in the second phase).

The three phases are: revolutionary, dogmatic, and non-dogmatic Communism. Roughly speaking, the principal catch-words, aims, and personalities corresponding to these various phases are: Revolution, or the usurpation of power—Lenin. Socialism," or the building of the system—Stalin. "Legality," or stabilization of the system —"collective leadership."

It is important to note that these phases are not distinctly separate from one another, that elements of all are found in each. Dogmatism abounded, and the "building in socialism" had already begun, in the Leninist period; Stalin did not renounce revolution, or reject the dogmas, which interfered with the building of the system. Present-day, non-dogmatic Communism is only non-dogmatic conditionally: it just will not renounce even the minutest practical advantages for dogmatic reasons. Precisely because of such advantages, it will at the same time be in a position to persecute unscrupulously the minutest doubt concerning the truth or purity of the dogma. Thus, Communism, proceeding from practical needs and capabilities, has today even furled the sails of revolution, or of its own military expansion. But it has not renounced one or the other.

This division into three phases is only accurate if it is taken roughly and abstractly. Clearly separate phases do not actually exist, nor do they correspond to specific periods in the various countries.

The boundaries between the phases, which overlap, and the forms in which the phases appear are varied in different Communist countries. For example, Yugoslavia has passed through all three phases in a relatively short time and with the same personalities at the summit. This is obvious in both precepts and method of operation.

Power plays a major role in all three of these phases. In the revolution it was necessary to seize power; in the building of socialism, it was necessary to create a new system by means of that power; today power must preserve the system.

During the development, from the first to the third phase, the quintessence of Communism—power—evolved from being the means and became an end in itself. Actually power was always more or less the end, but Communist leaders, thinking that through power as a means they would attain the ideal goal, did not believe it to be an end in itself. Precisely because power served as a means for the Utopian transformation of society, it could not avoid becoming an end in itself and the most important aim of Communism. Power was able to appear as a means in the first and second phases. It can no longer be concealed that in the third phase power is the actual principal aim and essence of Communism.

Because of the fact that Communism is being extinguished as an ideology, it must maintain power as the main means of controlling the people.

In revolution, as in every type of war, it was natural to concentrate primarily on power: the war had to be won. During the period of industrialization, concentrating on power could still be considered natural: the construction of industry, or a "socialist society," for which so many sacrifices had been made, was necessary. But as all this is being completed, it becomes apparent that in Communism power has not only been a means but that it has also become the main, if not the sole, end.

Today power is both the means and the goal of Communists, in order that they may maintain their privileges and ownership. But since these are special forms of power and ownership, it is only through power itself that ownership can be exercised. Power is an end in itself and the essence of contemporary Communism. Other classes may be able to maintain ownership without a monopoly over power, or power without a monopoly over ownership. Until now, this has not been possible for the new class, which was formed through Communism; it is very improbable that it will be possible in the future.

Throughout all three of these phases, power has concealed itself as the hidden, invisible, unspoken, natural and principal end. Its role has been stronger or weaker depending on the degree of control over the people required at the time. In the first phase, ideas were the inspiration and the prime mover for the attainment of power; in the second phase, power operated as the whip of society and for its own maintenance; today, "collective ownership" is subordinated to the impulses and needs of power.

Power is the alpha and the omega of contemporary Communism, even when Communism strives to prevent this.

Ideas, philosophical principles and moral considerations, the nation and the people, their history, in part even ownership—all can be changed and sacrificed. But not power. Because this would signify Communism's renunciation of itself, of its own essence. Individuals can do this. But the class, the party, the oligarchy cannot. This is the purpose and the meaning of its existence.

Every type of power besides being a means is at the same time an end—at least for those who aspire to it. Power is almost exclusively an end in Communism, because it is both the source and the guarantee of all privileges. By means of and through power the material privileges and ownership of the ruling class over national goods are realized. Power determines the value of ideas, and suppresses or permits their expression.

It is in this way that power in contemporary Communism differs from all other types of power, and that Communism itself differs from every other system.

Communism has to be totalitarian, exclusive, and isolated precisely because power is the most essential component of Communism. If Communism actually could have had other ends, it would have to make it possible for other forces to spring up in opposition and operate independently.

How contemporary Communism will be defined is secondary. Everyone who undertakes the work of explaining Communism finds himself faced with the problem of defining it, even if actual conditions do not compel him to do this—conditions in which Communists glorify their system as "socialism," "classless society," and "the realization of men's eternal dreams," while the opposing element defines Communism as an insensitive tyranny, the chance success of a terroristic group, and the damnation of the human race.

Science must use already established categories in order to make a simple exposition. Is there any category in sociology into which we can cram contemporary Communism if we use a little force?

In common with many authors who started from other positions, I have, in recent years, equated Communism with state capitalism or, more precisely, with total state capitalism.

This interpretation won out among the leaders of Yugoslav Communists during the time of their clash with the government of the U.S.S.R. But just as Communists, according to practical needs, easily change even their "scientific" analysis, Yugoslav party leaders changed this interpretation after the "reconciliation" with the Soviet government, and once more proclaimed the U.S.S.R. a Socialist country. At the same time, they proclaimed the Soviet imperialistic attack on the independence of Yugoslavia—in Tito's words—a "tragic," "incomprehensible" event, evoked by the "arbitrariness of individuals."

Contemporary Communism for the most part does resemble total state capitalism. Its historical origin and the problems which it had to solve—namely, an industrial transformation similar to the one achieved by capitalism but with the aid of the state mechanism—lead to such a conclusion.

If, under Communism, the state were the owner in the name of society and of the nation, then the forms of political power over society would inevitably change according to the varying needs of society and of the nation. The state by its nature is an organ of unity and harmony in society, and not only a force over it. The state could not be both the owner and ruler in itself. In Communism it is reversed: The state is an instrument and always subordinate exclusively to the interests of one and the same exclusive owner, or of one and the same direction in the economy, and in the other areas of social life.

State ownership in the West might be considered more as state capitalism than it is in Communist countries. The claim that contemporary Communism is state *capitalism* is prompted by the "pangs of conscience" of those who were disillusioned by the Communist system, but who did not succeed in defining it; they therefore equate its evils with those of capitalism. Since there is really no private ownership in Communism but rather formal state ownership, nothing seems more logical than to attribute all evils to the state. This idea of state capitalism is also accepted by those who see "less evil" in private capitalism. Therefore they like to point out that Communism is a worse type of capitalism.

To claim that contemporary Communism is a transition to something else leads nowhere and explains nothing. What is not a transition to something else?

Even if it is accepted that it has many of the characteristics of an all-encompassing state capitalism, contemporary Communism also has so many of its own characteristics that it is more precise to consider it a special type of new social system.

Contemporary Communism has its own essence which does not permit it to be confused with any other. Communism, while absorbing into itself all kinds of other elements—feudal, capitalist, and even slave-owning—remains individual and independent at the same time.

CHAPTER 23

Russia and the West:
A Comparison and Contrast

HENRY L. ROBERTS

Comparisons of Russia with the "West" have been a staple of historians and of contemporary observers for a very long time, and no end is in sight. A recent appraisal of Soviet developments in the decade after the death of Stalin was devoted in part to a consideration of the prospects for "a gradual convergence of the social and/or political systems of the West and the Soviet Union."[1] The variety of the contributors' responses—"very likely," "necessarily uncertain," "unlikely any meaningful convergence," "highly improbable," "depends on what is meant by 'gradual'"—suggests an ample range of disagreement, both in expectations for the future and in the characterization of the contrasts underlying these expectations.

The Russians themselves have, of course, been perennially preoccupied with this act of comparison. As Sir Isaiah Berlin has observed, in speaking of the nineteenth century: "Russian publicists, historians, political theorists, writers on social topics, literary critics, philosophers, theologians, poets, first and last, all without exception and at enormous length, discuss such issues as what it is to be a Russian; the virtues, vices and destiny of the Russian individual and society; but above all the historic role of Russia among the nations; or, in particular, whether its social structure—say, the relation of intellectuals to the masses, or of industry to agriculture—is *sui generis*, or whether, on the contrary, it is similar to that of other countries, or, perhaps, an anomalous, or stunted, or an abortive example of some superior Western model."[2]

Such concerns are not uniquely Russian. Americans and Canadians, colonial offspring of European culture, have spent a great deal of time meditating on their relations to the Old World; the inhabitants of the British Isles continue to have ambivalent feelings about the Continent; the Germans, though situated in Central Europe, have written at length about the significance of Germany's Eastern and

---

[1]*Survey: A Journal of Soviet and East European Studies*, No. 47 (Apr., 1963), pp. 37–42.
[2]Isaiah Berlin, "The Silence in Russian Culture," in *The Soviet Union, 1922–1962: A Foreign Affairs Reader*, ed. Philip E. Mosely (New York: Praeger, for the Council on Foreign Relations, 1963), p. 337.

Reprinted from Henry L. Roberts, "Russia and the West: A Comparison and Contrast," SLAVIC REVIEW, *XXIII, No. 1 (March, 1964), 1–12, by permission of the author and the publisher.*

Western "faces"; in Italy they say that Europe stops somewhere south of Rome; indeed, of the major European nations only the French seem not to have been much bothered by this particular problem of identification. Still, the relative intensity and persistence of the preoccupation in the Russian case, the fact that at times it has loomed as *the* question in discussions of Russian society and culture, would indicate a somewhat special problem.

It should be noted at the outset that what is involved here is not simply a nation-to-nation comparison but rather the relationship of one country, Russia,[3] to a more complex entity, the "West," by which is usually meant Western Europe. This latter entity, though comprising a number of nations, is assumed to have a degree of unity, the possession by its members of common features, against which Russia can be compared and contrasted. In other words, the question really means: does Russia belong to the West or not, is it a part of the West or is it somehow alien from the cluster of nations? Historically it is clear that the pathos and passion this question has aroused derive from the issue of participation or nonparticipation. And while, as we shall see presently, the historian might prefer to deal with it in different terms, this issue still lies at the heart of most discussions of Russia and the West.

We can take, as an example, two articles appearing recently in this journal. The one presented Russia as belonging to an East European cultural sphere quite sharply differentiated from, and opposed to, that of Western Europe: Eastern-Orthodox-Byazntine as against Western-Catholic-Roman. It urged, moreover, that the terms "East" and "West" in this setting "are so specific and meaningful that it would be unwise to introduce new concepts even as working hypotheses."[4] The second article, in contrast, was inclined to argue that while there have been periods, usually sterile ones, of Russian self-sufficiency and isolation, Russia and the West have "a common logic of development, a shared process of evolution. . . . Russian culture has no vital existence of its own apart from Europe."[5] Although the authors are addressing different themes, the trend of their thought is clear: one sees Russia as essentially distinct from the West, the other as linked to and dependent on it.

The disconcerting feature of this divergence—and both positions have respectable ancestries—is not simply their apparent incompatibility but their plausibility and persuasiveness when presented in the course of the authors' argument. From these, and other examples, we must suppose that in considering Russia's relation to the West we are not dealing with a simple question of fact—otherwise it would have been settled long since—but with a more subtle and troubling problem.

In the face of conflicting interpretations which do not appear to arise from crude errors of fact, one can explore at least three possible avenues of explanation: (1) One may look for the warping presence of animus or prejudice as the source of trouble; (2) one may attempt to achieve a more satisfactory "perspective" that can

---

[3]"Russia," of course, comprised numerous nationalities, and the term has occasioned much debate. In this piece, however, I shall not attempt to deal with this problem. By Russia I mean the Russian state or the culture and society of its Great Russian inhabitants only.

[4]Omeljan Pritsak and John S. Reshetar, Jr., "The Ukraine and the Dialectics of Nation-Building," *Slavic Review*, XXII, No. 2 (June, 1963), 224-26.

[5]Rufus W. Mathewson, Jr., "Russian Literature and the West," *Slavic Review*, XXI, No. 3 (Sept., 1962), 413 and 417.

somehow encompass or reconcile the conflicting interpretations; or (3) one may conclude that each interpretation is substantially correct in its own context but together they are not reconcilable because they are answering quite different questions and intentions and are, in fact, operating on different planes of thought. We shall look at each of these possibilities in turn.

That an enormous amount of passion and animus has entered into comparisons of Russia and the West is perfectly obvious. One thinks, for example, of Dostoevsky's painful encounter with Turgenev in Baden in 1867. According to Dostoevsky, Turgenev "abused Russia and the Russians vilely and terribly," and told him that the fundamental point of his (Turgenev's) book *Smoke* lay in the sentence, "If Russia were to perish, it would cause neither loss nor distress to mankind."[6] In his account of the meeting, Turgenev, while denying that he would have expressed his intimate convictions to Dostoevsky, allowed that the latter had "relieved his feelings by violent abuse of the Germans, myself and my latest book."[7] We cannot go into the roots of this particular clash, but the passion evoked here by the Russia-West controversy is intense and unmistakable.

When Poles or Rumanians, despite the presence of linguistic or religious ties with the Russians and a fair measure of common if hardly joyful history, argue that Russia is not of the West, whereas their own nations most emphatically are, one feels that this is more than an academic classification, that it is an argument born of fear or desperation, and that the extrusion of Russia from the "West" is at the same time a call for support and assistance on the part of the Western nations. When a German author contends that the Russians, from the very beginning of their history, have been quite incapable of scientific and technological advance and have had to borrow and steal such knowledge from the West, which they hoped to overrun, one can agree with his enthusiastic translator that "this book is part of the Cold War."[8]

Undoubtedly the advent of the Soviet regime has greatly intensified passion and prejudice by placing Russia in the most violent possible antithesis to the rest of Europe: Communist Russia versus the Imperialist West. Moreover, the search for communism's Russian roots or antecedents, a natural and perfectly proper inquiry, has led to heightened and perhaps inappropriate emphasis on those features of the Russian past that would seem to mark it off most sharply from Western Europe: the prominence that has been given in recent years to Ivan the Terrible's *Oprichnina*, the Marquis de Custine's animadversions, and the murky character of Nechaev is surely in good part a reflection of present concerns.

And yet, while we may grant that when passions are strong the door is opened to the tendentious selection and misuse of evidence, we may doubt whether this can be defined as the major source of our difficulty. For one thing, the presence of passion or prejudice itself requires explanation, and that may lead us back, in circular fashion, to tensions inherent in the Russia-West comparison. For surely the fact that Dostoevsky and Turgenev, whatever their personal differences, should have clashed

---

[6] Jessie Coulson, *Dostoevsky: A Self-Portrait* (London: Oxford University Press, 1962), p. 163.

[7] *Ibid.*, p. 165.

[8] Werner Keller, *East Minus West = Zero: Russia's Debt to the Western World, 862–1962*, trans. Constantine Fitzgibbon (New York: Putnam, 1962), p. 7.

so violently on this subject does point to a peculiar quality in the Russian society of the time that should have made such great artists so painfully self-conscious about the national identity. One might have reservations about certain Polish or Rumanian views of Russian-Western relations, but it remains true that these nations have had long and intimate exposure to Russia: their fear of Russia as an alien intruder is at least derived from immediate experience. As a Rumanian writer remarked not long ago, with some acerbity: "There are some who feel that personal experience of the things described, or the fact that the writer has personally witnessed the events discussed, throws a suspicion of bias upon the author. A writer, in other words, is suspect precisely because he has too great and too close a knowledge of his subject. For our part, we feel that ignorance is not a guarantee of objectivity."[9]

So while we may strongly suspect that when we run across that tired phrase "Scratch a Russian and find a Tartar" we are not likely to get much enlightenment about either Russians or Tartars, it does not follow that no problem exists.[10] More than that, when we ask such a question as whether the 1917 Revolution brought Russian history closer to that of the West by placing it in the sequence of the other great "modernizing" revolutions of the last three centuries, or whether, on the contrary, it increased the distance by destroying, or disrupting, some potentially important convergent lines of development, we find ourselves faced with a real and quite intricate problem of historical interpretation, one that is not reducible to animus or partisanship.

Turning now from the role of animus, which while making the subject more prickly does seem to be marginal rather than central, we may consider some of the efforts that have been made to overcome, modulate, or get around the antithetical "either-or" of the Russian-Western relationship. It is my impression that these efforts have been quite fruitful in new insights, although, as we shall see, they tend to blur the Russia-West comparison or at least remove it from the center of the stage.

The most obvious approach is to replace the Russia-West polarity (with its overtones of an even more extreme Orient-Occident opposition) by the conception of a European "spectrum" ranging clear across the Continent, with changes occurring by degrees and shadings. This conception has the distinct advantage of calling in question the picture of the "West" as a homogeneous unit, which comparisons of Russia with the West so frequently posit. For example, the much vexed question of the existence of East European "feudalism" is cast in a rather different light when we are told that "the existence of a hierarchy is no longer thought to be a prerequisite to feudalism in the West, largely because the neat hierarchy assumed to have existed in the West is found to have been virtually a phantom."[11] Once this simple unity of the West is dissolved and the tremendous variety of its historical experience and its

---

[9] Constantin Visoianu, in the introduction to *Captive Rumania*, ed. Alexandre Cretzianu (New York: Praeger, 1956), p. xvi.

[10] Happily, this question of what the Russians *are* if they are not Western is beyond the scope of this paper. I am informed that a discussion of Russia and the East has been prepared for the preceding issue of this journal.

[11] Oswald P. Backus III, "The Problem of Feudalism in Lithuania, 1506–1548," *Slavic Review*, XXI, No. 4 (Dec., 1962), 650.

institutional and cultural forms is taken to heart, then the way is open to a much more flexible and subtle series of comparisons: within and between regions of Europe, and on different levels—religious, social, institutional, and the like. Moreover, if Europe is seen as a spectrum, one can then attempt to locate the smaller nations of Eastern Europe in a more relaxed fashion; when the West, or Western Europe, is presented as a sharply identifiable unit then there always is the painful scramble to determine who will be permitted to slip in under the tent.

I have the impression that there is much to gain through comparative studies in this vein—studies that would include Russia in the spectrum. For one example, recent investigations comparing the recruitment and social composition of the higher bureaucracy in the Habsburg and Hohenzollern monarchies[12] could profitably be extended to include imperial Russia. For another, I should like to see a close historical study of the correlation, if any, between certain patterns of landholding and leasing and peasant unrest from France and western Germany eastward to Russia.

It must be admitted, however, that this picture of Europe as a spectrum, with Russia, say, at the red end, does not take care of several important problems. It does not overcome the subjective sense of sharp contrast and opposition, which, as we have seen, has played such a significant role in the making of Russia-West comparisons. Moreover, the existence of sovereign states, of political boundaries, does mark real breaks in the spectrum, which is not a continuum, as anyone who has crossed a frontier post in Eastern Europe well knows. Finally, the fact that in the important realm of power politics Russia is usually set off, not against its immediate smaller neighbors, but against great powers farther to the West has certainly had a polarizing effect, of which the Iron Curtain division of Europe after 1945 is only the most recent and violent example. The impact of this effect upon other spheres of life and politics is very great indeed, as is illustrated by the sad history of countries and individuals that at times have sought to play the role of "bridge" between East and West. Still, for the student of history or comparative politics the "spectrum" approach does have real attractions, not least in helping do justice to the enormous richness and multiplicity of the European scene.

A second device for tackling the Russia-West comparison has been that of the "time lag." For those inclined to seek similarities rather than contrasts the time lag is very convenient: features in the Russian scene that seem different from the West are shown to be the same, but corresponding to an earlier date in the West; opposing trends turn out to be merely tangents drawn at different points along the same curve. Thus, it is thought enlightening to say that the style of Soviet life today is Victorian or at the latest Edwardian. (Such resemblances or echoes do not, of course, necessarily imply a time lag. On a fresco from the Palace of Knossos there is a charming Cretan lady whom the archaeologists call La Parisienne: presumably the parallel, though attractive, is fortuitous.) The use of the time lag is valid only if a more or less identifiable sequence of stages is occurring and if more than one nation

---

[12]For example, Nikolaus von Preradovich, *Die Führungsschichten in Österreich und Preussen (1804–1918)* (Wiesbaden: Steiner, 1955).

or culture has come to participate in this sequence, usually by borrowing and adaptation. W. W. Rostow in his study of the stages of economic growth provides an analytical framework for the succession of stages and then places the different modernizing countries in their rank in this procession. Within such a defined setting he does show that Russia experienced a time lag vis-à-vis Western Europe in achieving the famous "take-off" and in reaching "maturity." At the same time the burden of his message is the general similarity of these stages: "In its broad shape and timing, then, there is nothing about the Russian sequence of preconditions, take-off, and drive to technological maturity that does not fall within the general pattern; although like all other national stories it has unique features."[13]

The time lag has its problems, however. As Thorstein Veblen pointed out some decades ago,[14] the latecomer to a historical sequence does not simply duplicate earlier performances; there is usually a foreshortening of the stages, a leaping over of certain steps, and a lumpy mingling of the old and the new. Among the Russian Marxists Trotsky had perhaps the best sense of this feature of the time lag; indeed it underlay his thesis of permanent revolution. Despite the Marxist predilection for a unilinear view of history and its stages, Trotsky was able to observe: "The indubitable and irrefutable belatedness of Russia's development under influence and pressure of the higher culture from the West results not in a simple repetition of the West European historic process, but in the creation of profound *peculiarities* demanding independent study."[15] In other words, the conception of the time lag, although serving to increase the comparability of nations by putting them on the same track, may actually, when refined, reinforce the appearance of individuality and uniqueness.

A study of the mingling of the foreign and the indigenous, for which the term "symbiosis" can sometimes be used appropriately, affords a third approach to the comparative study of Russia and the West. In my judgment this is probably the most fruitful of all, since it corresponds to the common-sense observation that, in modern times at least, all nations are increasingly taking over or being bombarded by external influences which they must digest, naturalize, or otherwise cope with as best they can.

Two examples can illustrate the utility of this approach to a comparative study of Russia and the West. It is certainly to the Slavophiles that we owe part of our sense of Russia's difference and uniqueness. Not only were they intent upon stressing the differences, but the way they wrote and the features of the Russian scene they chose to emphasize strike the Western reader as peculiarly Russian. And yet, as we know from their education and the intellectual currents that influenced them, the conceptual apparatus of the Slavophiles was borrowed directly from German idealism and romanticism.[16] Paradoxically, increased access to "Western" ideas was to sharpen the picture of a Russia-West antithesis.

[13]W. W. Rostow, *The Stages of Economic Growth* (London and New York: Cambridge University Press, 1960), p. 67.
[14]In his *Imperial Germany and the Industrial Revolution*.
[15]Leon Trotsky, *The History of the Russian Revolution*, trans. Max Eastman (3 vols.; Ann Arbor: University of Michigan Press, 1960), I, 464.
[16]See Nicholas V. Riasanovsky, *Russia and the West in the Teaching of the Slavophiles* (Cambridge: Harvard University Press, 1952).

Or, to take an instance from the eighteenth century, a recent essay on the education and upbringing of the Russian nobleman[17] first brings out certain "Russian" features in his childhood experience: "The Russian nobleman of the 18th century normally lacked strong roots in any particular area and had no real feeling of attachment to a specific locality and to a family estate on which his ancestors had lived for generations. . . . There is little evidence of the attachment to and the ties with the ancestral home which characterised the mentality of the western noblemen." The child was under the supervision of serf nursemaids and tutors who had no rights and very rarely any powers of discipline. From this very "Russian" setting the young nobleman was sent to a school, where he received "a completely western education which had practically submerged the Muscovite traditions of learning and education by the middle of the 18th century." The author suggests that the effect of the somewhat abstract Enlightenment education upon children with this particular background was to produce a distinct cast of mind, exceptionally rationalistic and didactic, that was to have important consequences for Russia in the next century. For our present purposes the most interesting feature of this analysis is the way in which Russian and Western influences are seen to combine to produce a personality that is neither the traditional Muscovite nor the French man of the Enlightenment but rather the forerunner of the nineteenth-century *intelligent*.

Such an approach to the historical evidence can be extremely productive in dealing with a number of major problems of Russian institutional and social history: the impact of the Mongol conquest in Muscovy; the effects of Peter the Great's adoption of the goals and methods of contemporary German *Polizeiwissenschaft*; the consequences of taking a peasant, the son of a serf, and dropping him into the large factory of advanced Western industrialism; or the particular combination of Russian and Western Marxist elements that went into Bolshevism.

This approach is hardly an exciting discovery; it is the familiar province of the historian. But in the present connection two points need emphasis. First, such an approach, if it is to be fruitful, must be closely related to the material at hand; the results are illuminating to the degree that they lead to a concrete historical picture. It is not an approach that yields sweeping generalizations. Second, while such study does look beyond Russia's frontiers for some of its evidence and insights, its central purpose is to advance our understanding of Russia. Comparative study is a valuable tool to that end, but comparison *per se* is not the goal.

Indeed, all these approaches that I have mentioned as methods of looking at Russia and the West move away from direct comparison, either by blurring the comparison through reference to a "spectrum" or by becoming an analysis of the various factors, belatedness or foreign influences, that have contributed to the formation of Russia.

Would this suggest that such a comparison is a fruitless enterprise, that we may be engaged in an impossible endeavor to answer a pseudo problem? In one sense the answer must be yes. If we are asked whether two objects are alike or different, we are immediately impelled to counter: "With respect to what?" or "In terms of what standard?" If we ask whether two maple leaves are alike, we can answer affirmatively

---

[17]Marc Raeff, "Home, School, and Service in the Life of the 18th-Century Russian Nobleman," *The Slavonic and East European Review*, XL, No. 95 (June, 1962), 295–307.

if it is a question of contrasting them to oak or elm leaves, or we can answer negatively if it is a question of their being congruent or having identical vein structures. We cannot make a comparison *sans phrase*, without reference to the setting and purpose of the question.

This rather simple but tricky ambiguity in the act of comparison was well analyzed by Kant in a section of his *Critique of Pure Reason*. As he observed, some scholars are interested in and attracted by the principle of "homogeneity," others by the principle of "specification." "Those who are more especially speculative are, we may almost say, hostile to heterogeneity, and are always on the watch for the unity of the genus; those, on the other hand, who are more especially empirical, are constantly endeavouring to differentiate nature in such manifold fashion as almost to extinguish the hope of ever being able to determine its appearances in accordance with universal principles."[18]

According to Kant these differences in attitude have nothing to do with questions of fact or with the nature of reality but with method. In his rather formidable vocabulary similarity and dissimilarity are "regulative principles"—working maxims, both of which are necessary and which describe diverse tendencies and interests of human thought. Difficulties occur when we mistake their function and take them to constitute reality. "When merely regulative principles are treated as constitutive, and are therefore employed as objective principles they may come into conflict with one another. . . . The differences between the maxims of manifoldness and of unity in nature thus easily allow of reconciliation. So long, however, as the maxims are taken as yielding objective insight, and until a way has been discovered of adjusting their conflicting claims . . . they will not only give rise to disputes but will be a positive hindrance, and cause long delays in the discovery of truth."

My mention of maple, oak, and elm leaves suggests the possibility that we might bring the Russia-West comparison into more manageable shape by establishing the criterion of genus and species, of making our comparison within a hierarchy of classification. The terms of our comparison—Russia, a country, and the West, a group of countries—would point to just such a classification. In some restricted but relevant areas this kind of classification can be useful. If we wish to compare the Russian language with those of Western Europe, we do have a linguistic structure locating Russian in the Slavic branch of the Indo-European languages, to which French and German, through their respective branches, also belong. Even in the more elusive field of religion we can, by tracing the course of theological disputes and schisms, construct a reasonably workable classification of the branches of Christendom and place Russian Orthodoxy in its appropriate niche.

But these classifications extend only to such relatively well-defined subjects as language and religion; we are here concerned with such vast complexes as national entities, of which language and religion form only a part. How are we to establish classifications that can enable us to make comparisons on this larger scale?

Max Planck, the originator of the quantum theory, remarked that while the

---

[18]See *Immanuel Kant's Critique of Pure Reason*, trans. Norman Kemp Smith (London: Macmillan, 1933), pp. 537–49.

introduction of order and comparison is essential to scientific treatment and that order demands classification, "It is important at this point to state that there is no one definite principle available *a priori* and enabling a classification suitable for every purpose to be made. This applies equally to every science. Hence it is impossible in this connection to assert that any science possesses a structure evolving from its own nature inevitably and apart from any arbitrary presupposition. . . . Every kind of classification is inevitably vitiated by a certain element of caprice and hence of onesidedness."[19]

Such a cold douche from the austere natural sciences should make us cautious about the absoluteness of classifications in our rowdy and disheveled political and humanistic disciplines. I am entirely skeptical of any claims for a system of classification that purports to be inherent in the structure of history itself and free of arbitrary presuppositions. I find none of the principles of classification, whether based on geography or geopolitics, religion, ethnic-racial categories, social structure, or political system to be persuasive in providing an *objective* basis for ordering and comparing such complex congeries as nations. For example, to take the familiar Orthodox-Roman Catholic division of Christendom, while granting the enormous importance of this division as a historical influence, it would appear to me, after periods of residence in Moscow, Bucharest, and Athens—all Orthodox capitals—that this religious factor is, at least in the twentieth century, altogether inadequate as a principle of classification, though obviously of value in helping to explain many attitudes.

If we concede that we are not likely to find a purely objective order of classification that will enable us to compare Russia and the West, then we are driven back to the view that difference and similarity, homogeneity and heterogeneity, are tools to serve our diverse intellectual interests. As such they are necessarily tied to and get their meaning from our purposes and concerns in making the comparison. This is not to say that they conveniently produce answers we feed into them, but that, depending upon the questions which we bring to bear in our comparisons of Russia and the West, we will get a multitude of answers, indicating widely varying degrees of similarity and dissimilarity, each perhaps valid in its own setting, but only there. An anthropologist interested in the whole range of humanity's social organizations would probably regard the Russia-West contrast as relatively narrow. The political theorist, working within the framework of highly articulated and sophisticated political systems, would find the contrast, say, between autocracy and democracy very great indeed, perhaps representing the extreme ranges of his particular scale.

If we could be satisfied with such a modest and circumscribed role for comparisons of Russia and the West, there would be much less acrimony and controversy on this subject. But there's the rub; as we have seen, the motives that have impelled both Russians and Westerners into such endless debate and wrangling are powerful and urgent. Although we can hold, with Kant as our guide, that attempts to make

---

[19]Max Planck, *The Philosophy of Physics*, trans. W. H. Johnston (New York: Norton, 1936), pp. 13 and 14.

absolute comparisons will produce intellectual confusion and error, I am afraid it is certain that efforts will continue to be made to find in a comparison of Russia and the West either support for normative positions on the *political* relations of the Soviet Union and the Western Powers or the basis for a prediction on the outcome of this relationship in the future.

With respect to the range of ideological, diplomatic, and moral issues that currently divide us from the Soviet Union, I should certainly not underestimate their reality and importance or question the need for us to defend our own positions. But while we have all become used to employing the term "West" as a kind of shorthand for "our side," it would be well if we based our policies on the preservation of values and principles because we believe in them and not because they are "Western."

As for the future, I do not believe that the outcome, whatever it may be, is prefigured in the comparison of Russia and the West. If we look back to the decade or two preceding the outbreak of the First World War, we have the impression that for a brief period the old debate over Russia's relationship to the West was losing its intensity and was perhaps beginning to appear irrelevant. These were the years of that profound intellectual and cultural eruption (Einstein, Freud, postimpressionism, etc.), the consequences of which are still jolting us and which, in a half century's retrospect, seems to have been one of the great historical watersheds. Russia in its "Silver Age" entered fully and immediately into that movement, its creative talents were at the forefront, there was no significant time lag. In this breakthrough, initially on a narrow front of thought and art, the traditional Russia-West debate seemed out of place, not resolved but overtaken by new challenges and horizons. The First World War and the Russian revolutions interrupted and in considerable measure obscured this development, and as we have seen the old antithesis reappeared in the harsh form of communism versus "imperialism."

While this antagonism has by no means played out, it is becoming increasingly evident that the new world adumbrated at the beginning of the century is coming on apace, and whether we prefer to symbolize it by $E=mc^2$, or automation, or the return of the repressed, or abstract expressionism, it is a strange world. While Russia and the West will probably respond to it in different fashions, there is a distinct danger that by keeping our attention focused on Russian-Western relations and comparisons we may be quite unprepared to meet the challenge of novelty. If we think of the Western tradition as a kind of comfortable interest-bearing inheritance that we can bank on for the future, we are in for serious trouble.

CHAPTER 24

Soviet Russia as a Model for Underdeveloped Areas

W. DONALD BOWLES

A wave of rising expectations has engulfed the underdeveloped areas of the world, and in its wake there remains an intense desire for economic advancement and national recognition. Until recent years the relatively luxurious level of living in the West, particularly in the United States, served as a natural focal point in the search for a pattern of economic life that would lead to growth in these areas. Since World War II, and particularly since about 1955, when the USSR adopted a stepped-up "trade and aid" program among the less-developed areas, increasing attention has been directed to the Soviet pattern of growth and the Soviet answer to "imperialism."

It is now being suggested by some that the USSR is a more appropriate model for underdeveloped areas than is the United States. Several obvious considerations may in part explain this shift. For example, the present rate of growth of Soviet gross national product is somewhere between 6 and 8 per cent per annum, or more than twice that of the United States. Individual levels of consumption in many items are now approaching levels comparable to those of Western Europe. The Soviet citizen can be seen going about his work with relative contentment and with pride in his country. On top of all this, the USSR is assuming a course of political independence that only a powerful nation can pursue fruitfully for long. Briefly, there seems to be ample reason why the Soviet Union might be considered as a model for underdeveloped nations today. However, it is the thesis of the present article that in fact the USSR provides a most unsatisfactory model for nations interested in rapid economic development.

In one sense it is surprising that the question of the Soviet model should even be raised. By 1913, Russia was a nation with a substantial industrial sector and a respectable rate of overall growth. It was a major trading nation and had already received relatively large amounts of foreign capital. Moreover, the area abounded in

Reprinted from W. Donald Bowles, "Soviet Russia as a Model for Underdeveloped Areas," WORLD POLITICS, XIV, No. 3 (April, 1962), 483–504, by permission of the publisher. I [author] am indebted to Lynn Turgeon for helpful comments on the manuscript. Responsibility for the views expressed below, of course, rests with me alone.

natural resources. In short, the country taken over by the new Bolshevik regime simply was not in the same economic class as the countries we today label "underdeveloped."

This qualification aside, in examining the relevance of the Soviet model we shall focus on four of the more important problems that typically emerge today as an underdeveloped country attempts to accumulate capital and begins to move forward economically: the problems associated with population size or increase, inflation, balance-of-payments difficulties, and social and economic dislocations associated with institutional changes. Our appeal here is neither to Russian area specialists nor to economists concentrating on questions of growth (to whom much of the material presented will be familiar, although further mutual understanding may be hoped for). Rather, the article is directed to that larger group of non-specialists in fields peripheral to these disciplines. For this reason, the use of technical jargon is minimized, long elliptical footnotes are generally avoided, and specific source references, except in special cases, are not indicated.

I. POPULATION PROBLEMS

Since Malthus, students the world around have compared the "arithmetic" growth of food supply with the more rapid "geometric" increase of population. The relationship cast in this crude mold has been abandoned by most modern demographers and economists. Yet, when this hoary proposition is criticized, a caveat is frequently uttered: perhaps Malthus really had something to say in relation to the underdeveloped areas of today. Ignoring possible traces of the concept of the "white man's burden" in this position, one can admit the existence of a phenomenon of fearful potential. For several reasons, there is a likelihood of substantial population increases in the foreseeable future in the very areas now attempting rapid economic growth. The tragic possibility exists in many such countries that the two conditions —rapid population increase and economic growth—may be mutually exclusive.

Two varieties of the "population problem" can be found in what are here characterized as the less-developed areas. Some countries, like Venezuela and Mexico (with an annual growth of about 3 per cent), do not presently have large populations in relation to resources, but these populations are growing rapidly. In others, like India and China, rates of growth are less of a problem than the already existing density of population. In either case, population pressures are a constant threat to the success of any attempt to carry out a development program. The essential nature of the difficulty may be illustrated by a simple arithmetic example. If the so-called incremental capital-output ratio may be assumed to approximate 4:1 (i.e., four dollars spent on new capital yielding one dollar of additional output), then a 1 per cent increase in per capita output requires a rate of saving of 4 per cent *if* the population remains stable. However, if the population is growing at, say, a rate of 2 per cent annually, a rate of saving of 12 per cent is required to achieve a 1 per cent annual increase in per capita output.

While this formulation is useful primarily to indicate the nature of the question raised rather than precise magnitudes, it does suggest two possibilities. The effort needed in the typical underdeveloped country to achieve an annual rate of saving sufficient to expand the capital on hand for future production is exceedingly burdensome and in many cases impossible to achieve. Similarly, there is the additional disturbing probability that a successfully initiated development program will soon encounter a rising rate of population increase (resulting primarily from a drop in the death rate), so that the rate of saving must be stepped up even more to retain the level of saving achieved initially. In brief, in countries with high rates of population growth, the rate of savings must be very high. In countries suffering from high density of population, while the rate of saving required to increase per capita levels of income may be somewhat less, it is all the more difficult to set aside *any* current income for capital accumulation.

The Soviet experience with the population issue is interesting on both counts. Officially in the USSR there can never be a "population problem," since this is considered to be a manifestation of capitalist infirmity.[1] Fortunately, the uselessness of this dogma for understanding Soviet reality is compensated for by the known facts regarding Soviet population. By 1928, the year in which the First Five-Year Plan was begun, production levels had recovered from the shambles of "war communism" (the period of civil war and foreign intervention) and climbed back to their prewar heights, stimulated in large measure by the adoption of the "New Economic Policy" in the early 1920's. In that decade, population increased by about 1.8 per cent annually, a rate that had characterized tsarist Russia from the turn of the century to 1914. With the initiation of rapid industrialization, collectivization, and the widespread and important institutional changes of the late 1920's and the 1930's, the birth rate decreased and the death rate rose, at least in several individual years. Over the period 1928–1939, the average annual rate of increase was only about 1.2 per cent. Following World War II, the rate of increase rose sharply, today averaging 1.6 per cent annually.

The rate of economic growth since 1928 compares favorably with increases in population since that year, and presently it is about 6–8 per cent. A particularly large share of the production pie is taken over for non-consumption purposes in the Soviet Union, but, even so, neither population size nor its rate of increase has ever become a "threat" to economic growth. By the late 1930's, the share of gross national product allocated to household consumption had been reduced rather sharply, although per capita consumption suffered less because total production was increasing rapidly. Preparation for war and the vast economic dislocations occurring during the war led to a general decline, so that by 1953 individual levels of consumption apparently were again comparable to those existing in the late 1920's.

[1] According to the official Soviet textbook on political economy, "The misanthropic Malthusian 'theory' was created for the purpose of justifying the social order in which the parasitism and extravagance of the exploitative classes thrive on the inordinate labor and the growing privation of broad masses of workers." *Politicheskaia ekonomiia* (Political economy), Moscow, 1954, p. 299.

Since the war, the level of living has improved continuously; over the period 1951–1955, for example, consumption by civilian households increased almost four times in excess of the rate of population increase.

In brief, the level of Soviet production increased at a sufficiently rapid pace to permit consumption levels to hold up fairly well (ignoring the war period), simultaneously with a relatively high rate of capital accumulation (the ratio of investment to gross product). Presently the rate of gross investment is 25–30 per cent (no allowance being made for replacement of capital). If defense is included, then the "rate of non-consumption" approximates 35–40 per cent of national income.

The crucial importance of these facts is clear. The typical underdeveloped area possesses an economy dominated by crude methods of agricultural production and a limited resource base. As a consequence, low levels of per capita production exist. In turn, it is impossible to allocate sufficient quantities of current production to the creation of capital goods by which to raise productivity and income in future years. Even in some of the less-developed areas possessing important raw materials that find acceptance in world markets, the sheer number of persons to be fed, clothed, and housed virtually exhausts gross national product. In contrast, once the Soviet economy had regained the levels of tsarist production by 1928, the regime was able actually to *reduce* the share of total output allocated to consumption, and to maintain the resulting higher rates of expenditures for investment and defense.

Any country that is able to set aside a third of its production each year for non-consumption purposes is a relatively wealthy country in terms of per capita levels of production. The implication is that such a country is not overpopulated in terms of its resource and technological potential, and that annual production increases are significantly well ahead of annual increases in population. Had World War II not intervened, Soviet population would today approach 300 million rather than the present figure of 216 million. Under that circumstance, it is conceivable that "population pressure" would have been felt by the regime and a different official attitude toward population problems would have evolved. As it turned out, however, the Soviet Union was (and is) in a class apart, in this sense, from the typical underdeveloped country.

II. INFLATION

"Pure inflation" is a term used to define a situation in which levels of production remain unchanged as prices rise. The more prevalent pattern, however, is a price rise associated with *some* production increase. Inflation is a characteristic of most underdeveloped areas as they attempt accelerated economic growth. While economists enjoy some agreement on desirable principles of monetary and fiscal policy under these circumstances (managing the money supply, taxing and spending, and changing the public debt), suggestions as to specific policy applications amount to about one per economist. A recent study indicates, however, that the difficulties encountered in this regard in less-developed areas are not centered about small, technical differences of professional opinion. Rather, they generally arise from the crude fact that a given government attempts to spend at a level too far above its income. In many years in the postwar period, the deficits in such countries often represented from 10 to 20 per cent of total expenditures. Thus, these governments, in the absence of a suitable taxing and borrowing policy, have tried to bid resources away from the private sector for developmental and other purposes. The following results of such a policy invariably occur: wages and the prices of other factors rise, and consumer goods' prices increase as the increased incomes (resulting from higher wages) are used to purchase a constant or possibly contracted supply of consumption goods.

The problem, phrased in this manner, implies its own solutions: either (1) lower government expenditures, thus probably reducing the rate of capital accumulation and growth; (2) raise taxes; or (3) achieve a rise in voluntary saving. The first solution lies outside the major premise that there *will* be accelerated economic growth. The other two solutions beg the question. Per capita annual income in 1954–1956, according to a U.N. study, approximated $50 in Burma, $65 in India, and $120 in Peru, to select some underdeveloped areas at random, while in the United States it was almost $2,000. Obtaining a significant amount of voluntary saving from such low income levels is a difficult practical matter, to put it conservatively. By the same token, but for other reasons also, increased taxes are often not possible. Thus governments, in their anxiety to move forward, allow their revenues to lag behind their expenditures and the inevitable result, ignoring the possibility of price control and rationing ("repressed inflation"), is inflation.

What is there in the Soviet experience that may provide guidance to the underdeveloped regions in this matter? Initially, it must be noted once again that Soviet experience is not strictly comparable to that of most underdeveloped countries. From the beginning of the Soviet planning era, per capita income was sufficiently high to afford at least the physical possibility of increasing the rate of investment and maintaining the higher rate. Despite the low propensity to save that characterizes Soviet citizens, the government has possessed several well-known sources of revenue that are available only to a totalitarian regime. Broadly speaking, such revenues fall into two groups: forced saving and forced taxes (forced in the sense of not having been levied by a democratic representative government). In the first instance, government bonds have been purchased largely on an obligatory basis, the payments in some cases being automatically deducted from the worker's wage. Although revenue from bond sales was less than 10 per cent of total budget revenue, in individual cases the value of the bonds purchased sometimes amounted to about a month's wage per year. In 1957, bond sales, which might more appropriately be labeled taxes, were abolished for the most part, and simultaneously a moratorium was declared on the payment of outstanding principal and interest.[2] In addition to forced bond sales, the Soviet Union has levied an income tax. This, too, has been a minor source of revenue. In his farewell speech in the United States in 1959, Premier Khrushchev stated, "In the near future we are going to abolish—and I mean

[2] Actually, until the late postwar period the price level was rising so rapidly that the effective rate of interest was negative.

abolish—all taxation of the people—and I think you will appreciate the significance of this step." What this amounts to, of course, is that the Soviet government will no longer toy with such minor sources of revenue as the income tax and bond sales, but will instead concentrate on bigger game—namely, the turnover (excise) tax and deductions from "profits" of state enterprises.

In the underdeveloped areas the complex problem remains of finding methods to transfer an increasing share of total national output from the consumer to the government without running afoul of inflation. Although control over the price level has been achieved since the late 1940's, it seems clear that the Soviet Union is perhaps one of the world's least instructive examples of how to achieve rapid economic development without inflation. To be sure, this is not to say that such development actually can or cannot be achieved without inflation, but merely to emphasize that the Soviet Union has not done so.[3]

Although there is yet much to learn on the subject, the broad pattern of Soviet inflation is relatively clear. With the onset of planning in 1928, high production goals were established, based on planned increases in the labor force and in labor productivity. The latter, in turn, were to result from making increased and improved plant and equipment available to the workers. As it turned out, many of the key production targets were not met, and productivity gains failed to materialize as planned. In the face of these facts, the individual enterprise managers, in their struggle to meet production targets, began to bid up wages in the hope of attracting a sharply increased supply of workers. If it seems surprising that in a "planned economy" managers should have excess funds for such purposes, the answer lies in the inadequate control possessed by central banking authorities over the wage funds in their charge. In many cases, the inflated supply of such funds made it possible for managers to bid against each other, so that wages rose considerably, perhaps by as much as 1,000 per cent over the period 1928–1953.

Until the late 1930's an "open market" in labor was maintained which meant that individuals sought out their own jobs and employers found workers by offering them favorable material inducements, primarily wages. ("Forced labor" obviously is not included in this group.) A good case can be made for the position that the central authorities tolerated—indeed, condoned—illegal wage payments even if they resulted in a rapidly rising price level. After all, the increases served to create the illusion of improving levels of living, in the face of sharp rises in living costs—thus keeping established industrial workers contented and attracting new workers from the countryside. Also, such wage increases had some propaganda value abroad in the absence of published statistics on price levels. In addition to wage increases, a second factor explaining the rise in prices was the increase in the rate of non-consumption expenditures by the state during most of the 1930's (investment, social security,

---

[3] This is an important qualification. In many underdeveloped areas, an investment program cannot be achieved except through government channels. This means that the government must find a way to mobilize a significant share of national income. Conceptually, the upper limit of the share is what the people can or are willing to do without. Some years ago, Colin Clark expounded the thesis that any country with a tax load of more than 25 per cent of its national income would suffer from inflation. His thesis, while provocative, yet awaits proof.

military expenditures), which created purchasing power without concurrent creation of purchasable consumer goods.

The factors noted thus far are, of course, familiar components of a "demand-pull" inflation in which, in the classic formulation, too much money chases too few goods. There was one additional feature of the Soviet scene, however, that contributed to the sharp rise in prices—namely, the system of taxation employed to generate savings, given the precondition that an open market for consumer goods was to prevail after 1935. Maintenance of an open market requires that the prices established must just clear the market. With purchasing power expanding rapidly, and not siphoned off to the state treasury through a direct income tax, prices were forced up on the limited supply of consumer goods. The major source of revenue was (and is) the turnover tax, which is levied at a primary phase of production or procurement (today, mainly on consumer goods). Prior to 1935, and again in the late 1930's, this tax as a percentage of consumer goods' prices rose rapidly on the average, indicating that the difference between consumer purchasing power and the value of consumer goods' production at cost was being absorbed through increased taxes. Inasmuch as the tax is included as part of the price, prices similarly rose. However, from the late 1930's until 1949, the year of the first and major postwar wholesale price reform, rising wage costs and stepped-up rates of non-consumption expenditures played a relatively more important role in the process of inflation.[4]

Now, for countries seeking a method of growth that will permit relatively stable price levels, the Soviet experience represents only a chaotic, apparently unplanned development which, in fact, was accompanied by severe inflation. It might be asked at this point: does this really matter? High growth rates were actually achieved, and the economy simply accommodated the inflation. At present, price levels are relatively constant or falling, which itself seems to indicate that there was, after all, always the possibility of bringing inflation under control when the central authorities believed it had served its purpose. It is impossible, of course, to give a clear-cut answer to this contention. At the same time, it should be noted that central authorities in most underdeveloped areas today cannot tolerate inflation of the Soviet type even for a year, and certainly not for the length of time permissible in the USSR.

The reasons are simple but fundamental. Inflation exacerbates many of the problems that need to be overcome if a low income level is to be raised over time. Saving, of course, is the heart of growth, and inflation reduces the level of voluntary saving, inasmuch as people are reluctant to see the value of their money claims depressed as prices soar. Because of the limited taxing and borrowing capacity of governments in the less-developed areas, the elimination of all voluntary saving

[4]Primary reference here is to prices of consumer goods sold in state stores, most of which originated in the non-industrial sector of the economy. The demand by Soviet enterprises for producer goods was controlled more directly by the state. Thus, the more than doubling of producer goods' prices in the prewar period was attributable to wage increases that outpaced the rise in industrial labor productivity. To some extent, the price line was held through the use of grants directly from the budget (subsidies) to make up the difference between cost and planned price.

would spell disaster for many development plans. Moreover, even if some voluntary savings persist, they tend to flow in the direction of assets that are resistant to value erosion (e.g., hoards of precious metals, land, materials, stocks, etc.), rather than into plant and equipment in search of high returns related to high productivity.

If these were the only difficulties encountered in the process of inflation—and they are not—then perhaps it might seem to leaders of underdeveloped areas that, while inflation is awkward, it is perhaps possible to muddle through, and that growth will still be possible. Regrettably, such is not the case. The Soviet economy could tolerate inflation because of the existence of a strong central government that could force the necessary rate of savings to a relatively high level. At the same time, dependence on the world economy was greatly reduced, owing to the potentialities for development found within the domestic economy. Neither of these situations exists in most of the economically backward areas today. Strong central governments are typically absent. Also, for better or worse, the underdeveloped areas are presently wedded to the rest of the world through economic necessity—the need for new capital, essential manufactured consumer goods, raw materials and, in some cases, food. Such ties—certainly such ties with the West—would disintegrate under the conditions of inflation described above, and along with them any hopes for significant economic advancement. Inflation caused no external problems for the Soviet Union, because early in its history a policy of economic isolation was adopted, a matter to which we now turn.

III. BALANCE-OF-PAYMENTS PROBLEMS

To understand the crucial and reciprocal influence of the balance of payments on the internal economic life of a country, it is useful to concentrate on the intimate relationship of the level of national income, the foreign trade balance, and the rate of exchange. These factors may be thought of as three marbles in a bowl—the movement of one will have an immediate effect on the position of the other two. Let us assume initially that a certain underdeveloped country embarks on a development program. Further, assume that the program is initiated by a grant from abroad, to be spent partly on domestic labor used in the construction of "social overhead capital," and partly on the importation of material and skills essential to these projects that are not obtainable locally. As domestic wage payments increase, total spending on locally produced goods rises. The supply of such purchasable goods, however, has increased only slightly,[5] with the result that prices are bid up. As prices increase, there is heightened impetus to import from countries abroad, since foreign prices become more attractive in direct proportion to the degree of local inflation. Similarly, but for reasons not related simply to price changes, as domestic incomes rise, imports tend to follow along, illustrating what the economist calls their functional relationship to domestic income.

Local inflationary pressure is thus transmitted to the balance of payments. Possibly a light industry established under the development program might produce for

---

[5] Construction of hospitals, water supplies, schools, etc., creates purchasing power, but the services thus available are provided free or at nominal cost.

export and thus provide a source of foreign exchange with which to purchase additional imports or repay the loan. Its effectiveness for this purchase may be partially or entirely offset by price rises that render the exports non-competitive in world markets. Under these conditions, local demand for foreign currency will exceed the supply, and pressure will begin to build up on the rate of exchange, resulting in a fall in the price of local currency in terms of other currencies. If the country has large foreign debts that must be serviced each year in foreign currency—and this is typically the case in underdeveloped areas—the depreciation means in effect that a larger quantity of real goods and services must be exported each year to facilitate the unchanged annual money payment. Another adverse effect of the depreciation is that it dries up new sources of foreign capital at the very moment when they are most needed—i.e., when imports are running high and exports are lagging.[6]

In an effort to control inflation and the associated balance-of-payments difficulties, it has been the practice of most trading countries of the world since the 1930's to impose direct controls over external economic affairs—namely, import restrictions, export subsidies, and exchange controls. The effects of the first two actions are, respectively, to reduce the local demand for foreign currency and to increase the local supply of foreign currency. The imposition of exchange control takes a different form in each country, but generally a central authority "mobilizes" all foreign exchange accruing to its citizens and "rations" it out to those who wish to import. The economic isolation permitted a country by virtue of its adoption of exchange control has the general effect of excluding or reducing the importation of commodities considered unessential to the development program (French perfume, for example), and of simultaneously encouraging exports to regions that provide badly needed quantities of foreign exchange.

Exchange control is a basic fact of life in the international economy today. At the same time, the General Agreement on Trade and Tariffs (GATT) contains provisions that encourage the elimination of such controls at the earliest possible date in all circumstances, for the reason, long recognized by Western economists, that world production is maximized by an international division of labor on the principle of comparative advantage. One of the exceptions permitted in the GATT regulation, however, is the case where such controls may be justified because of an economic development program in progress. This bow in the direction of reality by the framers of the provision is an indication that no magic formulas exist to exorcise these balance-of-payments difficulties; rather, they must be removed by what is typically a painful and protracted operation. In this light we may evaluate Soviet experience as a guide for the underdeveloped areas.

The period following the Russian Revolution was characterized by chaos and destruction, and over the years this situation was reflected in a heavy net import

---

[6]Long-term foreign capital, unless there are strong incentives to the contrary, is discouraged by the lack of profit predictability under such conditions. Short-term capital actually moves out of the country under inconvertible currency standards as local speculators sell their own currency, which further increases pressure on the exchange rate, as long as it is believed that the rate will continue to fall.

balance. During the period 1918–1922, imports exceeded exports by about six times. This deficit was financed by gold and the remaining foreign-exchange assets of the previous regime, and by relief from abroad. A catch phrase of the time was "mobilization of savings." In the international sphere this was carried out in several ways. A major step was the repudiation by the new regime of both external and internal debts, as well as the nationalization of large-scale enterprises, foreign-owned and domestic. Also, travel abroad by Soviet citizens was limited except for official purposes, and this was looked upon as a great step forward in husbanding precious supplies of foreign exchange. Similarly, so-called "luxury imports" were eliminated.

It was in this early period, allegedly, that Soviet leaders evolved the concept that economic development could derive solely from internal resources. Disregarding the inconsistency between this concept and the nationalization of foreign property and repudiation of foreign debts, it would seem that the powers of the Soviet economy to develop from internal resources were discovered only gradually and empirically, so to speak. Lenin, for example, believed that the Soviet Union might not even be able to recover, let alone expand, without foreign help. To this end he apparently took the lead in considering an expansion of foreign concessions in the USSR in the early 1920's. Also, many economists of the period felt that the financial requirements of economic development would be so great as to compel the use of foreign capital.[7] By the 1930's, however, the official position had crystallized about the concept of self-sufficiency, and economists were criticized for their earlier attitude.

During the second half of the 1920's, the Soviet Union again faced a balance-of-payments problem and an attempt was made to expand exports while simultaneously stabilizing imports at approximately their 1925 level. Writers of the subsequent decade give the impression that 1925 was an important turning point in Soviet external relations. From that time, greater selectivity was applied to imports, so that by 1928 the slogan "metals and machines" was said to characterize the import structure. Interestingly, the term "self-sufficiency" never had been much used in Soviet literature to describe the limitation and selectivity that were applied to imports. Rather, the honorific phrase "technico-economic independence" forms the basis of the rationalization of a policy which, when practiced by Nazi Germany, Soviet writers scornfully labeled "autarky."

[7]The international negotiations conducted by the Soviet Union in the 1920's indicate that foreign capital was welcome. After Lenin's overtures in regard to foreign concessions, the Soviet government in October 1921 expressed its readiness to open discussions on the tsarist foreign debt as a prelude to the receipt of new direct investment in the form of concessions. (Interestingly, the Soviet attitude at that time was different from that of underdeveloped countries today, which generally favor loans rather than direct investment so as to minimize "foreign control." One may speculate that the Soviet position was conditioned by the tactical consideration that foreign venture capital might be attracted in the direct form even while unconditional repudiation of the tsarist debt remained.) In 1922 at the Genoa Conference, and later at the Hague, Soviet delegates offered what they apparently thought were strong incentives for credit relations. For many reasons, delegates of the Allies did not accept Soviet terms, and the Soviet Union entered a period of what has been called the "credit blockage." Despite the breakdown of credit negotiations, foreign concessions became fairly numerous, although such enterprises never produced a significant share of national income. At the same time, from 1923 on, short-term commercial credits were extended to the USSR. Although no foreign government provided funds, after 1926 several governments established programs guaranteeing commercial credits extended to the Soviet Union.

In the First Five-Year Plan, foreign trade was assigned a rather ambiguous role. On the one hand, imports were to remain highly selective and limited, so that there would be a net export balance and the limited imports would provide the greatest possible stimulant to growth. In the same plan, curiously, there was provision for net receipt of foreign capital. The obvious inconsistency between the net receipt of foreign capital and a net export balance apparently was not immediately evident to Soviet planners.[8] Or, perhaps the political requirements of this period impelled writers and planners to prepare for the best of both possible worlds.

Plans to the contrary, substantial foreign credits did not materialize, and a stable net export surplus was not achieved until 1933. Herein lies the oft-told tale of the difficulty experienced by the Soviet Union in connection with short-term commercial debts, which totaled 1.4 billion gold rubles by the end of 1931. In the Soviet view, the heart of the difficulty was the "capitalist instability" that led to the collapse of the international economy in the 1930's, with its particularly disastrous effect on the prices of primary products—that is, on precisely those products that weighed heavily in the Soviet export balance. Apparently, the central planners had not considered world market prices in their foreign trade plan, and the planned supply of Soviet exports was initially rather inelastic. In the face of declining world demand, the terms of trade turned sharply against the USSR. By 1932 a unit of exports on the average would purchase in the world market only about three-fourths of what it would buy in 1929. Subsequent to 1932, Soviet terms of trade improved.[9] Lacking the possibility of short-term extensions or additional receipts of short-term credit, and to avoid defaulting, the Soviet Union took the only step possible—namely, it began to curtail imports and eventually, to avoid continued selling in a depressed world market, to curtail exports as well. After 1931, both exports and imports declined, but imports fell more than exports, with the result that a net export balance was finally achieved in 1933. By 1934 the physical volume of Soviet exports was 3 per cent above the 1929 level, whereas the volume of imports was about one-half that of 1929.

The restoration of the Soviet balance-of-payments equilibrium was achieved in this fashion—by curtailing imports, by the sale of gold abroad until as late as 1935, and by maintaining exports until the crisis had passed. The adjustment process was much less painful than it might have been for two main reasons. First, although it is statistically difficult to make accurate estimates, at the peak of foreign trade participation in 1930 Soviet exports accounted for less than 5 per cent of gross production. The Soviet Union was, in a word, lightly involved in world trade, whereas many underdeveloped countries regularly export far more than 5 per cent of their total production. Second, the adjustment process was greatly facilitated by sheer good fortune. The Soviet trade pattern in this period was essentially that inherited from the tsarist era, and luckily the traditional Soviet export surplus was with countries that did not resort to foreign trade controls, or that did so very late (e.g.,

---

[8] A net export balance in effect means that the rest of the world is incurring a debt to the exporting country, i.e., that the exporting country is granting, not receiving, credit abroad.

[9] The greatly reduced prices of Soviet exports at this time led many Western observers to label Soviet shipments abroad as "dumping," although sufficient evidence never existed to support such a charge technically.

England). In contrast, the large Soviet import surplus was with Germany, the first to resort to severe limitations of convertibility. The result of this distribution of the trade balance was that the Soviet Union was able to apply net earnings on current account to meeting its German commitments. If its net current-account earnings had occurred in countries with exchange controls, it would have been hard pressed to meet commitments arising elsewhere.[10]

In summary, the Soviet answer to the difficulties encountered in finding both capital and markets abroad was to withdraw from the world economy. Obviously, industrialization has proceeded briskly in the Soviet Union. This does not mean, however, that self-sufficiency was the best policy, inasmuch as the economic costs of such a policy were high. If one ignores official "Marxist" explanations, the Soviet policy of autarky can be understood simply as one imperative that followed from the decision of the regime to industrialize quickly at all costs. For many reasons, some of which we shall explore in the following section, the task was so large and the changes required so massive that it could be carried out only in a closed society, one whose entire range of foreign contacts—economic, political, cultural—was controlled by the regime. In this light, Soviet foreign economic policy is seen to have developed from a set of values that casts the individual in the role of a servant of the state, curiously reminiscent of early mercantilism.

The luxury of following an independent political course and severing one's ties with the world is scarcely conceivable for underdeveloped areas today. Low incomes prevent significant capital accumulation. The primitive nature of technology and the typically limited resource base interact with the lack of capital to cause low incomes to persist. The institutional structure frequently is such that the innovation required to compensate for the lack of capital and a weakly developed technology is impeded. Given these circumstances, *and* the unwillingness of the leadership in the underdeveloped areas to live with the poverty of the past, recourse must be had to aid from abroad. The particular form this assistance takes will vary from country to country, but aid there must be, as well as two-way trade. The Soviet Union, in terms of production, resources, and technology, was wealthy by current standards in many of the less-developed areas and hence could "afford," for whatever reason, to adopt a policy of self-sufficiency. In this light, the practical significance of the Soviet Union as a model in this respect becomes exceedingly remote.

IV. INSTITUTIONAL CHANGE

Between man and his physical environment in any society there is a set of institutions, a pattern of regularized ways of thinking about things and of acting, in accordance with which man provides for his wants. Along with such obvious factors as resources and technology, institutions in large measure determine how well a given society will take advantage of the surroundings in which it finds itself. While traditional economic theory ordinarily treats institutions as one of the "other things

---

[10] While a 1936 regulation prohibited exports to countries whose currencies were controlled, the pattern of Soviet trade reflected at this time the inherent features of the world economy, so that the regulation perhaps may be best viewed as a bargaining device.

being equal," there is nevertheless growing awareness of their importance in the matter of economic development. One textbook on economic growth, for example, lists the following changes (fostered by the Soviet regime) in the institutional pattern of Russian life. These changes, the author suggests, are perhaps more important than the actual material goods produced since the Revolution: ". . . the spreading of education and the very idea that education is desirable (even though it may be warped by wrong economic and world-political ideas); the overthrow of the superstitious element in Russian religion and education; the animation of the population to mobility, social and territorial; the sponsoring of the idea of careers open to talent; the inflaming of economic acquisitiveness; and the adoption of the principle of pay graded to production."[11] Although one may argue about the necessity or desirability of individual items on this list, there seems little doubt that many of these changes are "permanent and essential ingredients of economic development."

One might suppose that the Soviet Union is therefore, at least in this respect, a suitable model for underdeveloped areas to emulate in their quest for a formula by which to achieve rapid economic growth. The evidence again, however, discourages such a conclusion. It is useful to consider institutional change—change in the way people think, act, and interrelate—in terms of the costs incurred in the process of change, and in terms of alternative ways in which necessary institutional change might be stimulated and guided. In viewing the USSR in this respect, the alterations associated with the very early moves toward industrialization will be discussed. While institutional changes obviously have continued down to the present time, these more recent innovations seem relatively minor when compared with the societal convulsions experienced in the 1930's.

In 1921, the sown area in the Soviet Union was roughly one-third below the pre-revolutionary level, and the grain harvest was about 40 per cent below. Also, livestock levels were generally below prewar levels. Under the New Economic Policy there was a general recovery, and by 1927 the total area sown to grain, the mainstay of the Soviet diet, was about the same as before the Revolution, and livestock levels were generally above the pre-revolutionary mark. These developments would appear to be auspicious for the initiation of an industrialization drive, but in fact this was not the case.

The recovery in agriculture was based on peasant farming. While the government had nationalized land at the time of the Revolution, the peasants in fact operated the land pretty much as if it were their own, and the government recognized their right to the produce of the land. By 1928, only about 2 per cent of all farms were collectivized, and it was the resulting private nature of farming that was to cause difficulty. For one thing, the leaders suspected that the peasants were the last remaining stronghold of capitalist sentiment in the economy, and that they had an adverse political and ideological effect on the country. While these peasant households remained outside the planned sector of the economy, there was also a serious

---

[11]Herman Finer, "The Role of Government," in H. F. Williamson and J. A. Buttrick, eds., *Economic Development: Principles and Patterns*, New York, 1955, p. 416.

question as to the very success of planning itself. At this time, for example, the Russians used a planning year beginning on October 1—i.e., a harvest year—which in effect meant that it was necessary to plan on what the peasants were *expected* to do, not on what they could be counted upon to do.

But the more serious aspect of the situation was the fact that, in 1926-1927, grain marketings were only at about one-half their prewar level. Several factors account for this decline in the marketed share. The grain harvest did not fully recover its prewar level, so that out of a reduced production the quantity the farmer kept for himself amounted to a larger percentage of total production. Inasmuch as the total sown area was about the same as that of the prewar period, the conclusion follows that the decline in production was due primarily to a decline in yields. This, in turn, was perhaps greatly conditioned by the agricultural reorganization through which land previously owned and managed as relatively efficient large estates was broken up into many small, relatively inefficient farms.

The most important reason for the decline in the marketed share was the dominance of the middle and small peasants in agriculture by the late 1920's. Prior to World War I, this group accounted for about one-half of the total grain production and marketed about 15 per cent of its output. By 1926-1927, however, it produced about 85 per cent of total output and its marketed share was only 11 per cent. In contrast, the so-called "landlords" (who disappeared following the Revolution) and the "kulaks" (rich peasants) together produced about half of total output in the prewar period, while the remaining kulaks in the late 1920's produced only about 13 per cent. In the earlier period the landlords and kulaks together accounted for about three-fourths of all marketings, whereas the kulaks by the late 1920's produced only one-fifth of total marketings. Interestingly, the small number of collective farms in existence on the eve of industrialization marketed about one-half of their total output, approximately the same share of the harvest as the wealthy landowners had marketed prior to World War I.

In the context of peasant dominance of agriculture, the so-called "scissors crisis" was instrumental in reducing the marketed share. Agricultural prices in 1927-1928 were about 50 per cent above their level in 1913, while prices of manufactured consumer goods were almost double those of 1913. In effect, in the 1920's the terms of trade turned against the peasant, with the most adverse terms experienced in 1924. This problem was discussed openly in the Soviet press at the time, and Party statements alleged that it endangered the "union" between worker and peasant thought to be so necessary to the success of the Revolution. The important result of the shift in the terms of trade was that the peasant now chose to market less grain in the face of his deteriorating "parity" situation.[12]

---

[12] An interesting question exists on this point. The typical American farmer, in the face of declining prices for his agricultural produce and rising prices for the products he must buy, is thought to expand his production to compensate for this price squeeze. Had the Soviet regime felt that the peasant supply of produce would have been equally responsive to such price changes, conceivably the collectivization drive need not have occurred at that time, if ever. In fact, at least two factors account for the response of the Soviet farmer to the "scissors crisis." First, the peasant by 1928 was paying very small taxes in relation to previous

In brief, this was the problem of the marketed share of grain. Marketings represented supplies made available for cities and for export, and, presumably, stockpiles and industrial uses. Because the landlords and kulaks had been replaced by people who were both producing and marketing less, the regime was confronted with a situation that was intolerable on the eve of a high-tempo industrialization drive. Faced with peasant reluctance to part with their produce, the Politburo appears to have had several alternative courses of action. A do-nothing policy in agriculture—i.e., relying on the forces already operating—was out of the question, of course. Similarly, giving the peasantry greater incentives to increase both production and the marketed share would have meant, in effect, diverting consumer goods away from the cities or stepping up the development of light industry. The first alternative would have been politically embarrassing, while the second would have retarded the development of heavy industry and industrialization, the original intent of the program.

In this light, gentle persuasion is seen as precluded by the goals and nature of the regime; thus, the question resolves itself into one of selecting the most effective form of compulsion. In principle, increased agricultural taxes might be considered here inasmuch as such taxes would have meant forcing about 80 per cent of the population to act against their will, as evidenced by their decision *not* to market a significant share of output. The form of compulsion adopted was rapid and widespread collectivization of agriculture, through which the peasant farms lost their identity and the farmers became members of a *kolkhoz* under the leadership of a representative of the regime (though such a leadership was nominally elected by the farm membership). The main effect of the transformation was that, although production remained relatively unchanged, the quantity of marketings increased significantly. Doubtless, the ideological preference of the regime for this "socialist" form of agriculture played a role in the decision. Also, it was felt that collectivization would make for efficiency and raise yields. But the strategic consideration in the collectivization drive was something quite apart. By this policy the government abandoned indirect methods and substituted direct controls over agricultural activity, including particularly the marketed share of agricultural production.

The Soviet Union in this fashion eliminated one of the chief impediments to rapid industrialization—that institutional pattern of life which gave the peasant private property rights to the produce of his farm. Naturally, the peasantry as a group did not instinctively grasp the beneficial significance of this change. On the contrary, to many the Man from Moscow or Kiev with the briefcase, roaming about rural Russia, represented the enemy and he was frequently treated accordingly. Resistance to collectivization took many forms, including the killing and eating of approximately one-half the livestock herds, the burning and destruction of farm buildings and

---

levels, perhaps only one-third as much as before the war, and rent payments as such were virtually non-existent. As a result, the peasant did not need the cash that previously had been derived from marketings. Second, a change in the price structure within the agricultural sector itself made it more profitable over time to use grain for livestock feed. While the amount of grain used for feed was about the same in 1927–1928 as in the prewar period, in percentage terms this meant an increase because total output was less.

implements, and the consumption of existing grain and other produce. The regime fought back with the secret police, the militia, and the army. Estimates of the number of people killed, starved, or deported in this period run into the millions.

The point illustrated in this brief description of agriculture in the first years of the USSR is a simple one. Soviet leaders at an early date committed themselves to rapid industrialization and then eliminated or transformed those institutions that stood in their way. Similar examples from other spheres of Soviet life abound: the suppression of labor unions, as that term is understood in the West; the use of "convict" labor in areas where workers were otherwise unavailable; the harassment of the Church, and so on. In agriculture, the change was so massive and so rapid that the associated social and psychological costs, which increase in a kind of exponential relationship to the rate of change, were enormous. Whatever motives may have guided Soviet leaders in their collectivization decision, the actual operation was achieved by the clumsy hand of force.

The immediate problem of the time was solved. The peoples of the underdeveloped areas might well ask, however, if the Soviet model minimizes the cost of the changes that may be required. The extended family system, a nomadic economy, and paternalism all inhibit the introduction of institutions judged important to the development of a highly interdependent, complex, productive economy. These and other dominant institutional patterns of life abound in various forms in most of the less-developed areas. Whatever their intrinsic worth in terms of human, non-material values, it is recognized even by the leaders of such areas that the institutions must be seriously modified, if not eliminated, if the fruits of modern technology are to be shared by their people.

In the light of these required institutional changes, which include changes in social and moral values as treated here, the Soviet model appears to be largely irrelevant. First, only through a highly centralized, terroristic campaign was the Soviet regime able to maintain the important institutional changes carried out in the Russian economy after 1928. In the case of agriculture, the rapid abolition of the collective farm system in the Ukraine in the wake of the German occupation during World War II testified to the shallowness of peasant acceptance of this institution. The point is that most underdeveloped countries today lack precisely the type of strong government necessary to make such drastic changes stick.

Second, for many reasons the moral climate of the world today inhibits the cruel excesses of the early phases of the Soviet industrialization, just as British methods of industrialization in the early nineteenth century are now judged harshly. While an ambiguous standard is applied (e.g., the death of a political enemy on a battlefield is approved, the domestic execution of a political prisoner is disapproved), in a world of great interdependence such a moral climate is important to each country. Thus, as leaders in the less-developed areas begin to wrestle with institutional problems in the course of economic development, they must seek a model in which a relatively humanistic, largely persuasive (not coercive) leadership guides the economy, while at the same time avoiding the appalling human costs that characterized many past economic transformations. The experience of the USSR gives no clue to such a model.

V. CONCLUDING STATEMENT

In terms of the values that have evolved in Western civilization since the times of the Greeks and Romans, the Soviet pattern of economic growth must be rejected as a suitable model for underdeveloped areas to emulate in the twentieth century. In terms of the Aristotelian syllogism: unless human life is conserved in the process of economic development, that process must be judged unacceptable; the Soviet industrialization pattern did not conserve human life; therefore, the Soviet pattern must be judged unacceptable. While the reader might insist with Plato that we define our terms, the gist of this syllogism at least does not jar Western sensibilities.

At the same time, this logic would be judged by Soviet theoreticians as both inane and vicious. To understand why this is so is to understand what is perhaps the most important aspect of the Soviet appeal to the underdeveloped areas today. Stated briefly, it may be that the areas of the world in which attempts at development are occurring are precisely those areas in which Aristotle, the liberal culture and, in effect, the whole complex of the Western value system are of little or merely academic interest. In some countries the effort to hold on to life is such a difficult chore that the liberal tradition—that is, the tradition of free men—has never achieved root. In others, where many of the important citizens may have received a Western education, the ruling group may attach much importance to the way of life we are here terming free. Even in this instance, however, the press of circumstances—the passion not just for growth but for rapid growth, and the insistence upon national recognition and prestige—often force the hand of the government so that steps are taken away from a liberal society.

Under such pressures, Marxism, or the Russian brand thereof, conceivably can have great attraction. If all values are sacrificed to nationalism and economic growth, a totalitarian regime that serves these ends might find approval. The Soviet experience may indeed appear to offer some guidance in the matter of techniques vital to the establishment of a police state controlled by an "elite" or even a "vanguard of the working people." In the absence of democracy in many underdeveloped areas—democracy in the sense of possession of civil liberties by an articulate and informed electorate—a Soviet-type regime might not be criticized for taking from the people what they never possessed. Similarly, it might be held that the Soviet Union has shown the way for a reordering of social values deemed essential to accelerated economic growth. Finally, if abject poverty, pestilence, and death have dominated the lives of persons in a certain area for centuries, many of the moral values to which we in the West attach importance may be ignored in the desperate scramble for material improvement.

In a particularly lucid way, Professor Adam Ulam has outlined one possible view of this type of appeal. He argues that, as it evolved, the body of thought called Marxism incorporated, unlike any other school of thought, two characteristics of the period. As a result, it contains today, a "passion for material improvement and fanaticism in the service of industrialization." At the same time, Marxism displays an "insight into the psychology of large masses of the population in a society that undergoes the process of transition from a traditional, pre-industrial phase to industrialization."[13] Given these two characteristics, Marxism is seen as alluring to a people who place development and national power above other values.

Although Professor Ulam indicates that the time span during which these countries remain susceptible to Marxism may be relatively brief, to persons of a liberal persuasion the brevity of such a period seems small compensation for the human costs that may be incurred. The Soviet experience, with all its technical shortcomings, hardly appears to be a model of perfection. Nevertheless, it is there to be followed if a country so chooses. Indeed, if economic growth were the only goal, it would be difficult to "prove" via the syllogism or any other figure of logic the natural superiority of the Western system as a pattern for the underdeveloped areas.

The crucial point is, however, that economic growth has never been and probably never will be a nation's one and only goal. One need merely recall such names as Thomas Jefferson, John Adams, and Alexander Hamilton to conjure up a picture of a growing nation struggling to conquer the material forces of production while simultaneously taking steps to ensure the enhancement of what we in the West have traditionally called the "dignity of man." It is in this simultaneous achievement of material comfort and a free society that the uniqueness of the West lies. There is little evidence to indicate that the Soviet pattern of industrialization or, more accurately, the Soviet political and economic institutional structure offers any hope of this dual accomplishment. The underdeveloped areas of the world in their search for methods to increase their material level of living ignore this fact at their own peril.

[13] Adam B. Ulam, *The Unfinished Revolution*, New York, 1960, pp. 285 and 284.

CHAPTER 25

The Comparison of Soviet and American Law

HAROLD J. BERMAN

"The science [of law] has been degraded to a national legal science. A humiliating, unworthy form for a science!" So wrote the great German jurist von Jhering more than a century ago.[1]

Are we seriously interested in developing a science of law—not, of course, in the narrow sense of the term science, but in the broader sense of a coherent body of knowledge about the nature of law, its principles of development, its social

[1] Von Jhering, *Der Geist des römischen Rechts*, I, 15 (1852).

Reprinted from Harold J. Berman, "The Comparison of Soviet and American Law," INDIANA LAW JOURNAL, XXXIV, *No. 4 (Summer, 1959), 559–70, by permission of the publisher.*

functions, its relationship to political, economic, religious and other ideas and institutions? If we are, we must study more than one legal system.[2] If we are not, the study of foreign law will be superficial and even dangerous, for without legal philosophy, without legal history, without legal sociology, we shall misunderstand foreign legal institutions. In the case of our own law, we acquire some feeling, at least, of its philosophy, its principles of development, its functions, its social and historical background, from our general education and from our daily experience; though we may be unable to articulate its major premises we nevertheless have a sense of them, an awareness of their existence. The danger in studying a foreign legal system is that we will uncritically transfer to it the assumptions which we make about the underlying foundations of our own system; therefore we are compelled, in studying a foreign system, to undertake the difficult task of analyzing its foundations.

It is because the study of diverse legal systems compels the articulation of their underlying premises that comparative law is both important and exciting. To observe that American judges attach more authority to previous judicial decisions than do French judges is merely interesting, at most; but to relate that fact to the social and political role of the judiciary, the method of selection of judges, and similar factors in the two countries, is to make what Whitehead calls "inert" knowledge come to life. Similarly, a comparison of the Continental European system of indictment of crimes after extensive preliminary investigation by an examining magistrate with the Anglo-American system of indictment is an empty academic exercise until it is related to the role of the official in Continental European countries, the inquisitorial tradition, attitudes toward crime and criminals, the political significance of the writ of habeas corpus in English and American history, and like matters. When such relationships are drawn, they illuminate the entire law.

It is true that where two countries have similar legal systems, similar political, economic and social structures, and similar traditions the mere comparison of legal rules and institutions, without explicit consideration of the entire legal and social structure which surrounds them, may serve an important practical function. No better example of this can be found than in the comparative study of judicial decisions and statutes in our own states—undertaken every day in our law schools as well as in our courts and legislatures. Thus we may profitably compare the personal injury law of Indiana and the personal injury law of Massachusetts, taking for granted that social conditions, habits, values, and the like, as well as the technique and theory of the law, are basically the same in both states. By the same token,

---

[2] One of the earliest and best statements of this point is that of *Feuerbach, Kleine Schriften vermischten Inhaltes* 163 (1833): "Why does the legal scholar not yet have a comparative jurisprudence? The richest source of all discoveries in every empirical science is comparison and combination. Only by manifold contrasts the contrary becomes completely clear; only by the observation of similarities and differences and the reasons for both may the peculiarity and inner nature of each thing be thoroughly established. Just as from the comparison of languages the philosophy of language, the science of linguistics itself, is produced; so from the comparison of the laws and legal customs of nations of all times and places, both the most nearly related and the most remote, is produced universal jurisprudence, the pure science of laws, which alone can infuse real and vigorous life into the specific legal science of any particular country."

however, such comparative studies yield relatively meager results for the development of legal science in the broad sense in which I have used that phrase.

On the other hand, when two countries have legal and social systems which are entirely different in character, it becomes extremely difficult to find appropriate criteria of comparison. Llewellyn and Hoebel, in their study of the law of the Cheyenne Indians,[3] were able to derive significant insights regarding the nature of law, but those insights were necessarily on a very high level of generality and hence are not readily applicable to the specific problems of more mature legal systems. Their demonstration, for example, that it is a basic function of Cheyenne adjudication to "kill off the grievance tensions" which are engendered by dispute suggests an important avenue of inquiry for the study of any system of law. But the way in which that function is performed in a primitive tribe which has no professional lawyers and no written records, which uses horses rather than money as the chief measure of value, which makes no explicit distinction between criminal and civil cases, and so forth, offers only a few guides to how such an inquiry should be conducted and by what standards it should be tested in dealing with a complex legal system such as our own.

Clearly, then, the particular value to be gained from the comparative method of study of law depends upon what legal systems are chosen for comparison. The study of primitive law helps us to find a broad anthropological and psychological basis for analysis of more complex legal systems. The comparison of diversities among states in a single federation helps us to test the merits of alternative solutions to relatively narrow (although often quite important) questions. In between these two extremes is the comparative study of legal systems within Western Civilization. This has been the traditional focus of comparative law as studied and taught in our law schools. It offers a very good basis for broad comparisons of a sociological nature as well as for narrower analysis of alternative solutions to specific practical and theoretical legal problems. Yet it also has certain limitations from both points of view. The legal and social systems of the various countries of the West are sufficiently different from each other to make direct borrowing of legal rules and techniques very difficult; they are, on the other hand, sufficiently similar in their basic structure and in their underlying philosophy to confine sociological analysis to what is, in many respects, a single culture. For despite the inroads of nationalism in recent centuries, there remains a strong family likeness in the legal systems of the various nations of the West; in the words of Edmund Burke, "the law of every country of Europe is derived from the same sources."

FEATURES OF SOVIET LAW

In analyzing the common foundations of Western legal systems, it is important, and perhaps even necessary, to view them from the perspective of a non-Western culture; and here the study of Russian law can be extremely valuable. For the law of

---

[3]*Llewellyn and Hoebel, The Cheyenne Way* (1941).

Soviet Russia has many links with Western legal systems, and yet it has its roots elsewhere. It deals with problems very similar to those which confront our law—housing, automobile accidents, workmen's compensation, mass distribution of goods, family disorganization, inheritance, theft, homicide, and many others; and its legal solutions to these problems are in many respects similar to ours. Yet its roots are in a Communist political and social system, in a centrally planned economy, and in the historical tradition of Russia, with its Byzantine, Mongol, and Russian Orthodox heritage, on the one hand, and its window to the West, its adoption of Western institutions, on the other.

Let me speak first of some of the features of Soviet law which are more or less common to Western legal systems. Soviet law is found in constitutions, in legislation, in administrative regulations, and in judicial decisions—although the boundaries between these different forms of law are not as precise as in Western systems generally. The Soviet judicial system includes trial and appellate courts of general jurisdiction within each of the fifteen union republics, with a single all-union court, the Supreme Court of the U.S.S.R. Also there is a separate system of permanent military courts in the nine military districts into which the country is divided, culminating in the Military Division of the Supreme Court of the U.S.S.R. The basic instruments of Soviet criminal and civil law and of judicial procedure are in the form of codes (in addition to the criminal code, code of criminal procedure, civil code, and code of civil procedure, there is also a family code, a labor code, and a land code); these codes are republican, rather than all-union, but the differences from republic to republic are slight, inasmuch as the codes of the Russian Republic served as a model for the other republics and inasmuch as all the codes are subject to all-union legislation. There are a substantial number of lawyers in the Soviet Union—some 60 to 70 thousand—as well as law-trained notaries who draw contracts, wills and other documents. A substantial amount of litigation is conducted in the courts over such matters as housing, divorce and alimony, workmen's compensation, discharge of workers, property rights in houses and other property which may be personally owned, inheritance, author's royalties, personal injury, and other similar matters. In addition, a special system of courts called Gosarbitrazh (literally, state arbitration) decides some 400,000 cases a year involving contract disputes between state business enterprises, which are juridical persons with substantial powers of disposition of goods and a limited power of disposition of capital.

Soviet criminal and civil procedure follow the pattern set by the Western European systems, a pattern which came into Russian history with the judiciary reforms of 1864. In contrast to Anglo-American law, the Soviet system of indictment of crimes—like the French, German, Italian and others—provides for investigation by an examining magistrate, who questions the suspects and witnesses at length over a period of time and presents the indictment. Trial both of criminal and civil cases is by a court consisting of a single professional judge and two lay judges, called people's assessors, who are chosen from the population to sit for ten days out of a year. In a criminal case the burden of proof of the allegation of the indictment is upon the prosecution, and the accused is entitled to a defense counsel.[4] There is a right of appeal in civil and criminal cases; the appeal is heard by a court consisting of three professional judges. Review of the appellate court's decision by a still higher court is discretionary.

What I have been describing thus far are certain surface features of Soviet law. Of course we are concerned with what lies behind the surface. But the surface is also important, for it gives us a key to unlock the first door to what lies behind. An American law student does not have to feel entirely lost in starting out, at least, to understand Soviet law. He can read opinions of Soviet judges in decided cases—many of them have, in fact, been translated into English by American scholars. He can read legislative materials and codes which speak a language familiar to him—a language of property, contract, negligence, unjust enrichment, intent of the parties, fault, statute of limitations, right of appeal, burden of proof, damages, criminal intent, criminal negligence, right of self-defense, right to counsel, and, in general, the entire apparatus of concepts and institutions of Western legal systems. He can read treatises and textbooks by Soviet law professors—provided he knows Russian—written for the benefit of the Soviet legal profession (including judges and prosecutors) and of law students in the thirty-two Soviet universities, expounding and commenting on the various branches of Soviet law in a manner in many respects similar to that of treatises and textbooks with which he is familiar. The American student will confront serious problems in using these materials—including problems arising from their paucity as compared with ours, from the fact that some Soviet laws are unpublished, from the fact that much of Soviet legal literature is colored by Party jargon and from the fact that writers and courts are bound by Party policies which are not always explicitly stated. One must learn to read between the lines of Soviet writings. Yet it is of considerable value that the lines are there.

TECHNIQUES OF ADJUDICATION AND LEGISLATION

What lies behind this first door? Let us turn first to techniques of adjudication. Here the American lawyer does not feel nearly as much at home as would the French or German or Italian lawyer—though they, too, would encounter some surprises. Soviet adjudication is distinguished by its informality. A civil action is commenced by a written complaint which need state very little and may state as much as the plaintiff wishes. No written reply is required. There is no set procedure for formulating issues in advance of trial and no pretrial procedure for taking depositions or procuring documents. As in Continental European countries, the trial

[4]Until recently the accused was not entitled to defense counsel prior to trial. *The Fundamental Principles of Criminal Procedure* adopted by the Supreme Soviet of the U.S.S.R. in December 1958 provide for defense counsel after the preliminary examination is completed, but before the indictment is finally ready to be presented to the court. This means that counsel may petition the investigator to conduct further inquiries or to amend the first draft of the indictment and may contest it in the preliminary session of the court prior to trial. Also the *Fundamental Principles* provide that the investigator has discretion to admit counsel at earlier stages of the investigation. See *Fundamental Principles of Criminal Procedure in the USSR and the Union Republics*, Pravda, December 26, 1958, Arts. 22, 23, translated in 11 *Current Digest of the Soviet Press*, no. 4, p. 7, March 4, 1959.

may continue at intervals over weeks or even months. The judge and, if they wish (which they seldom do), the two assessors take the leading role in questioning the parties and the witnesses, although the lawyers (and also the parties themselves) may interrupt with questions. There are no rules of exclusion of evidence; however, the judges are supposed to be guided by rules of evaluation of evidence. In criminal cases the complaining witness may also claim his damages for losses which he suffered as a result of the crime; again, the pattern is European. Upon appeal the case is, in effect, retried. The informality of the proceedings is emphasized by the fact that in a great many civil cases the parties prefer not to be represented by counsel.

The informality of civil trial procedure is bound up, of course, with the absence of the jury; both are connected with the conception of the trial primarily as an official investigation of the truth of the claims and defenses presented, rather than primarily as a forum for achieving justice through the clash of adversaries. This, too, is European—to a much lesser extent, English—and quite un-American. But the informality of Soviet civil and criminal trial procedure is also connected with a special feature of Soviet law, the paternal relationship of the judge to the parties; the Soviet judge is supposed to guide and discipline the parties and to help inculcate in them the attitudes and beliefs which the State seeks to encourage. (About this I shall have more to add later.)

Soviet judges are not bound by precedent and indeed it is prohibited by Soviet Law for the lawyers to cite precedents. Statements of law by higher courts in decided cases, and especially by the Supreme Court of the U.S.S.R. are, however, binding upon lower courts; therefore Soviet lawyers can and do cite such statements. The difference is a subtle one. The distinction between the development of law by analogy of previous decisions and the development by analogy of code provisions, statutes, and doctrine may be of little practical importance to the outcome of a particular case. The absence of a doctrine of precedent in Soviet law is nevertheless significant in that it reflects the absence of a strong sense of the growth of law over time. There has been some organic development of legal doctrine in the Soviet Union, but on the whole Soviet legal development has been more fitful, and in many areas of law there have been wide swings back and forth. One wonders in any event whether the doctrine of precedent, in its strict form at least, could possibly be introduced into a system such as the Soviet, in view of the informality of pleadings. Without precision in the formulation of issues, it is impossible to achieve precision in the distinction between holding and dictum.

Finally, with regard to Soviet techniques of adjudication, it must be noted that Soviet judicial opinions reflect a rather mechanical mode of reasoning, a conceptual rather than a pragmatic logic. All flavor of "sociological jurisprudence" is missing from them. Law seems to be conceived in terms of fixed rules; its application is viewed as requiring accuracy, not policy. The opinions are short. Occasionally the facts of the case are developed at length, but rarely is there any elaborate discussion of the law. The opinion has more the form of a decree, and indeed is entitled a "decree" (postanovlenie). Typically it is in the following style: "This case is governed by Article Such-and-such. The contention that that provision is superseded by the statute of such-and-such date is unfounded since the statute does not refer to that provision." And so on. On the other hand, where it is conceived that there is a gap in the law, the Soviet courts often do not hesitate to fill it. In fact Soviet codes and statutes are full of lacunae, and as a result much of Soviet law is avowedly judge-made. For example, the Soviet courts, without any code or statutory provision on the subject, early decided that damages awarded in tort cases for loss of earning capacity should be paid in the form of annuities, and should cease to be paid if the plaintiff subsequently recovered his earning capacity. We know this practice in workmen's compensation, of course, but the Soviet judges introduced it in tort actions generally. Indeed, most of Soviet tort law has been spun judicially out of a few broad code provisions. To cite an even more striking example, workers found guilty under a law (repealed in 1956) imposing two-to-four months' imprisonment for quitting a job without permission of management, were held by the courts to be *not* obliged to return to the former job after serving their sentence, on the ground that by quitting, though unlawfully, they had dissolved the labor contract. This was surely "mechanical jurisprudence," though it made new law. Incidentally, the decision provided a strong incentive for otherwise reluctant managers to give insistent workers permission to quit.

Mechanical jurisprudence—in the sense of a logic which tends to read legal propositions literally, rather than in terms of their social purpose—is connected in Soviet law with the movement toward stability which began in the early 1930's and which has gathered considerable momentum since Stalin's death six years ago. Earlier Soviet jurisprudence had taught that law and policy are virtually indistinguishable, and that Soviet law in the period of construction of socialism should have maximum flexibility. Article 1 of the Civil Code of 1922, providing that rights are to be protected only to the extent that they are exercised in accordance with their social economic purpose, and Article 16 of the Criminal Code of 1926, providing that a socially dangerous act, though not specifically proscribed, may by analogy be punished under a provision proscribing similar acts, were reflections of the earlier conception of law as policy science. Since the mid-1930's these provisions have become obsolete. Article 1 has been severely restricted and the doctrine of analogy has been expressly repudiated. The reasons for this shift toward what Stalin in 1936 called "stability of laws" and "strict socialist legality" are complex. They are connected with a stabilization of social life generally, the stregthening of the family, the restoration of military traditions, the emphasis on the achievements of pre-revolutionary Russian history, and, in the economic sphere, a recognition that efficiency in an industrial society requires the kind of calculability and predictability which only a stable legal system can provide. Under Stalin legal stability did not extend to the political area and to a lesser extent the same is true under Stalin's successors. Where the top leadership has felt its own security threatened, it has resorted to terror—terror by abuse of law and naked terror without law. But in areas of social and economic life in which the supremacy of the Party leadership was not challenged, Stalin discovered that his dictatorial powers could be maintained more effectively with a system of law than without it. The dictator can easily enact statutes; his problem is to establish a system for their interpretation and application in a manner consistent with his will. For this purpose mechanical jurisprudence is safer than a more creative approach—safer for the leadership and safer also, and perhaps primarily, for the judges and lawyers.

Thus Soviet techniques of adjudication lead us to Soviet techniques of legislation. As in France after the French Revolution, so in the Soviet Union, restrictions on the doctrine of precedent as well as on liberal judicial interpretation of statutes are seen in terms of legislative supremacy. The difference is, however, that the Soviet legislature, the Supreme Soviet of the U.S.S.R., is hand-picked by the top leadership of the Communist Party, and its freedom of debate is extremely narrow. Only recently has there begun to function a committee system; there are now some thirteen standing committees for drafting legislation in various fields. But the initiative is with the Party leaders, and it is they who set the limits of possible action. It is the absence of due process of legislation, the absence of a genuine adversary system of enacting statutory law, which is the chief formal distinction between Soviet legal system and the legal systems of the West. Here we find one of the principal links between Soviet law and Soviet autocracy.

The de facto legislative power of the top Soviet leader or group of leaders reflects a fundamental difference between the nature and function of law in the Soviet Union and in the West. It is a basic postulate of all Western systems that law is essentially a means whereby society exercises control over the political leadership; it is a basic postulate of the Soviet system, on the other hand, that law is essentially a means whereby the political leadership exercises control over society. We must remember, in this connection, that Russia has never had a deep-rooted tradition of a system of law which binds the ruler himself.

SOVIET LAW AND SOCIAL ORDER

We must not confuse the rule of law, however, with the role of law. Despite their initial hostility to law both in theory and in practice, the leaders of the Russian Revolution learned, as leaders of the western revolutions in earlier centuries had learned, that principles for which the Revolution was fought cannot be preserved unless they are institutionalized in law. In particular, the Soviet leaders learned that the system of planned economy cannot function without law; and they learned that a collectivized—one might better say, a mobilized—social order requires law.

Let me speak briefly first about the law of a planned economy. Soviet industry and commerce are largely in the hands of state agencies; and Soviet agriculture is under stringent state control. The plans under which the state officials who are thus in charge of Soviet economic activities operate come from higher administrative authorities and ultimately from the Council of Ministers of the U.S.S.R. and its State Planning Committee. Yet plans are not self-executing. It is all very well for Moscow to say, or for the new regional economic councils to say, that so much coal, so much steel, so many pairs of shoes, so many television sets, should be produced; but where are the managers to get the necessary resources to produce them? Labor must be paid; goods must be procured; the product must be distributed. The bottlenecks of planning are insuperable unless a rational system of incentives, of criteria which will guide individual decisions, is worked out. The Soviet leaders in the late 1930's worked out such a system—worked it out in orthodox terms of monetary rewards, contract law, corporate autonomy of individual enterprises, and in the case of the collective farms, cooperative sharing in the profits of the collective as well as personal ownership by the peasants of their household plots.

I have mentioned that there is a great deal of contract litigation between state business enterprises. Gosarbitrazh adjudicates disputes arising over failure to deliver, breach of warranty, and similar questions. The individual state enterprise has the same incentive to sue for breach of contract that a private firm in this country has: its losses will come out of its profits, a portion of which goes into bonuses. In addition there is an added incentive: if it can show that its failure to fulfill the plan is due to the fault of another enterprise, it can escape measures of administrative discipline. Finally, the State itself has a strong interest in an impartial determination of responsibility, as well as in the correction of the causes of breach of contract. Indeed, one of the most interesting aspects of the work of Gosarbitrazh is its educational role in so-called pre-contract disputes. If the parties cannot agree upon the terms of a contract which, under the general plan, they ought to conclude, they may resort to Gosarbitrazh which will determine how the contract should be written. It is apparent that here contract law and administrative law are intertwined, and that although Soviet contract law is often quite similar to our own in terms of the rules of formation, breach, damages, and the like, it fulfills a quite different function.

Finally, I should like to say a few words about the functions of Soviet law in maintaining a collective social order. In comparison with the United States, Soviet society is more highly mobilized, more highly directed, not only for military purposes but also for social and economic purposes. It is the idea of Soviet society that each person in it must have a place, a job to do. The legal system is more immediately concerned than ours, therefore, with the allocation of duties, and although rights are of course correlative with duties, it is implicit in Soviet law that duties precede rights, that rights arise out of duties.

How is this manifested concretely? In the first place, it is manifested in the increased role of criminal law. The concept of "official crimes"—intentional or negligent abuse of authority by persons in responsible positions—plays a very important part in the control of the economic behavior not only of managers of state enterprises but even of persons quite far down in the official hierarchy, including, for example, waiters in state-owned restaurants. From 1940 until 1956 workers who quit their jobs without authorization of management were subject to two-to-four months' imprisonment. Recently in several of the smaller republics a law has been enacted whereby a person who is considered an "anti-social and parasitic element" may be exiled from two to five years to another part of the republic by an *ad hoc* body of neighbors with review by the executive authorities only.[5]

---

[5] See Berman, *Soviet Law Reform—Dateline Moscow 1957*, 66 YALE L.J. 1191, 1208–09 (1957). The "parasite law" has been widely discussed, pro and con, in the Soviet Press, since 1957. It has not yet been adopted, as of this writing, in the largest Soviet republic, the RSFSR.

The role of law in maintaining a mobilized social order is seen also in the powers of the Procuracy not only to prosecute crime but also to supervise generally the legality of administrative acts. The general supervisory powers of the Procuracy make it the watchdog of the central government; it may protest any illegality of management, of local government, or of individual officials, to the next higher administrative authorities. Although the institution of the Procuracy is known also to other countries, only in Russia has it played so crucial a role in maintaining social order in general.

The Soviet courts, too, play an important part in maintaining the sense of collective unity and collective purpose in the society. Their task is conceived to a large extent in terms of educating people—the parties, the spectators, and the whole society—to be the kind of "new Soviet man" which the state is seeking to develop: hard-working, honest, resourceful, cooperative, and above all conscious of his membership in a socialist society and loyal to its leadership. Both substantive and procedural law manifest a jurisprudence in which the balancing of interests yields place to the shaping of interests. It is not merely that Soviet judicial decisions are thought to have an educational function, but rather that the educational purpose of judicial decisions becomes a central factor in the determination of rights and duties in particular cases. An American parallel may be found in the laws governing juvenile delinquency: here we are expressly concerned primarily with the effect of procedures and of decisions upon the person before the court. Something of the spirit of the law governing juvenile delinquency is carried over into all branches of law, civil and administrative as well as criminal, in the Soviet Union. This is not to say that Soviet law is benign. It is benign in some respects and ruthless in others. And it is not necessary to stress that the whole conception of a parental relation between the court, as an arm of the state, and the people as its wards, so to speak, lends itself to arbitrariness, to political interference in adjudication, and to a philosophy of historical relativism under which rights may be withdrawn altogether if circumstances so require.[6]

Yet it is a mistake to view Soviet law as merely an instrument of dictatorship. It is a genuine response to the crisis of the 20th century, which has witnessed the breakdown of individualism—in law as well as in other areas of spiritual life. "Where today are the economically selfsufficient household and neighborhoods, where is the economically selfsufficient, versatile, restless, self-reliant man, freely making a place for himself by free self-assertion . . .?" asked Roscoe Pound in 1930. "Where, indeed, but in our legal thinking in which it is so decisive an element."[7] Since 1930 it has gone out of much of our legal thinking as well. The excesses of totalitarianism can teach us the dangers of rushing to the opposite extreme; nevertheless this is not a matter where we can afford to be smug. Our own law may be seen in the light of Soviet experience, as seeking a level which combines the virtues of individuality and collectivity, while avoiding the mythology of either "ism."

[6]*Id.* at 1214-15.
[7]Pound, *The New Feudal System*, 19 KY. L.J. 1, 6 (1930).

Augsburg College
George Sverdrup Library
Minneapolis, Minnesota 55404